Tiny Nature

Discovering Nature's Hidden World

Through the Lens of Macrophotography

JAMIE ROSENCRANS

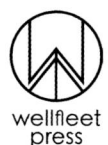

wellfleet
press

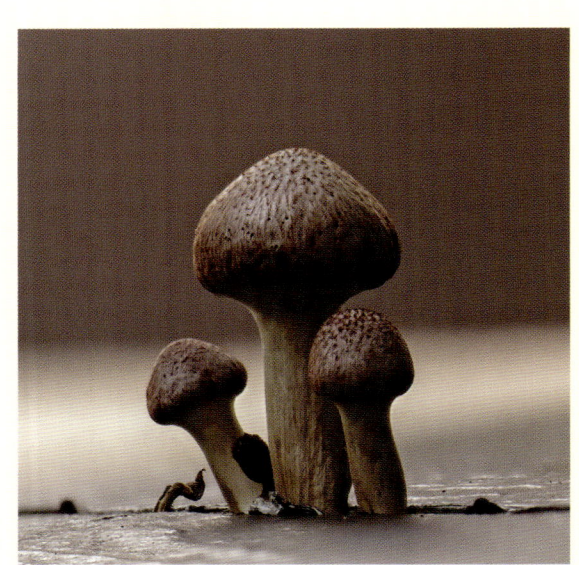

ABOVE: UNIDENTIFIED FUNGUS IN NASHVILLE, TN, US.

To Poog and Tiny Horse for joining me for so many wonderful woodland adventures—you are my heart.

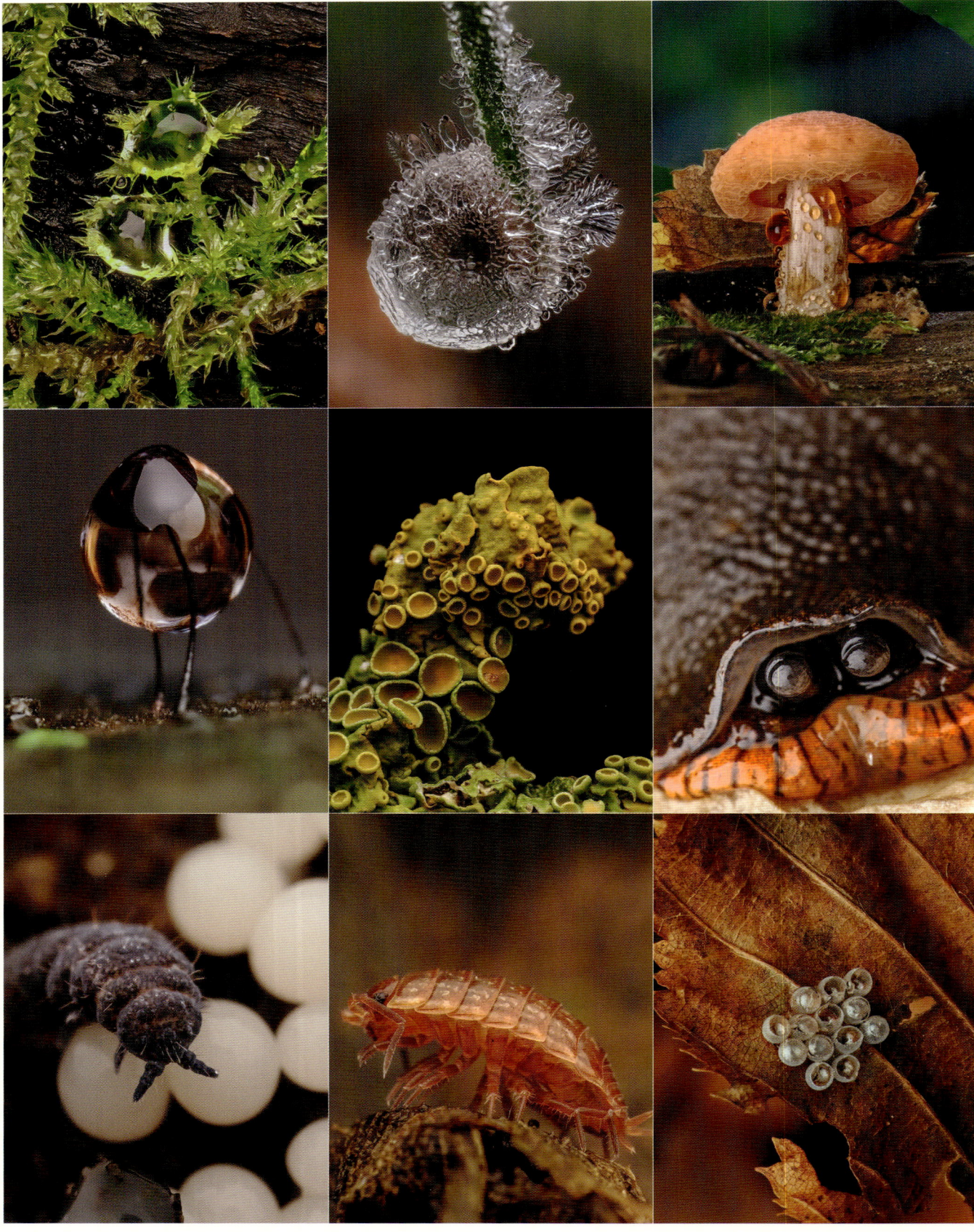

Contents

Introduction

It's a bit sad and strange how so many of us (myself included) can wander through life for so many years, blissfully unaware of the things not starkly visible or right in front of us. I grew up in the lush and leafy nature reserve–filled suburbs of Nashville, Tennessee, yet there was so much in the natural world that I never noticed.

I'm embarrassed to say that a hike through the woods was usually just a means to an end. (Yes, fresh air! Yes, nature! Isn't this all lovely? Look at me getting my steps in!) I *liked* being in nature, but I had no understanding of what nature *was*.

However, I always felt connected to trees. I was drawn to the forest, especially during the changing seasons with falling leaves and blossoming flowers. But it felt all so surface level. I walked the same woodland paths week after week, completely oblivious, listening to a podcast and looking straight ahead in a dreamlike state, with an occasional glance up at the tree canopy. I didn't know about the tiny ecosystem right under my feet. Its existence simply hadn't occurred to me. I think I viewed the forest as a safe and magical place where I could daydream and get some exercise as well. I didn't really need to know the details of the magic; I was just there for the view!

All this is to say, I had the *foundations* of a nature person. Moreover, I am artistically inclined—I enjoyed drawing and painting at an early age and wrote some little fiction novels. That then transformed into a fixation on becoming an actress or maybe a singer. No, no, a poet. All those careers seemed realistic to me.

As a romantic and imaginative kid who felt things very acutely and largely existed in a fantasy world, I was terrible at sports, bored in school, and longed to grow up and see what the future held for me. It was also pretty clear to everyone around me that I'd end up in a creative field of some kind because my feet were certainly not attached to this earth.

When I was about nine years old, my dad (an avid hobbyist photographer) gave me one of his old disk film cameras to experiment with. I started taking it to school with me, snapping photos of my friends on the playground, and on the bus. The first time I held developed photos in my hand, something clicked. This was the moment my love for photography began and from that point I was hardly ever without a camera. I had a near compulsion to capture every moment. I still have stacks of physical photo albums full of candid shots and large online libraries of digital photos that span my whole life. I was the official paparazzo of my friends' circle, for summer picnics, wild parties, weddings, and even a birth.

During the blogger boom, I had a brief and impassioned stint photographing food and being a food blogger (didn't we all?). But then, in my late twenties, I expatriated to the United Kingdom and transitioned to mostly travel and street photography. There was a period when I delved into some

6

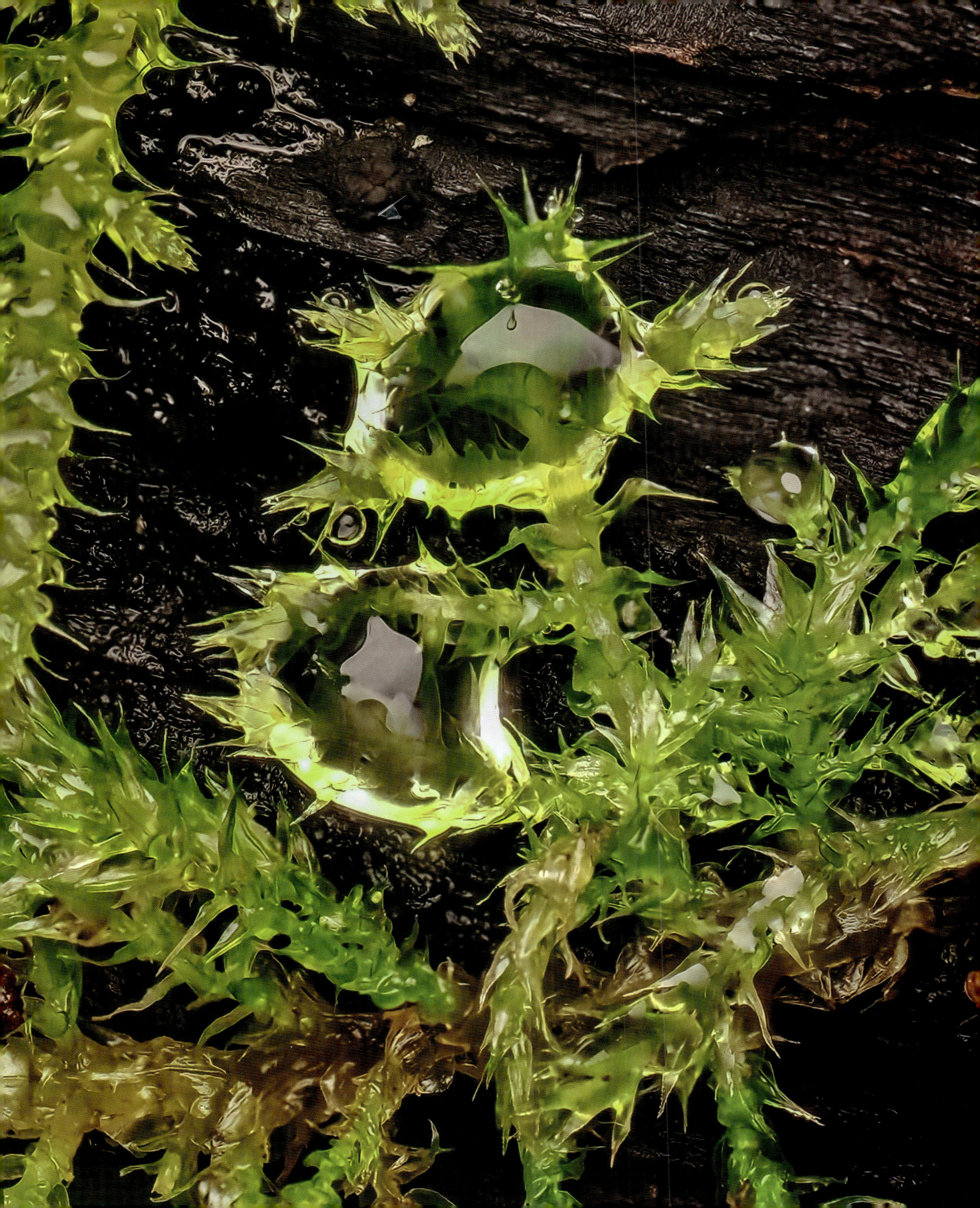

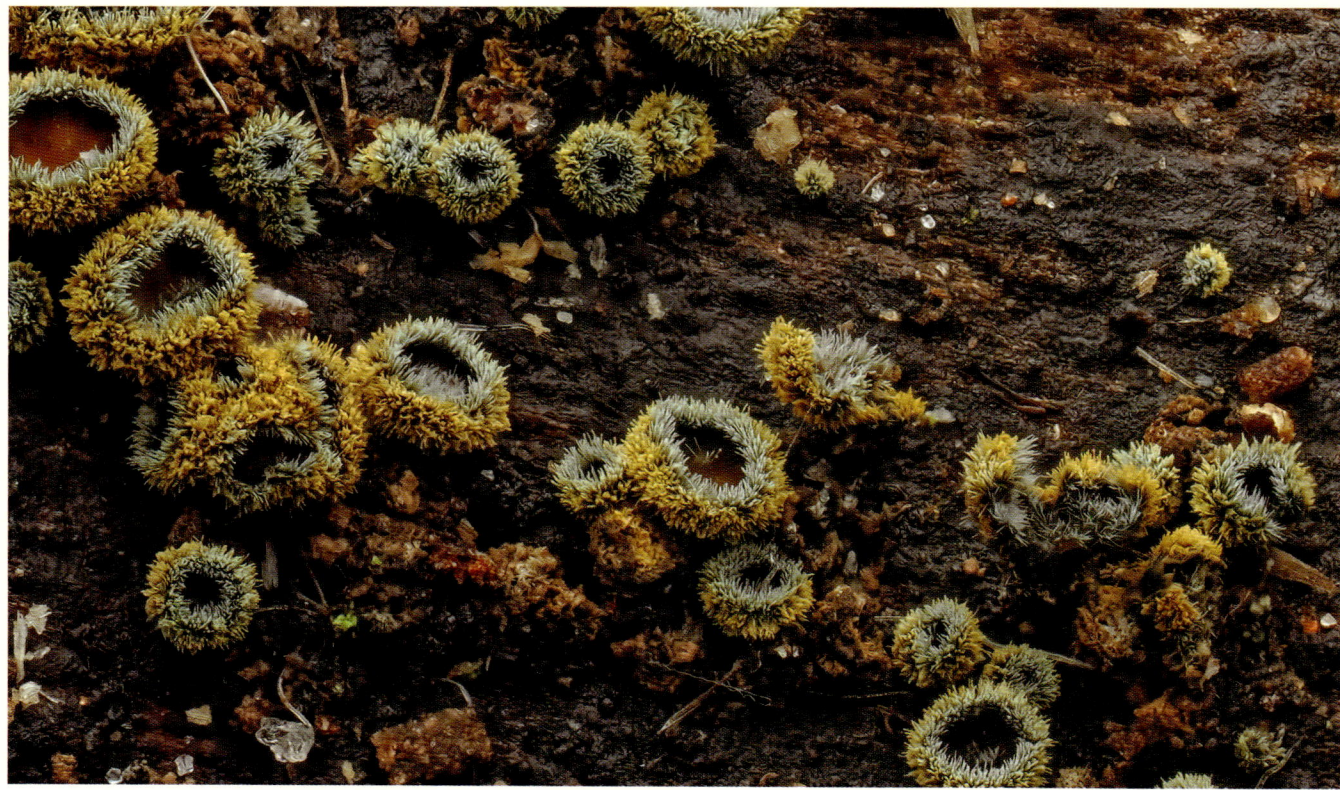

8

commercial photography, but by then, my enthusiasm had waned. Now I had a fancy phone with a fancy camera that made it easy to get a "good enough" shot for my everyday purposes. I wasn't as inspired as I had been all those years ago. The artist inside me was in some sort of coma.

Then, 2020 happened and suddenly the only thing we were allowed to do for a year (aside from whatever was deemed "essential") was to be alone and go for walks outside. The gardens across the street from my house became a haven for me, and in my daily solo excursions, I started walking slower, looking around, and *noticing* things. I began taking photos again with my mobile phone camera of flowers and ducklings, tree bark texture, water droplets on branches, or a pollen-dusted bumble bee nestled in a fresh blossom.

These daily walks became a lifesaver—a ritual that kept me sane during one of the strangest years of my life. And I was suddenly so intrigued by nature on a much deeper level. I was enraptured by all the tiny things that had suddenly revealed themselves to me. Without knowing it, I was developing my *macrovision*. Luckily, I relocated to the edges of an amazing urban nature reserve in London in the late summer of 2020. By that time, I had started using my camera again instead of my phone.

I began following nature photographers on Instagram in droves and kept on seeing these unbelievable "macro" photography images. Until then, this type of specialized photography had never been on my radar. I couldn't wrap my brain around some of the close-up insect portraits with unreal clarity and startling details. Winged creatures (that through

ABOVE: *NEODASYSCHYPHA CERINA.*

the wizardry of a macro lens) now had *personality*. Critters that I once ignorantly categorized as "gross" or "scary" had transformed into beautiful and friendly extraterrestrials that intrigued me.

Delicate mushrooms studded with the most surreal guttation droplets, and neon lichen with textures that could only be described as otherworldly. Something called *slime mold* that wasn't quite fungi yet wasn't an animal but was intelligent—it was like equal parts alien invasion and fantasy fiction. So rad! I had never noticed or even heard of most of these before, so I was shocked to find out that they were apparently very common. Where had I been? Had I just gone my entire life sleepwalking through a vast universe of complex and incredible wonders? How is this possible? I wanted to be part of it. I *needed* to be part of it.

So, I bought a macro lens, upgraded my camera, and decided to spend all my free time trawling through the understory, desperate to be part of this hidden world. I spent hours on the forest floor practicing technique, experimenting with light, and cursing my dying batteries and the fading sun. I lost track of time and space from concentrating so hard. But I learned something new about photography and met fascinating critters on every single forest walk. Knee deep in mud, soaked through with rain, sitting patiently stalking a damselfly in the scorching heat while being devoured by mosquitos, stabbed with thorns, stung by nettles every day, I've honestly never been happier than those first years of discovery. Macrophotography excursions became a nearly daily occurrence and swiftly unlocked a new period in my life where I finally felt creatively activated and fulfilled again. I amassed photos in a quantity that became overwhelming and exhilarating and I wasn't sure what to do

with them other than share them with others on Instagram—which is what I did.

Sometimes when I'm alone, sitting among the leaves, deep in the forest just listening to music on my headphones with no human in sight for hours, it really feels like I have journeyed through a portal to some secret leafy galaxy. All the humans are gone and it's just me and Radiohead and these glistening mushrooms and fluffy caterpillars.

Through macrophotography, I have rediscovered such an important part of myself that had laid dormant for ages. Somewhere beneath the layers of all the fun and terrible experiences in my life, I was a deeply curious individual, desperate to connect with something greater than myself. I had grown into a cynical and pragmatic adult and had finally found the antidote to my complacency.

So here I am, softened and humbled by the vastness and the mystery of nature. I am a real nature person. I am an artist again. I've found a community of fellow enthusiasts on the internet and together, without knowing, we've built a little supportive space to share and explore our love of nature and macrophotography. It's a tiny oasis somehow insulated from all the other toxicity and trash on social media. It's been a beautiful experience and I'm deeply thankful to still be a part of it.

I'm fully aware that this may all seem super self-indulgent and nauseatingly earnest. However, I'm not capable of fully expressing the magnitude that the discovery of the forest floor ecosystem and macrophotography has had on me without the risk of waxing poetic in a syrupy and obnoxious way. It's just the truth. I'm typically a sarcastic, glass-half-empty nonbeliever, but let me assure you that nature is magic, and I stand by that.

What Is Macrophotography?

Please note that this isn't a how-to-master-macrophotography book. Enough of those already exist! However, for anyone stumbling into this niche world for the first time, a greater appreciation of my work and the work of other macrophotographers might be achieved by me at least giving a brief overview of what macrophotography is.

As you have probably already gathered, macro is a type of photography that helps bring to life the microscopic details of a subject (in my case, nature). The term "macro" comes from the Greek word "makros," meaning *large* or *long*. With macrophotography, you are magnifying small subjects to appear larger than life in the final image.

One of the biggest challenges in macrophotography, given the proximity to the subject, is sufficient depth of field. This is where focus stacking can save the day! Focus stacking is a technique I often use to overcome shallow depth of field issues. Essentially, this means capturing a series of images, with each frame focusing on different points along the subject. Once you have all these different frames, you can then blend them into one seamless composite photo. This can be achieved in post-processing using specialized software (or directly in-camera if you have an OM SYSTEM camera like me). Focus stacking enables me to create a final image with an extended depth of field, ensuring that all parts of the subject are in sharp focus.

But this is not to say that I don't take single images. Most of my photos are single shots, but I like to employ a variety of different techniques and compositions depending on what I'm shooting. I want my photography to continue evolving, and I try to lead with an *art-first* approach. Even if a photo of an insect is technically perfect, with regards to clarity, focus, and detail, if it's not also beautiful and interesting, I'm honestly just not into it.

Why Isn't It "Micro" Photography?

I get asked this a lot. My understanding is that those with microscopic-level magnification are considered microphotographers. Through a macro lens you basically get visual superpowers. It reveals previously unseen complexities with a visual articulation you just can't get otherwise—even with perfect vision. With macrophotography, I use specialized macro lenses designed to achieve high magnification ratios engineered with optical elements that allow them to focus sharply on subjects at very close distances, often achieving a 1:1 magnification ratio or higher. This means that the subject appears life-sized or larger on the camera's sensor, capturing even the smallest details with stunning clarity. With Lasik surgery or not, your eyes just aren't that good.

Equipment Overview

I have always used Olympus cameras (now rebranded as OM SYSTEM). My first digital

OPPOSITE: WATER DROPLETS FROZEN ON A BLADE OF GRASS.

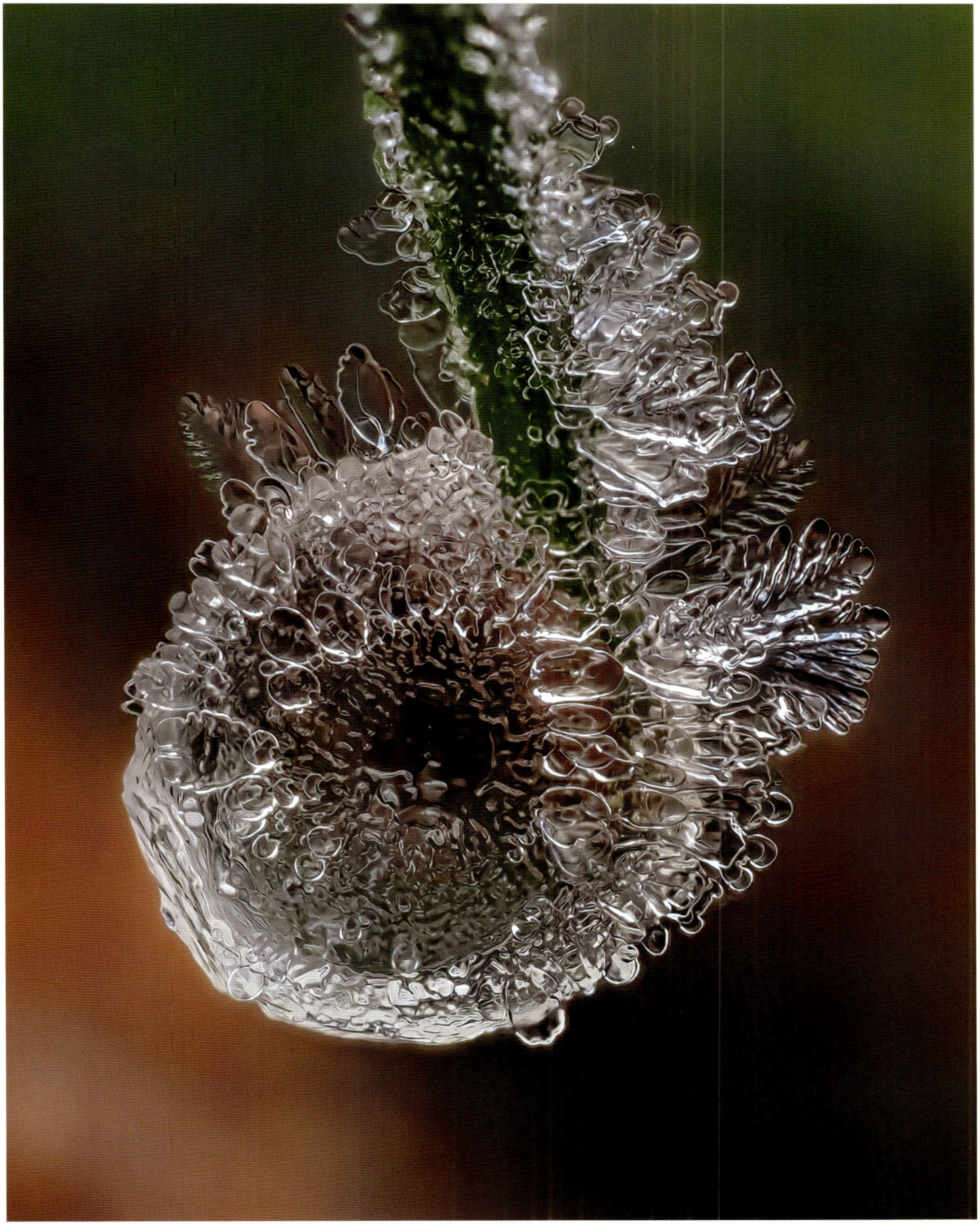

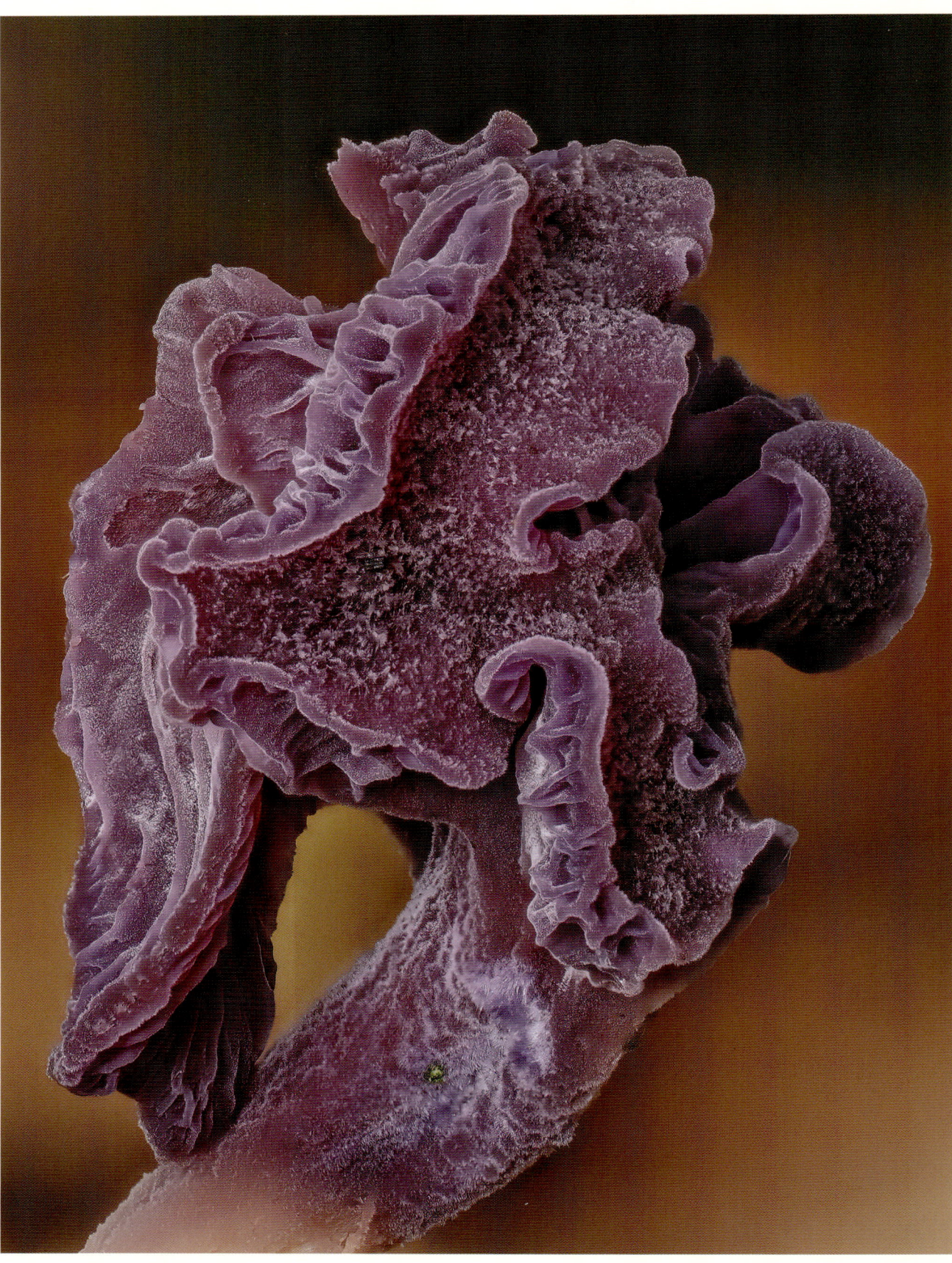

12

camera was an Olympus PEN series, and I fell in love and never looked elsewhere. This company offers incredible cameras and lenses (not just for macrophotography) but at this point, I'm completely biased beyond belief as I'm also an ambassador for them. But seriously, even if they dropped me tomorrow (please don't!) I'd still swear by their system as being absolutely phenomenal for macrophotography!

I have a few different OM SYSTEM cameras and currently switch between their 60 mm and 90 mm macro lenses. Some of the photos in this book were taken with flash and diffusion, and some are purely natural light, some are taken with long exposure, some on a tripod, and some handheld. Some are focus stacks of five to fifteen images, but most are single-frame shots.

You do not need a ton of expensive specialized kits to achieve incredible macro photographs. As I mentioned, when I got started with macro, I was using my phone camera and its "macro" settings. For a cost-effective entry into macrophotography, you can get macro attachments that clip onto the lenses of your phone and will still result in some absolutely incredible images. Photographers, like David Joseph who's based in Nigeria (@abcdee_david on Instagram), started with only a camera phone and has produced some of the best macrophotography images I've ever seen. David's work is an absolute masterclass in what can be achieved when just using a phone.

Embrace Your Artistry

Up close, mushroom gills and guttation transform into organic architecture. Abstract sculpture. A fantasy world. Zoomed in on the unearthly details, you can easily lose sense of what you are even looking at. Some of my favorite images have been a fluke—a fleeting shot that turned out half-decent by some miracle. On the converse, many shots were a struggle to get right, and I spent hours attempting different techniques and pulling every trick out of the bag to achieve the composition and outcome I desired.

A lot of hard work, dedication, and irritation has gone into learning and growing as a macrophotographer. I never thought I'd be that great at it; it seemed painfully complicated and overwhelming when I first started. I don't enjoy failing, although I hear it's important for growth (I'm working on that!). However, I embraced my mistakes and absorbed as much information as I could from experimenting and exploring. I stopped fixating on what I thought I was expected to do and instead leaned into the joy and fulfillment of just creating art again.

Macrophotography is my favorite way to see the world. At times when you don't know what you are looking at, but you are moved by it; when you have a visceral reaction to the colors and textures and appreciate the beauty of the unknown, you open yourself up to endless possibilities of connection and appreciation. I am my own person, artist, and photographer. I can do whatever I want because it's *my* photography and *my* process.

To be honest, I probably don't know that much; maybe more than some and less than others. I'm not formally trained in photography or science. I'm an artist who fell in love with photography at a very early age. I am not a scientist, and I can't

13

OPPOSITE: AMETHYST DECEIVER (*LACCARIA AMETHYSTINA*).

always recite every formal name of every springtail species I photograph because I'm more interested in capturing the fascinating things first, and then looking them up later. Most of the forest ecosystem knowledge I possess is all anecdotal and has been absorbed over the years through conversations with smarter people and by learning on the job through firsthand experiences. Yes, I've read some books, some blogs, and seen some documentaries, but I spend a lot of time chatting with other nature lovers and have gained knowledge that way too.

Discovering the Forest Floor

Developing the skills to find mushrooms, slime molds, and bugs to photograph is truly half the battle for a macrophotographer or budding naturalist. Nearly every species has a peak season and figuring out where and when to find your target subject is part of the excitement and wonder of being in the woods.

If you are just getting started with macrophotography, or even if you aren't a photographer at all and simply want to locate and appreciate the tiny hidden life of the forest, you'll need to first build a foundation of knowledge through time and exposure to the same woodland locations season after season to know what grows or thrives near you. You will, through repetition and perseverance, slowly become an expert on where and when to find things. (And you will probably become obsessed—a fair warning!)

I still live in the United Kingdom, and most of the photos in this book are from urban London forests. I say this primarily so that there's a clear understanding that my tips and advice in this book are limited to my specific location and experience.

Depending on where you are in the world, you will have a completely different forest ecosystem, climate, and likely an even greater diversity of species to photograph! Once you start finding things, I suggest making notes. From my many repeated trips to "check on that log," I've learned that in certain weeks and months, specific species are abundant in particular areas. I mark these spots on Google Maps and add notes to my photos in the Lightroom app. Returning to these jackpot locations year after year sometimes works like clockwork, but other times there's nothing. No matter, though, because the forest is constantly changing, and you'll always find new hot spots and species, so it never gets old.

The best macrophotographers and naturalists develop "macro vision." (I should trademark that.) Basically, you must train your eyes to spot certain shapes and colors, even things less than a centimeter in size. Scan the ground for signs by walking slowly and focusing intensely. In late spring, there's a necessary shift from scanning the forest floor to looking higher for insects. This transition from stationary fungi to moving insects can be strange and a bit jarring at first, but in my case, my posture certainly appreciates the change.

Many people get a camera and macro lens, head out, and often complain they can't find anything. I especially get lots of messages from city dwellers asking me how I find so many different species in London. A lot of it comes down to the amount of time I've devoted to being in the forest and improving my observational skills. When I started, I wasn't reliably finding subjects. It was pretty much always pure luck. I highly recommend just going out into the woods without expectation and just paying attention.

The Best Time for Macrophotography

No matter where you are, early mornings and late evenings are ideal for most subjects. Mushrooms and slime mold will be popping up and likely not eaten yet; flying insects will be sleeping, often covered in dew droplets; understory creatures normally hidden in the nooks and crannies will be out and about in full, making the forest an unrecognizable nightclub of activity. You can certainly go out in the day and easily photograph stationary subjects at any time. But flying insects and soil animals that rarely show themselves will prove challenging unless you flip logs and rifle through decaying leaves.

For beginners keen on locating fungi and slime molds, autumn and winter are where it's at. Autumn creates a unique type of magic in the forest. It's the changing leaves setting the trees ablaze in warm colors and the crisp air that allow a diversity of mushrooms and slime molds to be found! Gently rifle through decaying leaf litter to reveal tiny mushrooms sprouting from the ground, little bits of wood with beautiful, tiny fungal cups, or even juvenile slime molds. Beneath the leaves is a microcosm teeming with soil animals providing countless opportunities for photography. Pick a spot in the woods, sit down, be patient, and you'll find enchantment. Autumn, being a captivatingly romantic season, has so much beauty to capture.

In the winter months, I'm still finding lots of slime molds and soil animals as well as lichen and tiny fungi in the urban forest. I don't find winters terribly harsh or long in London—it's really a milder but colder and wetter version of autumn. Although we rarely get snow, we do always get frosty mornings, and I love photographing ice crystals on the grass.

If you are an insect lover, spring and summer are your months. In spring, I tend to find lots of slime molds absolutely thriving in damp conditions. The gastropods come out in full force and the bluebells serve as great hiding spots for baby soil animals. I tend to take lots of droplet photography, thanks to the constant rain creating really lovely abstract scenes in the grass and leaves.

Finally, in the summer months, the forests come alive. Everything that has laid dormant during the colder months is buzzing and crawling around. Lush and vibrant colors create incredible backdrops for photos. The tree canopy is so thick you get the most perfect dappling of bokeh in the background.

In the temperate climate of London, I find fungi, slime molds, and insects year-round but every season brings its own special array of species to seek out and capture. I also strongly suggest night photography. A summer forest after dark feels like another planet where all the tiny critters come out to play.

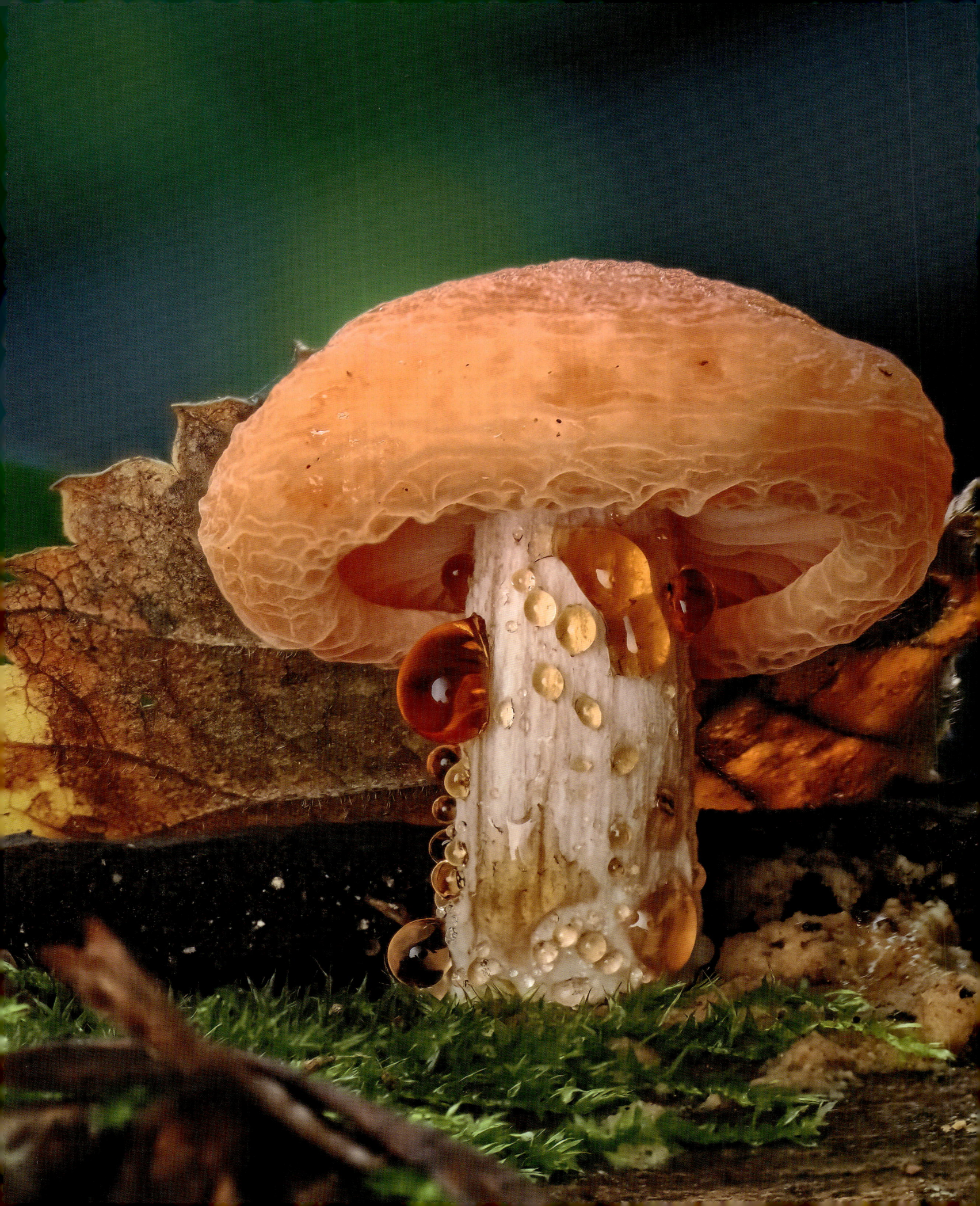

Fungi

With as much time as I've spent in the woods over the years, it seems improbable to not have seen and noticed fungi before macrophotography entered my life. Yet, I can't remember seeing any! There is just a black hole where those mushroom-shaped memories should fit.

Although embarrassing to admit, given what I've just said, I probably would have still described myself as a nature lover. Safe to say I was an *aspiring* nature lover! Because I liked lovely weather and pretty trees, I must've thought that qualified me, but I was afraid of most bugs and had no real knowledge or deep interest in them. It probably goes without saying that I don't relate to that person anymore.

This lifelong oversight has most certainly influenced my affinity for fungi. They are both poetically sculptural and indispensable to the planet's functioning. Fungi came before us as the stewards of the earth and will certainly outlive us all. ⌒⊃

OPPOSITE AND FOLLOWING PAGES: WRINKLED PEACH (*RHODOTUS PALMATUS*).

Although autumn is typically "the season" for mushroom lovers, as I mentioned previously, I find all different types of fungi year-round. As long as there is rain and there are fungal spores about, eventually fungi will pop up. The variety of fungi that can be found—even in bustling city parks—is unbelievable.

As a baby naturalist that has simply absorbed whatever knowledge I can manage over the years, and is certainly not an expert, my understanding is as follows:

Fungi start with spores. These spores germinate and send out *hyphae*, super thin, noodle-like branching filaments. The hyphae spread out and form a network beneath the soil known as *mycelium*. This mycelial network extends beyond the tree roots, decomposing and absorbing nutrients along the way. Fungi are the masters of decomposition but do much more than decompose. They sort of team up with plant roots, forming an underground system of nutrient exchange. Through this symbiosis, fungi give plants the nutrients they need, and plants give fungi the sugars they need for survival. It's a mysterious world that scientists are just now unraveling.

Fungi serve as ideal subjects for nature and macrophotography as the range of shapes and colors in the fungi kingdom creates endless compositional possibilities. There are so many types of fungi, and they are an important food source for animal life. On the forest floor, small creatures such as slugs, snails, and an array of insects consume fungi regularly. The act of eating mushrooms also serves an ecological function by helping with additional spore dispersal. Creatures that feast on fungi often travel long distances before excreting the spores in new locations, helping new fungi grow in different areas of the forest.

The various colors, shapes, and smells of mushrooms also attract insects. Some mushrooms like Stinkhorns emit odors that mimic decaying protein, capturing the attention

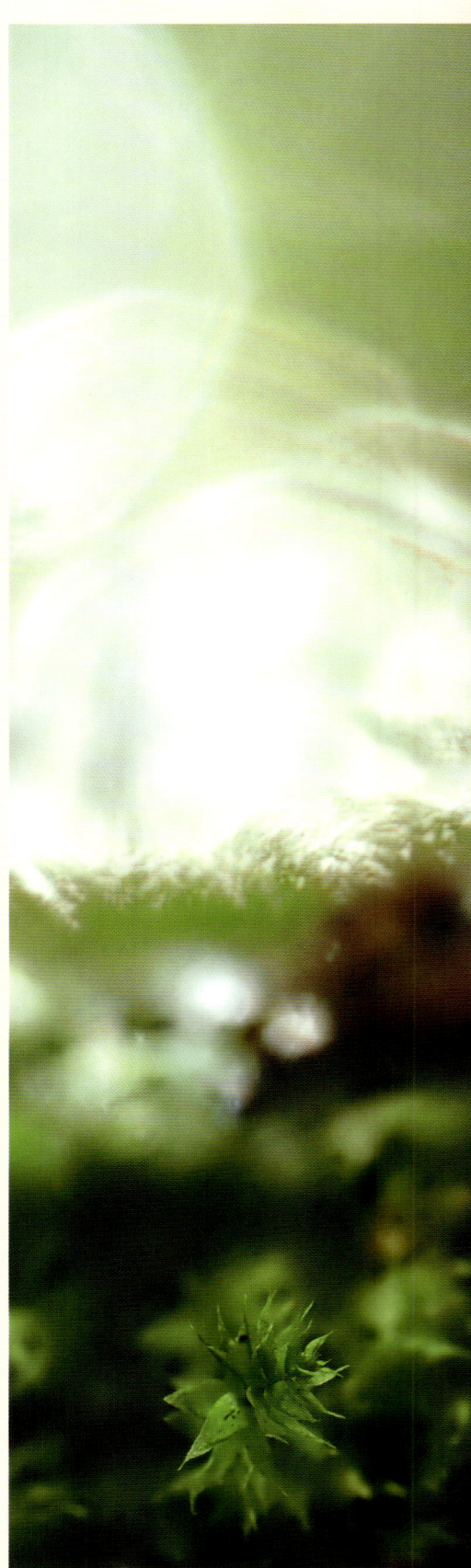

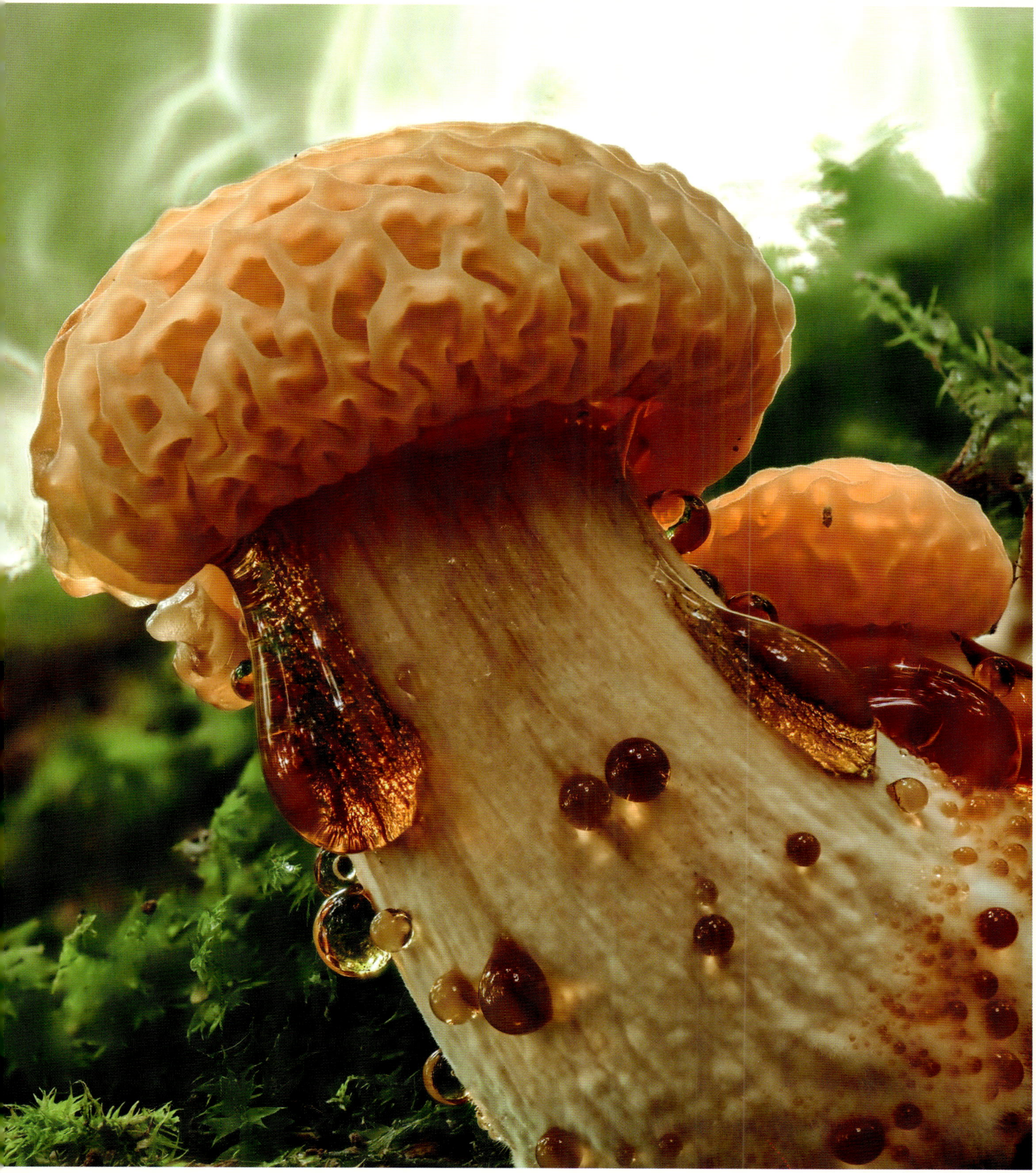

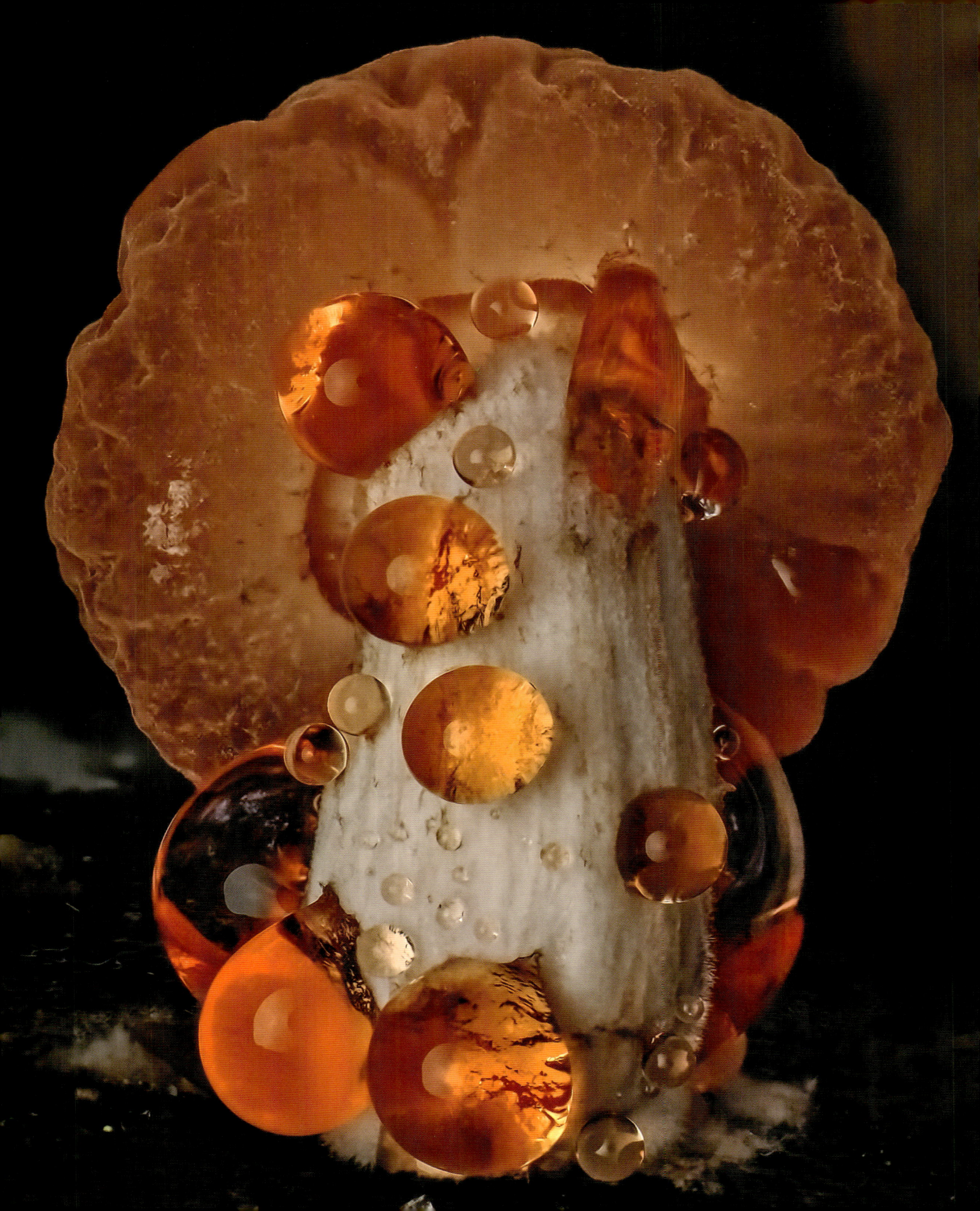

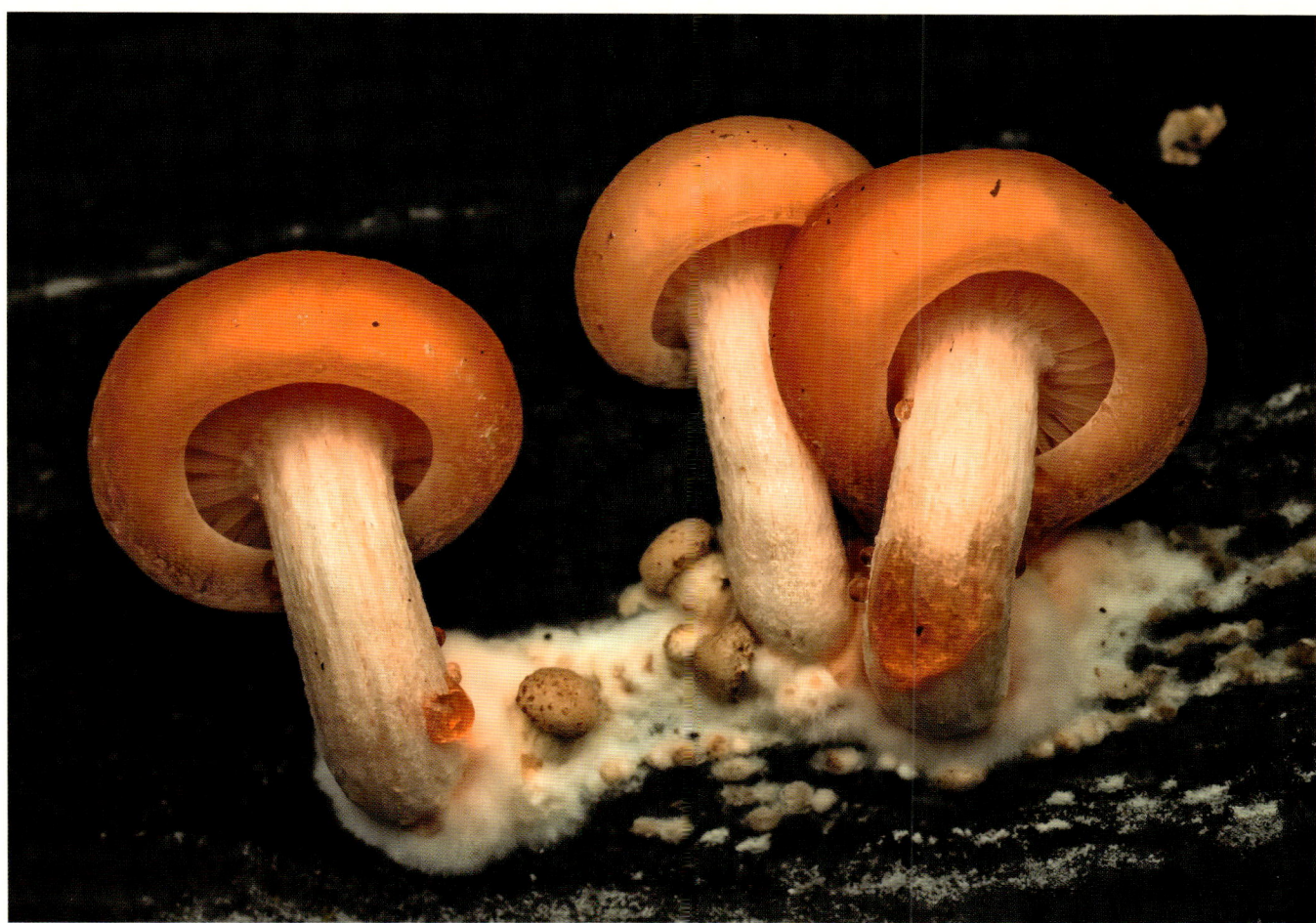

of flies and beetles that then further assist in the dispersal of spores through their activities.

Mushrooms also serve as shelter for the tiniest of soil animals. I often find springtails and small arachnids tucked into the gills of mushrooms. And then there is guttation, a process that results in the formation of small water droplets on the surface of fungi. I often refer to it as "mushroom sweat" and it creates some of the most beautiful abstract imagery.

My favorite mushrooms to photograph are *Rhodotus palmatus* (wrinkled peach mushrooms) and *Chlorociboria* (green elfcups). I've taken thousands of shots of these incredible fungi.

Wrinkled peaches are exquisite, complex, and slow to grow. Every part of their growth journey provides an entirely different aesthetic. They start out as fuzzy white patches, then develop ruby droplet–covered nubbins. Within a few weeks, they'll transform over several very different and highly photogenic stages. They are famous for their wrinkled, peach-colored, web-like caps and the breathtaking amber guttation droplets that cling to the stems and gills. They are rare and the last time I checked, are still red-listed in the UK, due to Dutch elm disease eradicating much of their habitats. I'm low-key mushroom-blessed to live near several hot spots for these fungi. I protect their habitats as if they belong to me, often camouflaging the path to reach them with sticks and holly, ensuring the dead wood they grow on stays safe and unbothered.

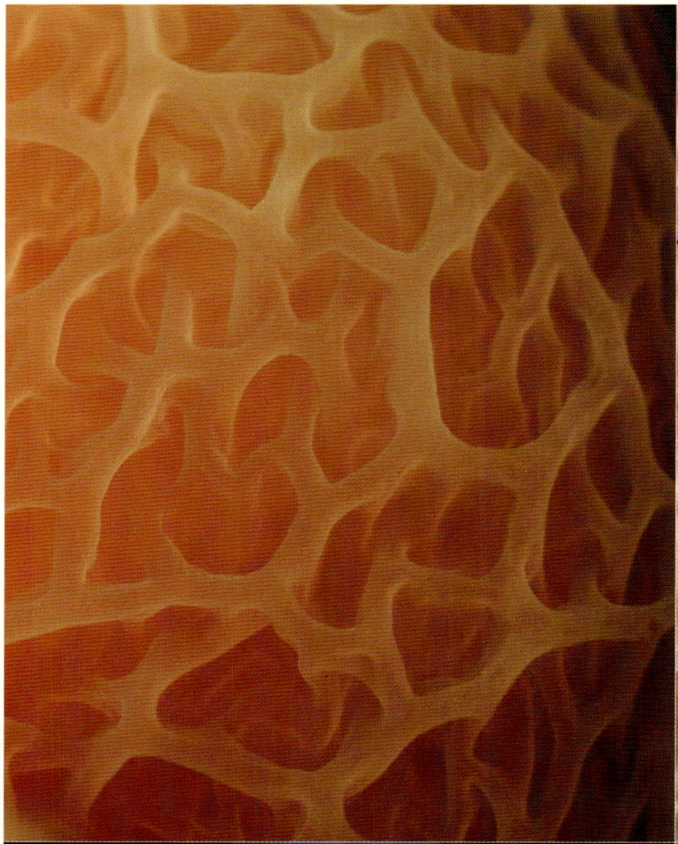

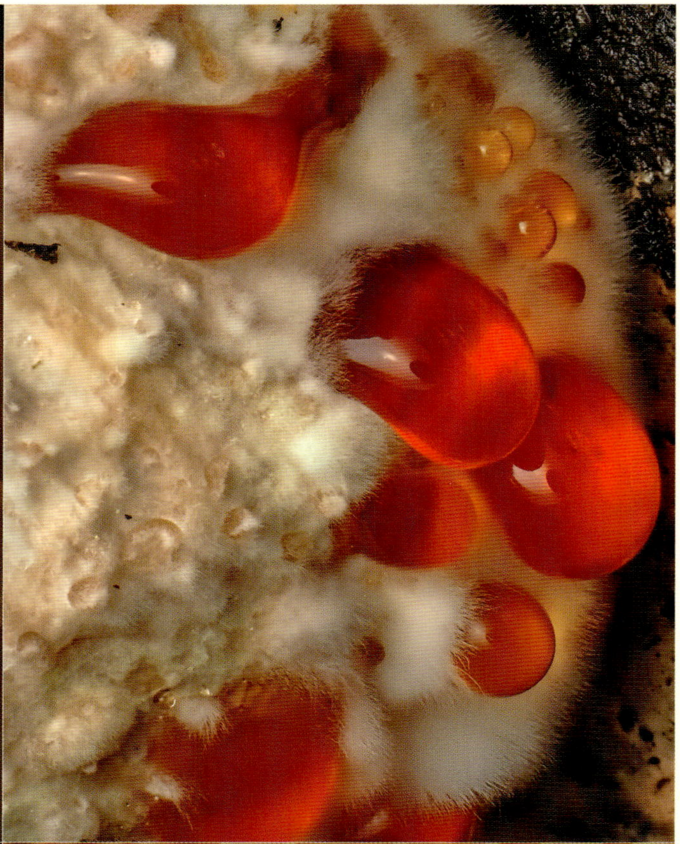

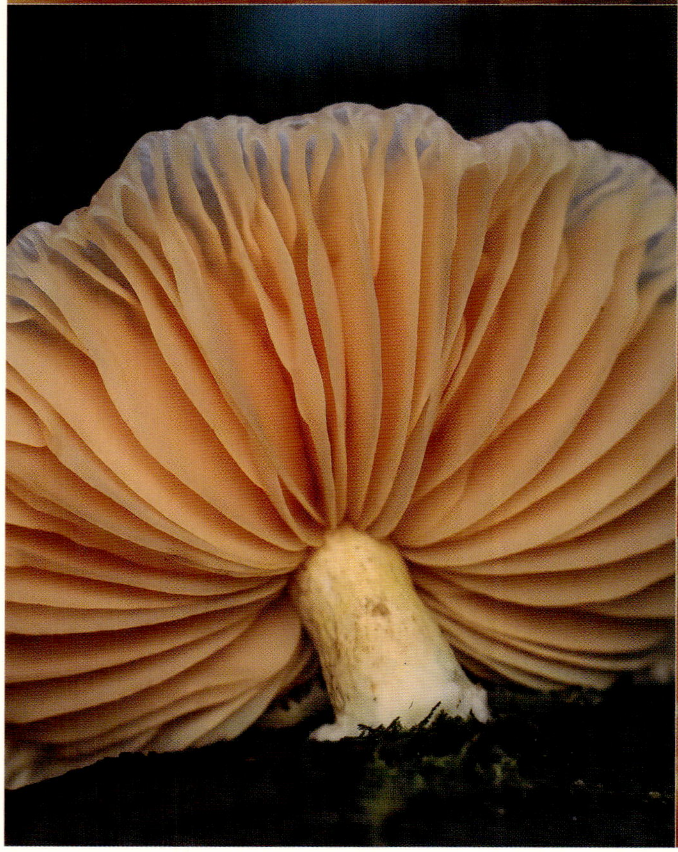

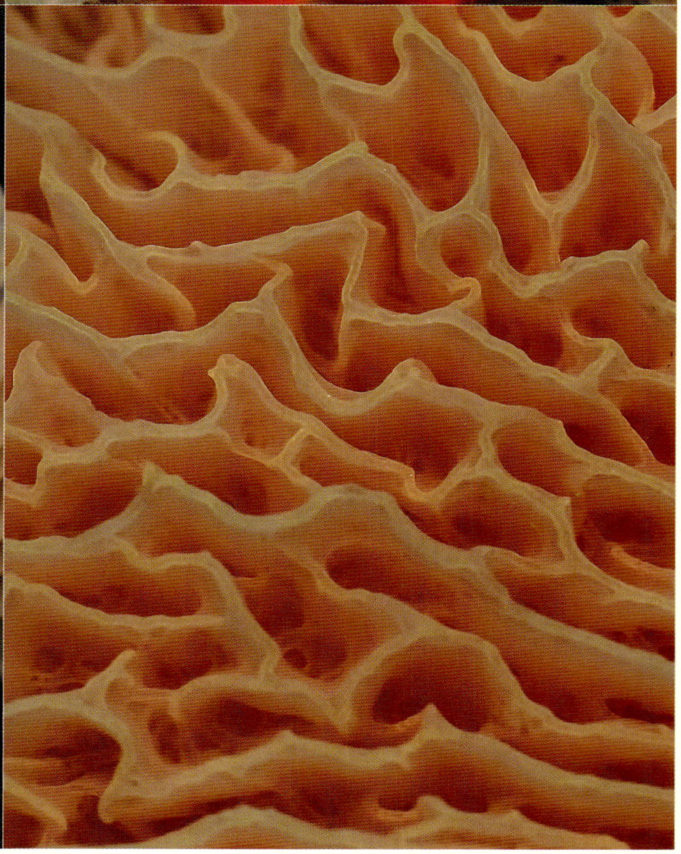

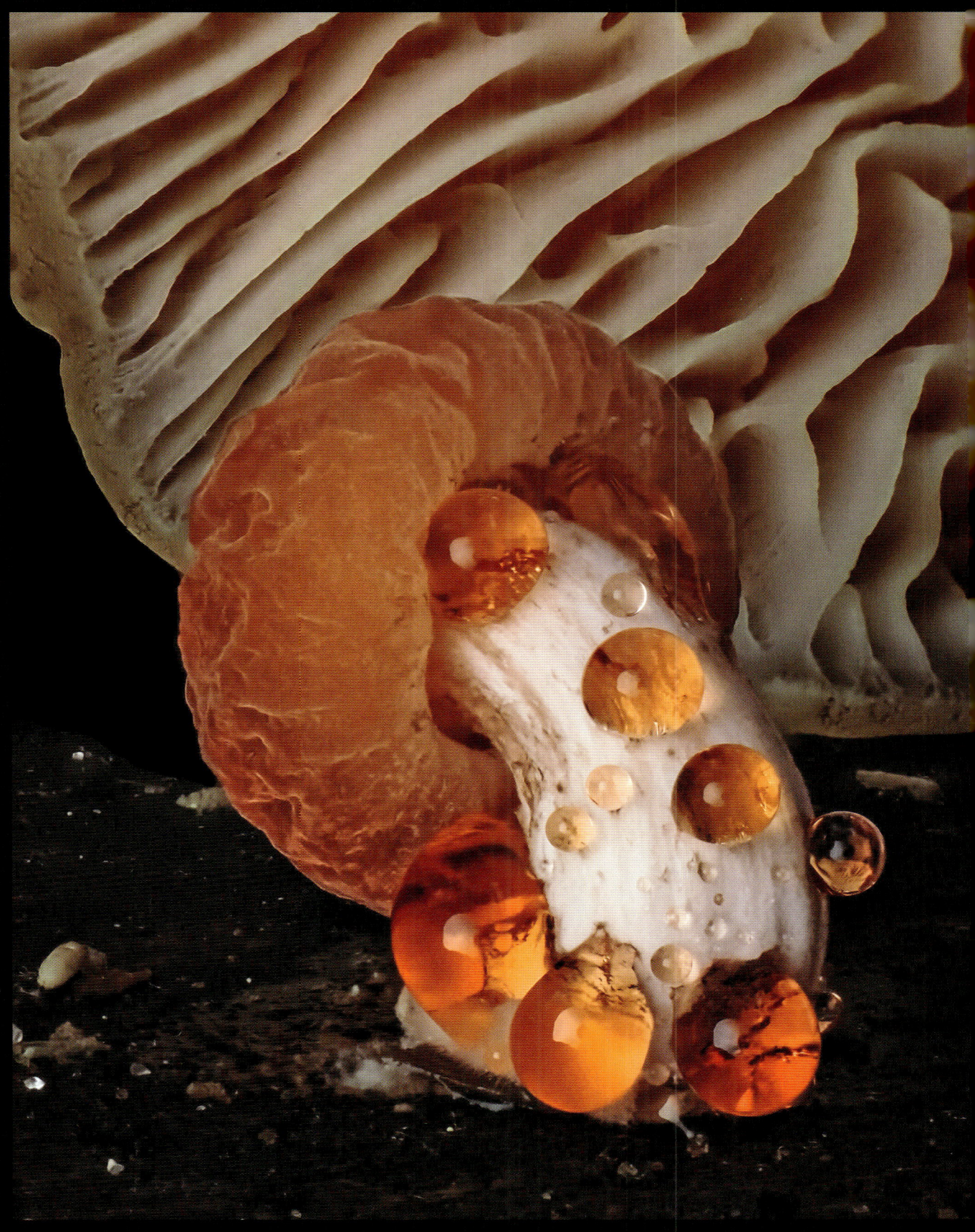

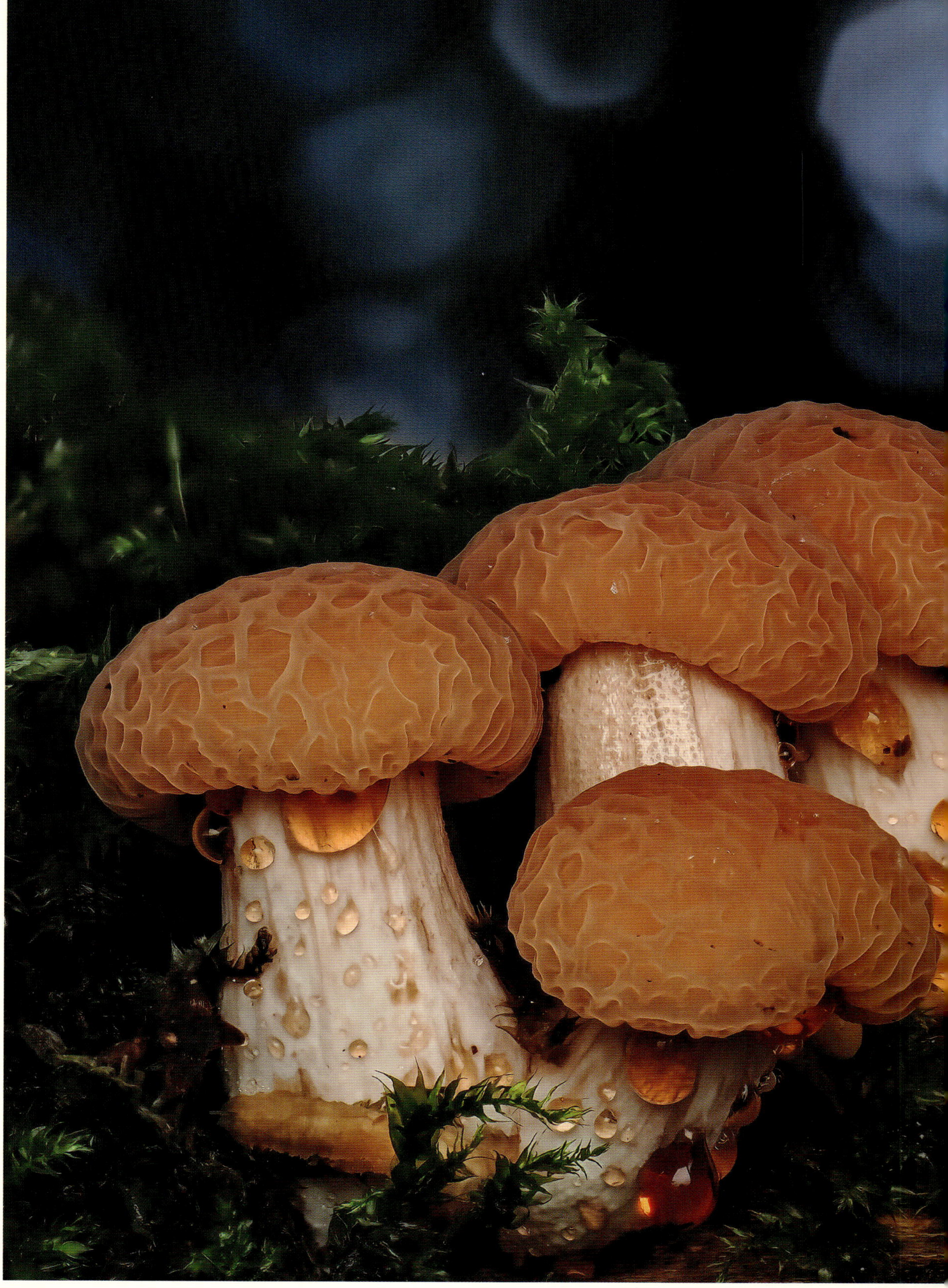

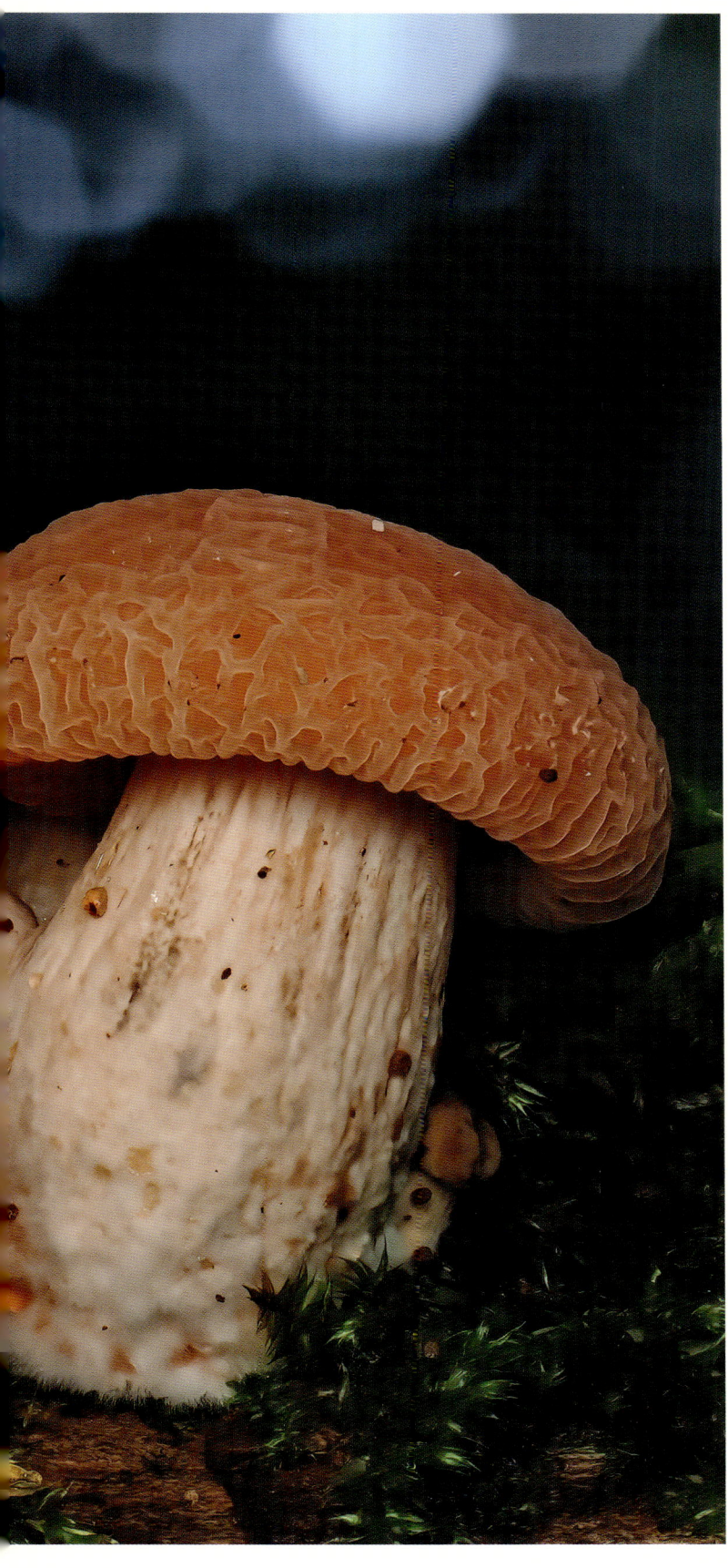

It took me a long time to find *Chlorociboria* fungi—mostly because there isn't much in my local park. However, other urban forests in London that are magical wonderlands chock-full of dead birch and beechwood have an overwhelming supply of this spectacular turquoise blue–cup fungi. They not only show up as gorgeous blue goblets but also stain the wood they grow on a delightful cerulean that makes finding them quite easy once you find an area where they are abundant.

As cup fungi, they are the ultimate water droplet holders. Unless it's been dry for ages, almost all the photos I take showcase the fungi cups bejeweled with rain and dew droplets.

I'm still out here learning, constantly mesmerized by new information (I can't say it enough: I'm not an expert on fungi. I just have a respect and reverence for the fungal kingdom and that unquestionably lends itself to how I approach photographing them). I try to be careful, respectful, and protective of the habitats. And while I don't know how many macrophotographers are out there wandering around the urban woodlands of London looking for fungi (because I never see any of you; where are you?!), since I'm not picking the mushrooms I photograph, you'll hopefully have a greater chance of photographing them as well! You are welcome!

25

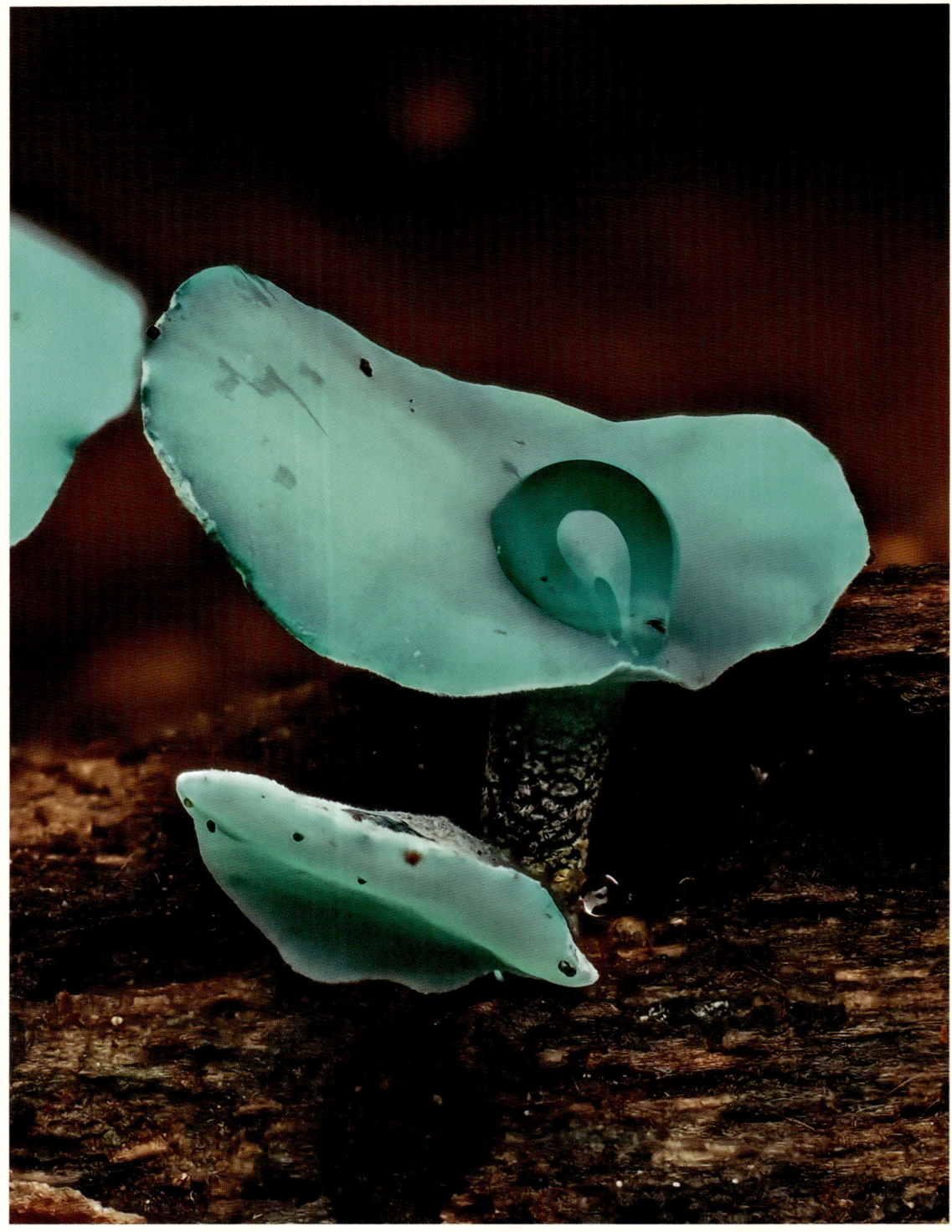

THIS SPREAD AND FOLLOWING SPREAD: GREEN ELFCUPS (*CHLOROCIBORIA* SPP.).

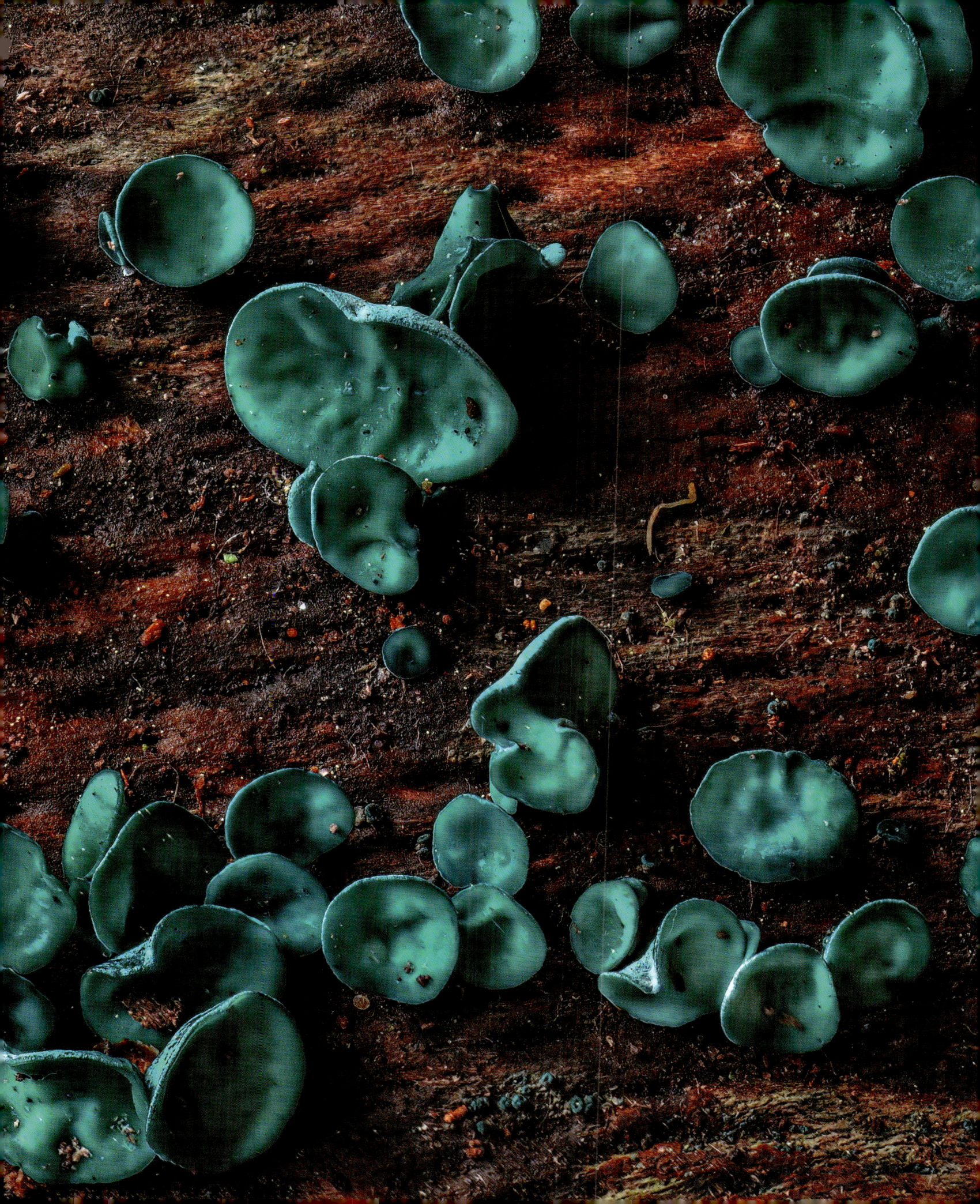

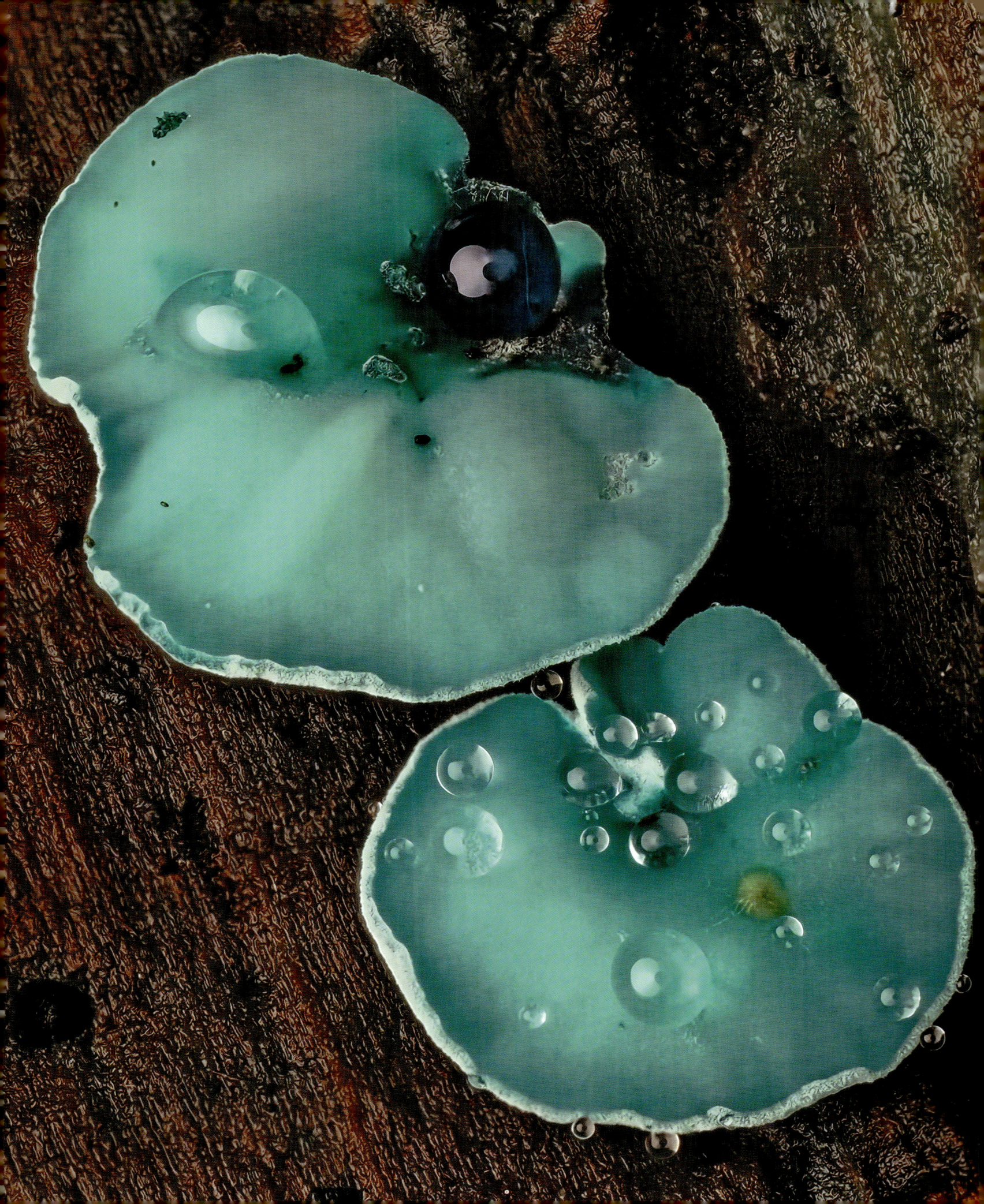

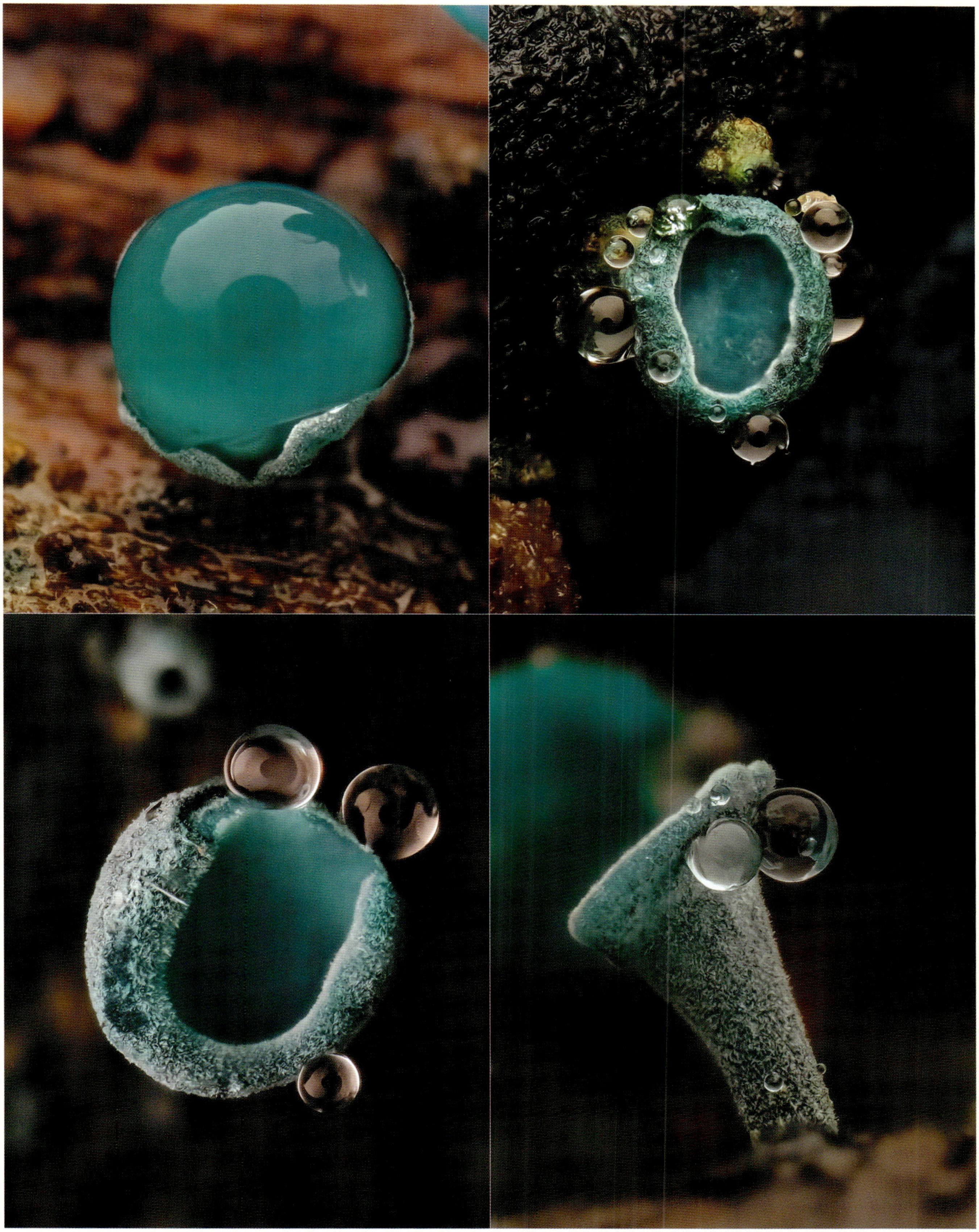

30

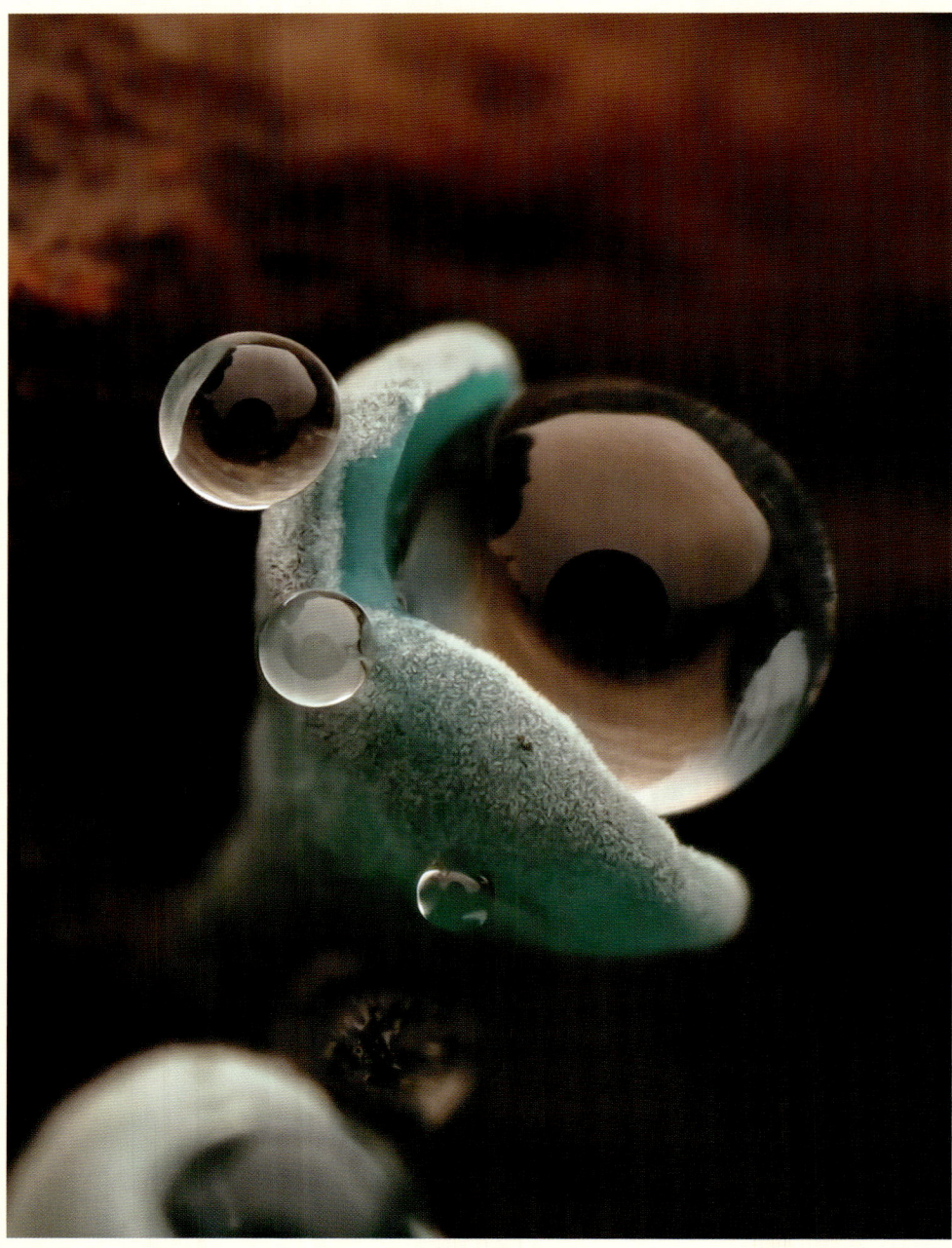

THIS SPREAD AND FOLLOWING SPREAD: GREEN ELFCUPS CATCHING WATER DROPLETS.

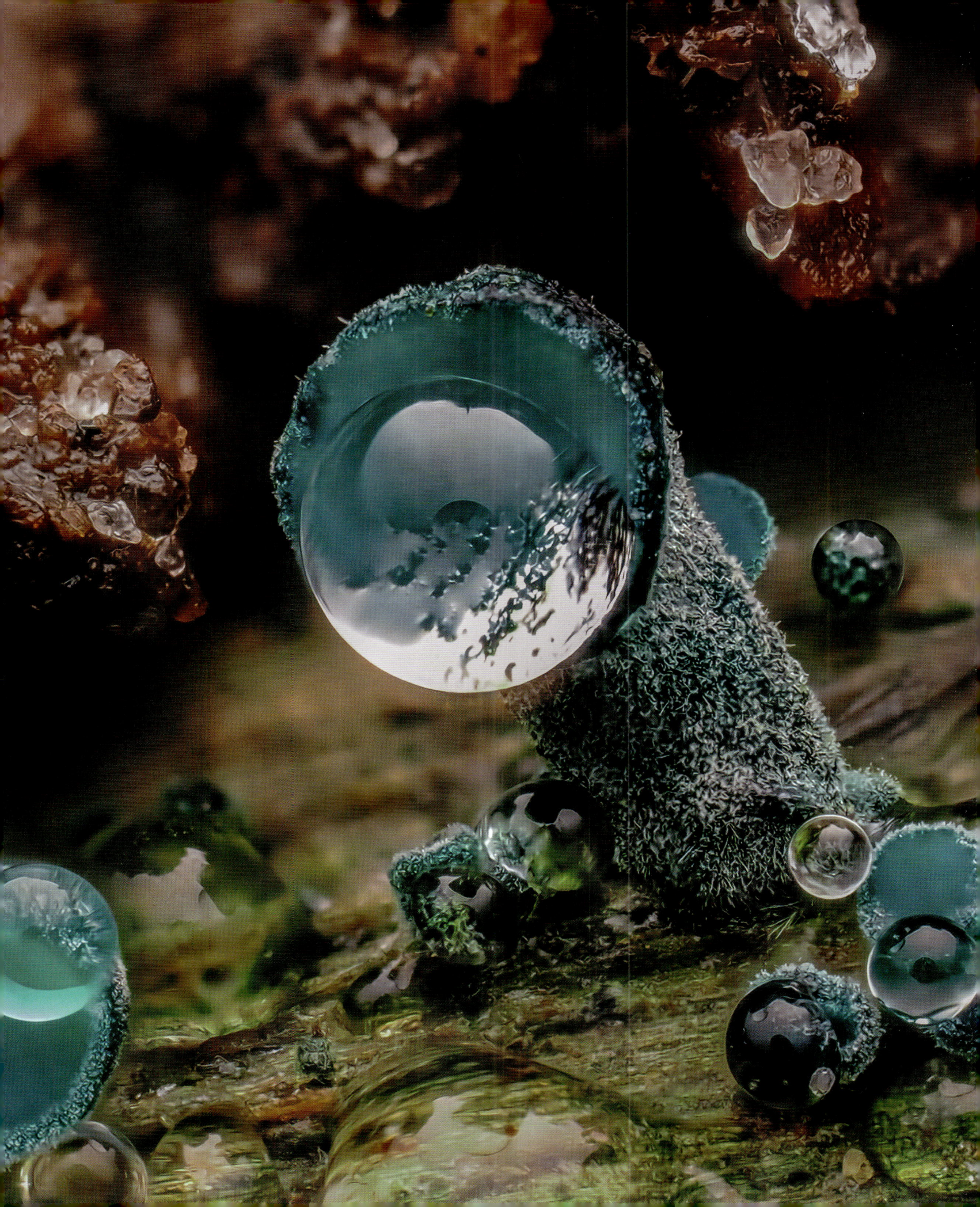

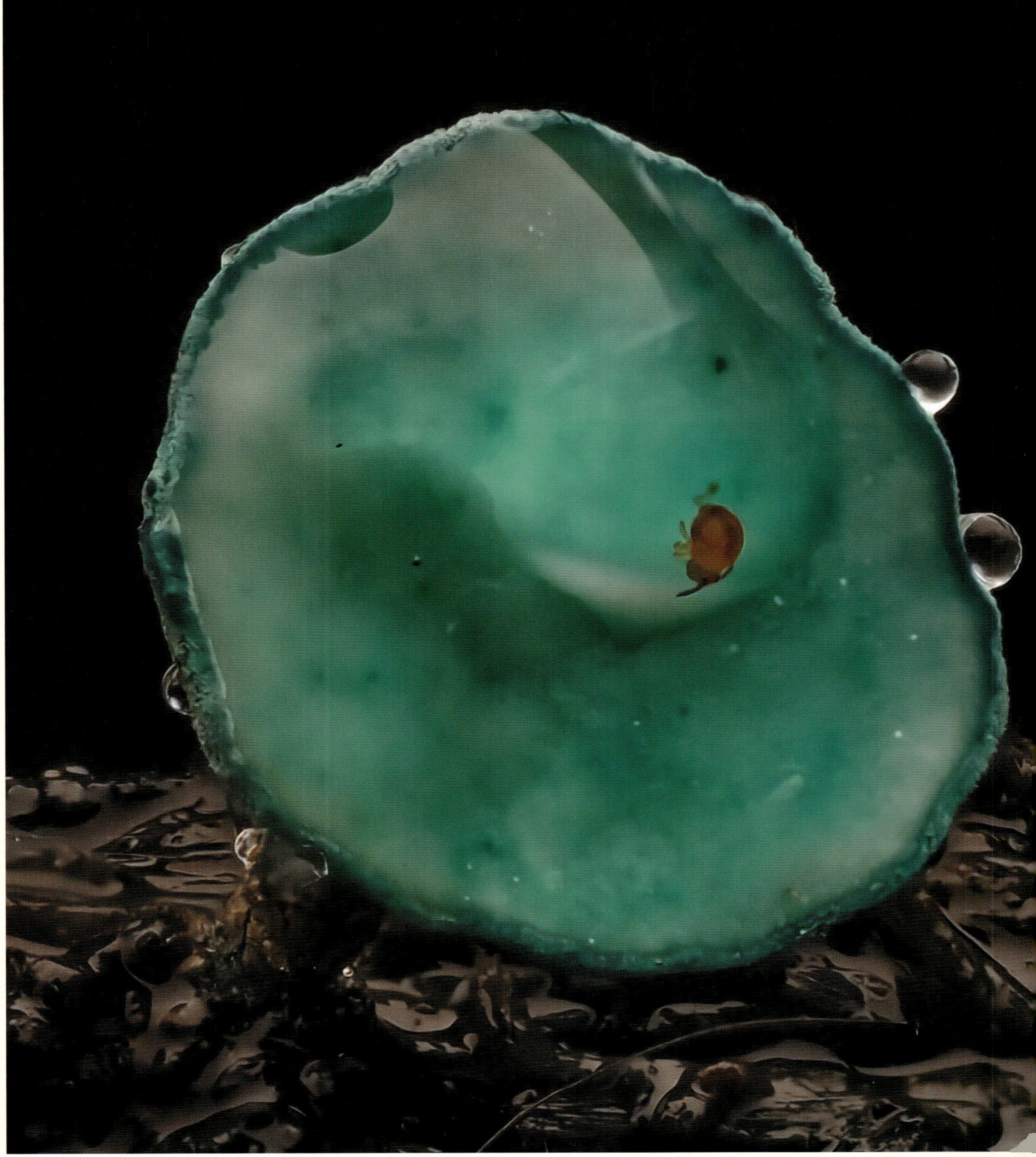

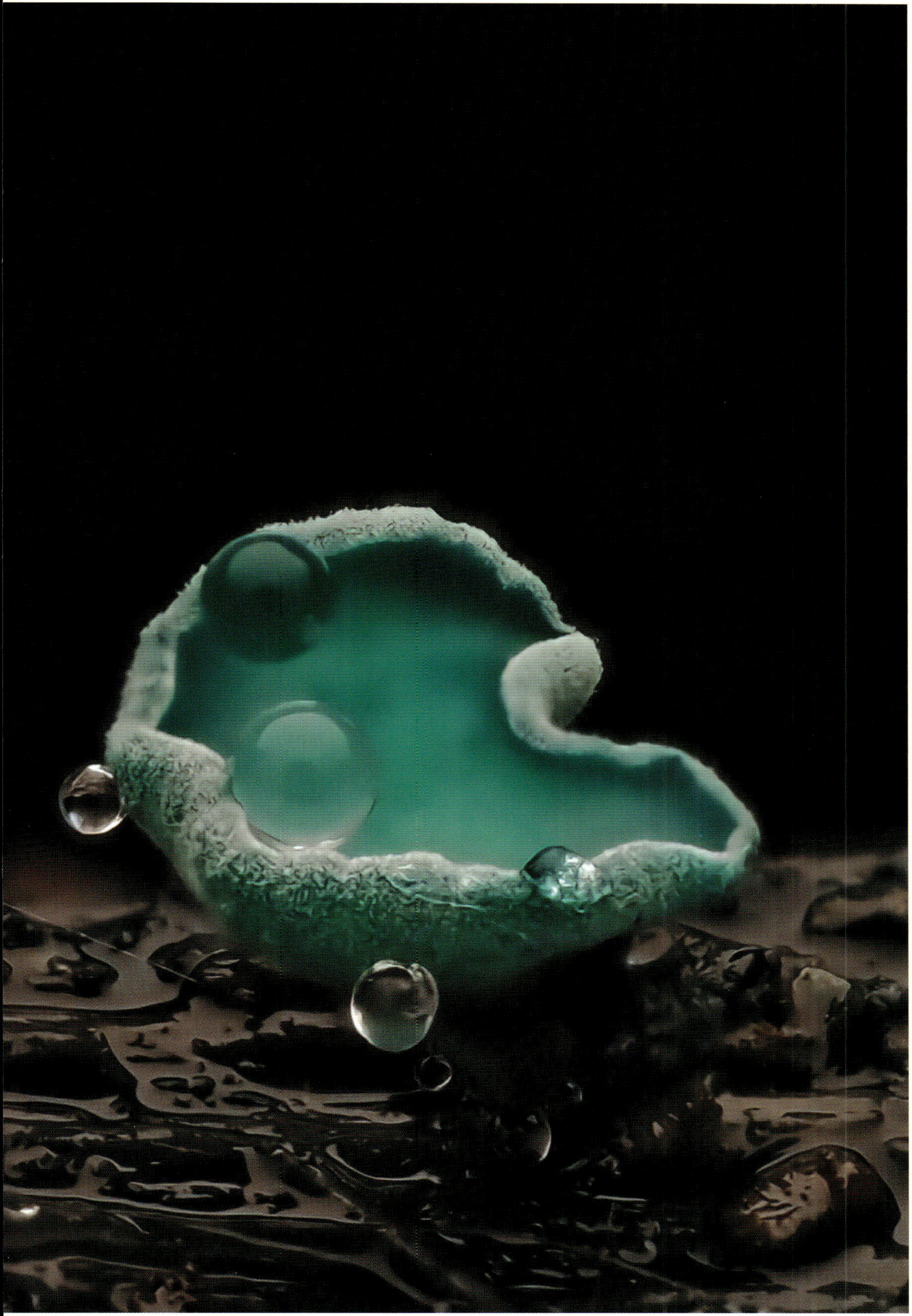

33

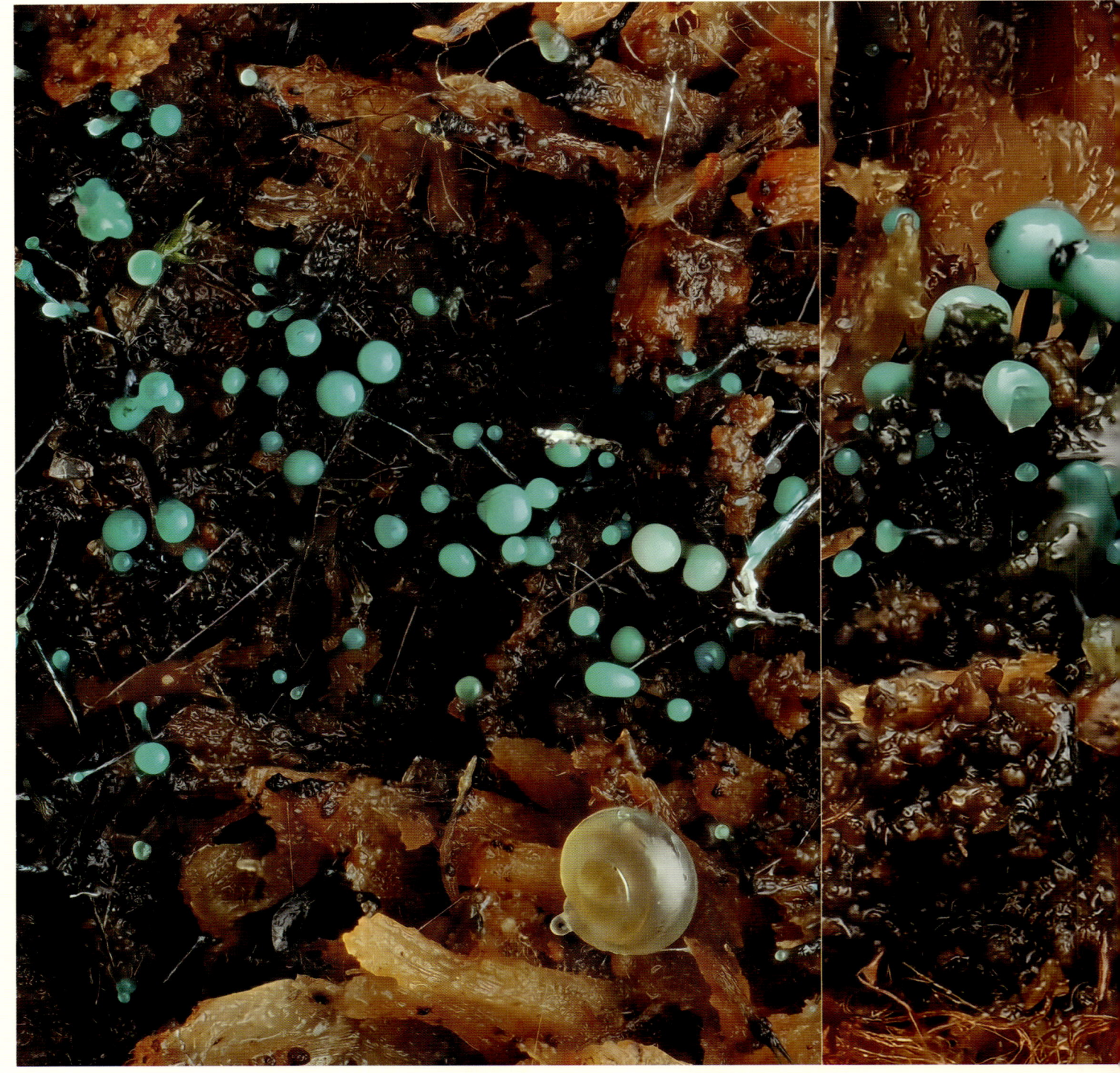

34

ABOVE AND OPPOSITE: TEAL CONIFER-PIN (*DENDROSTILBELLA SMARAGDINA*).

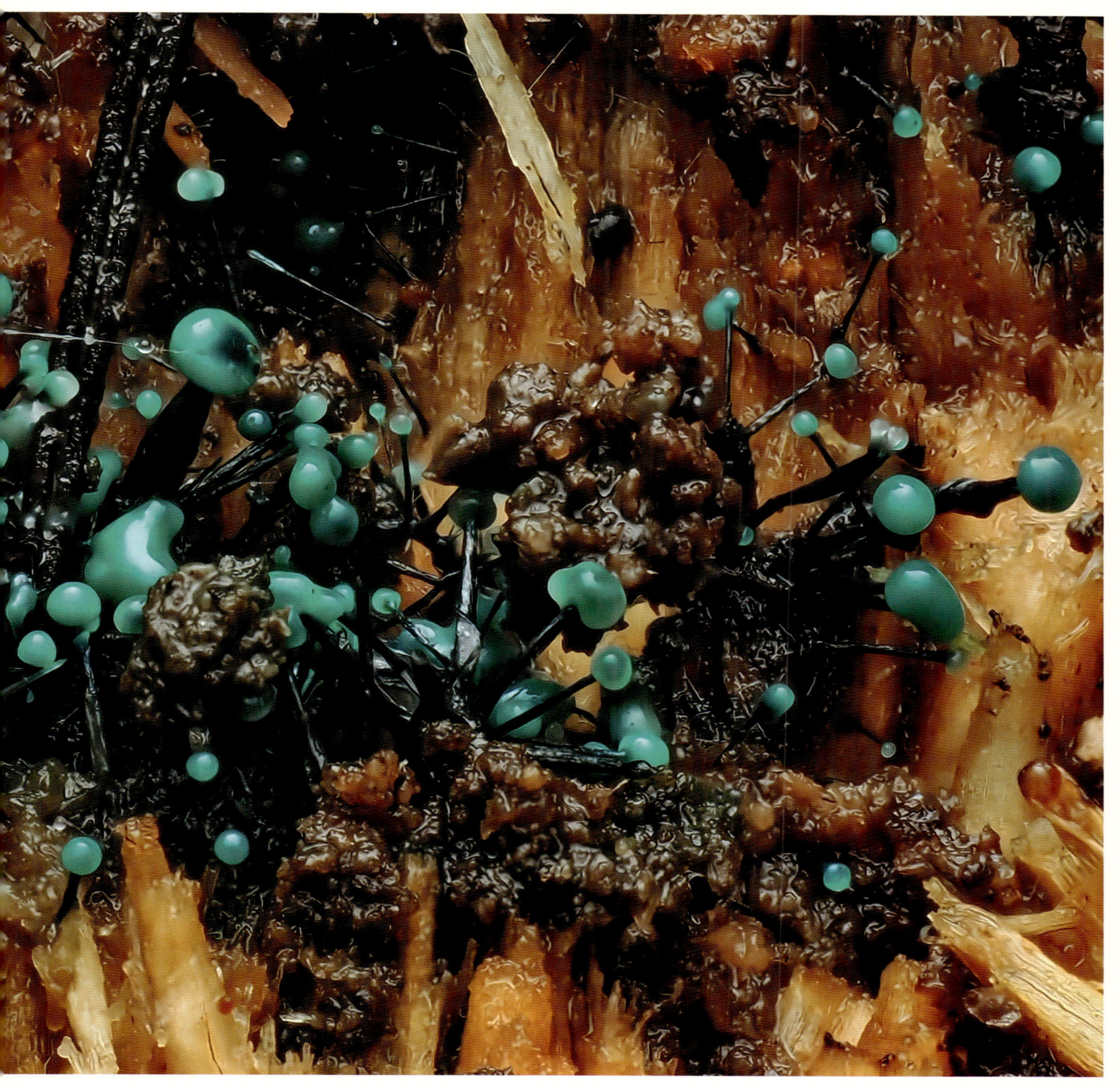

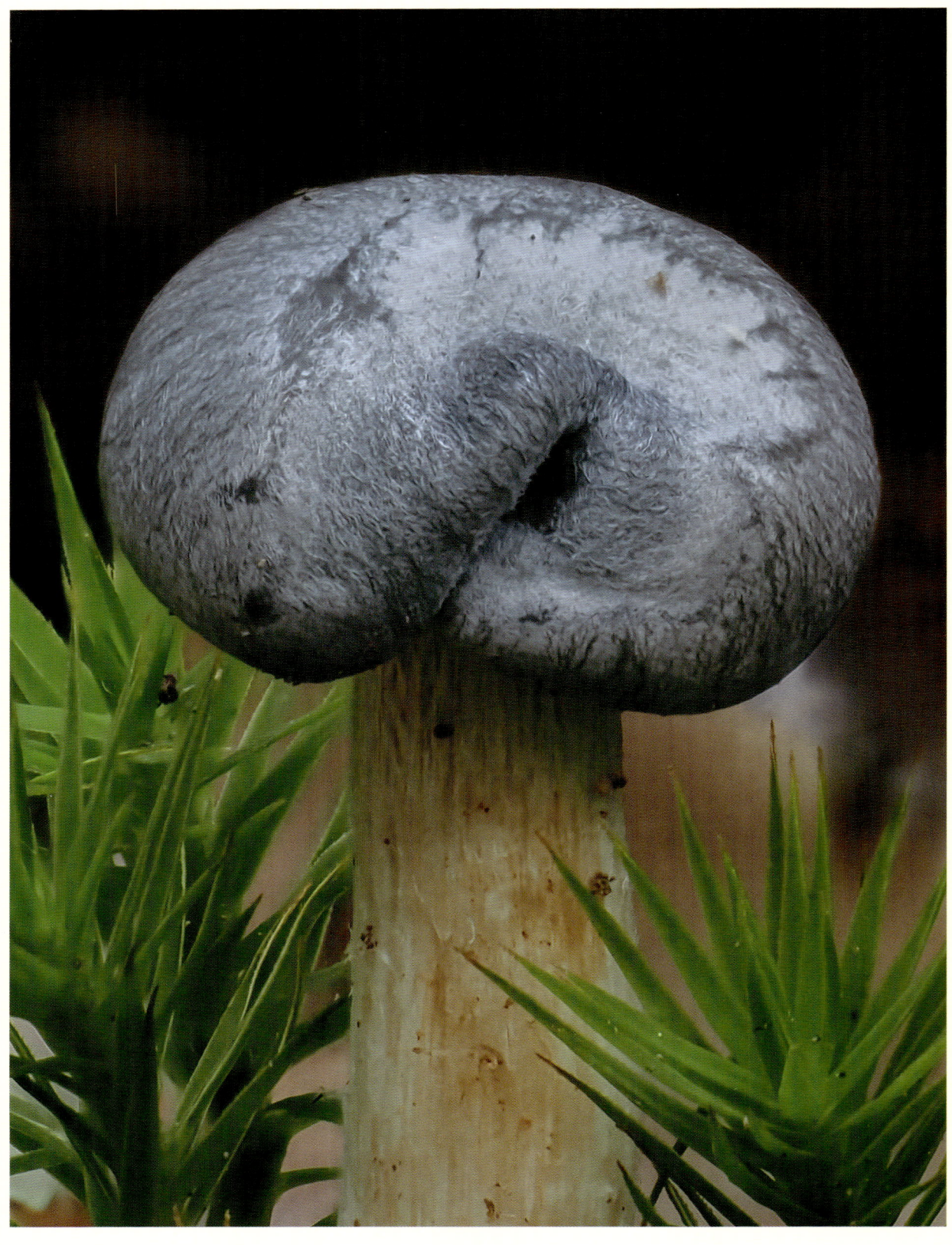

ABOVE AND OPPOSITE: ANISEED FUNNEL (*COLLYBIA ODORA*, PREVIOUSLY NAMED *CLITOCYBE ODORA*).

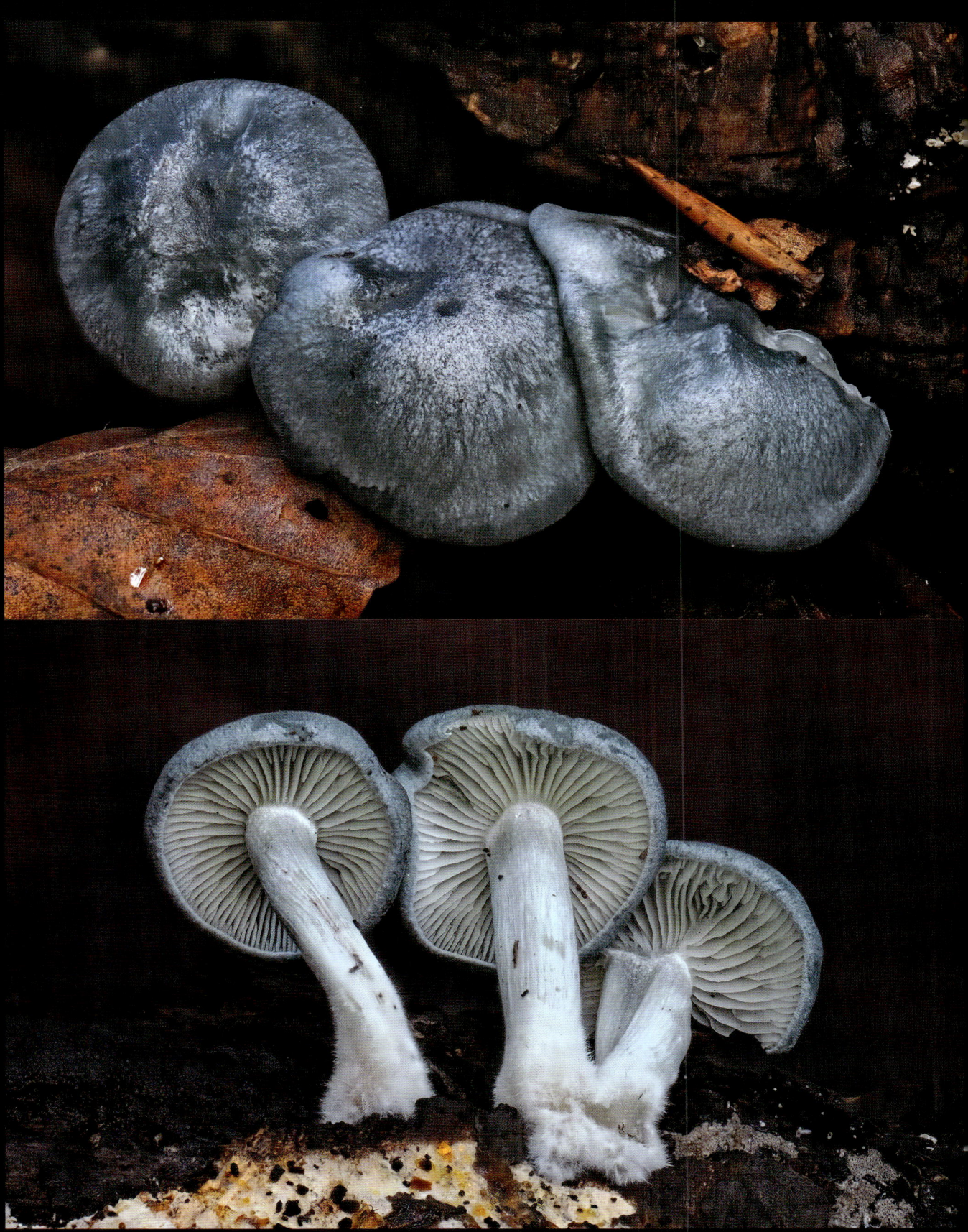

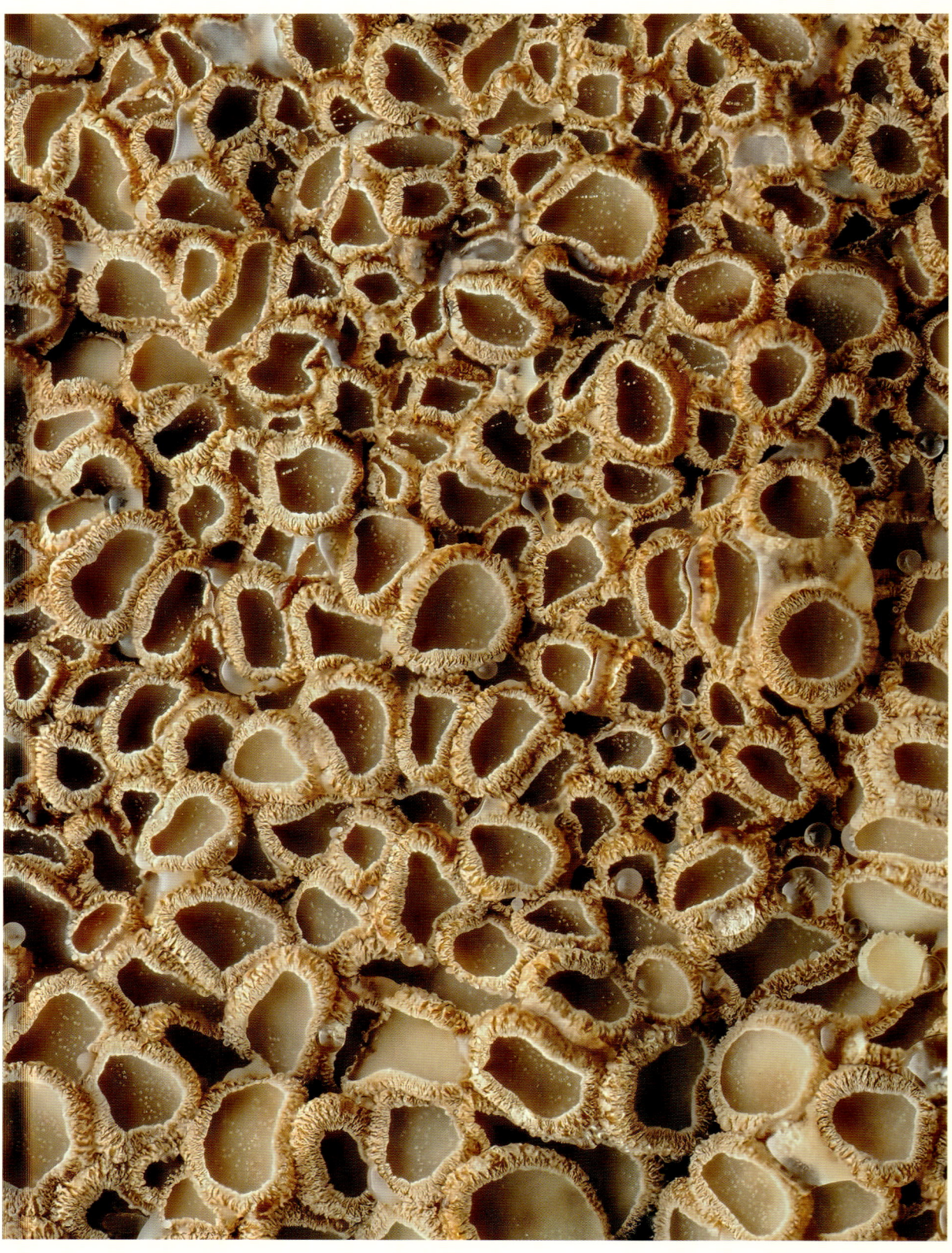

38

ABOVE: CROWDED CUPLET (*MERISMODES FASCICULATA*). OPPOSITE: LEMON DISCOS (*CALYCINA CITRINA*).

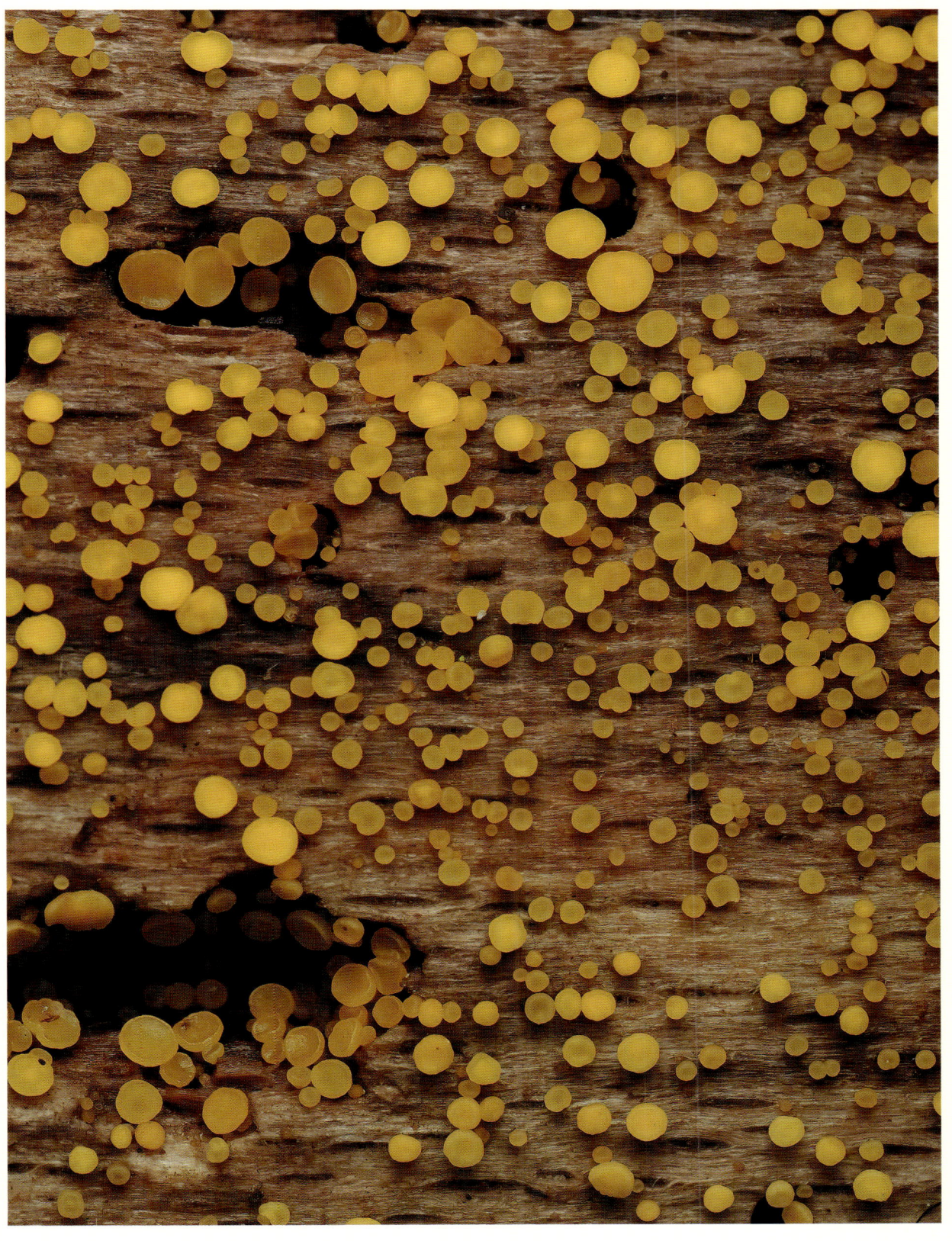

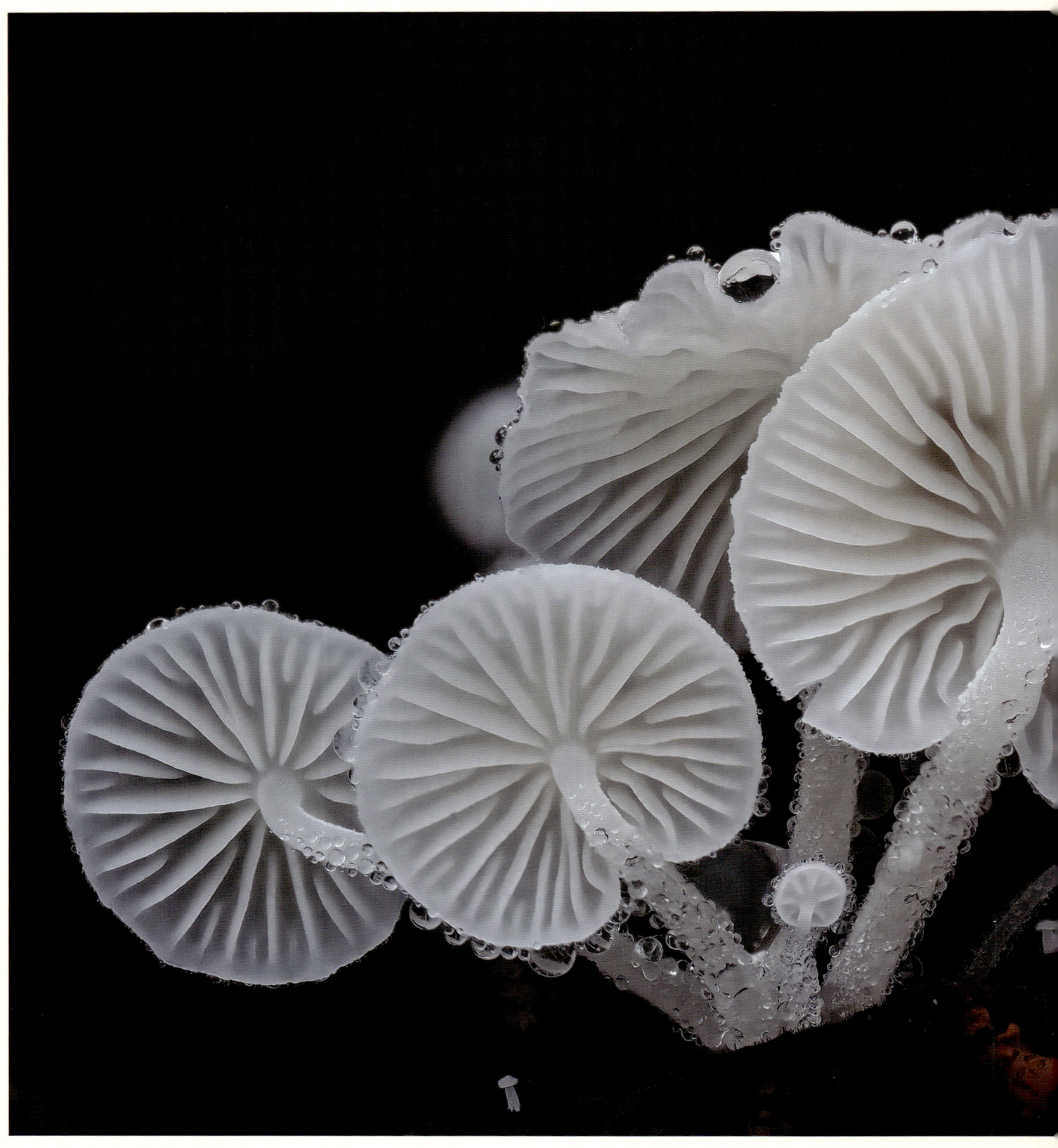

ABOVE: DEWDROP BONNET (*HEMIMYCENA TORTUOSA*).

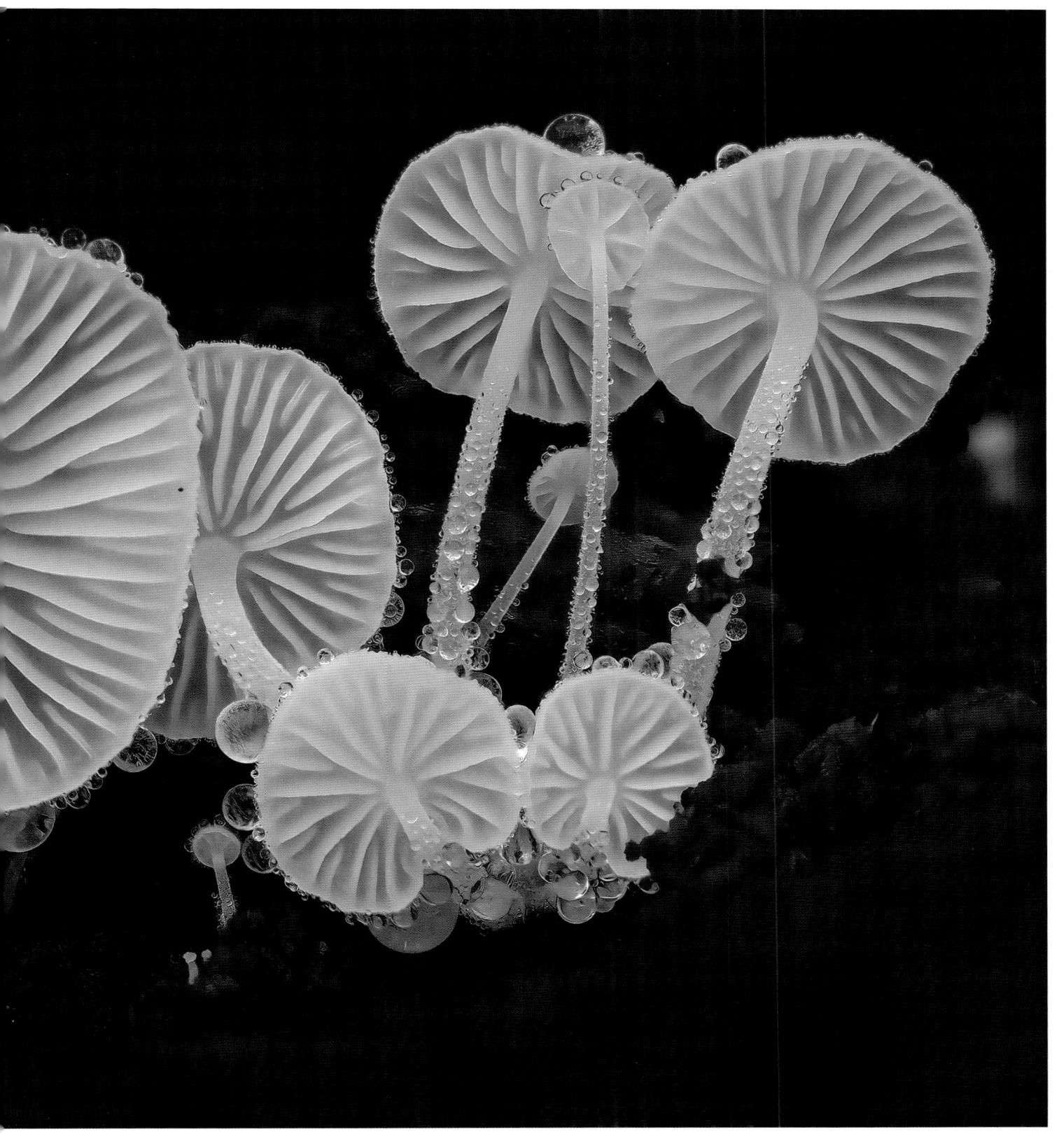

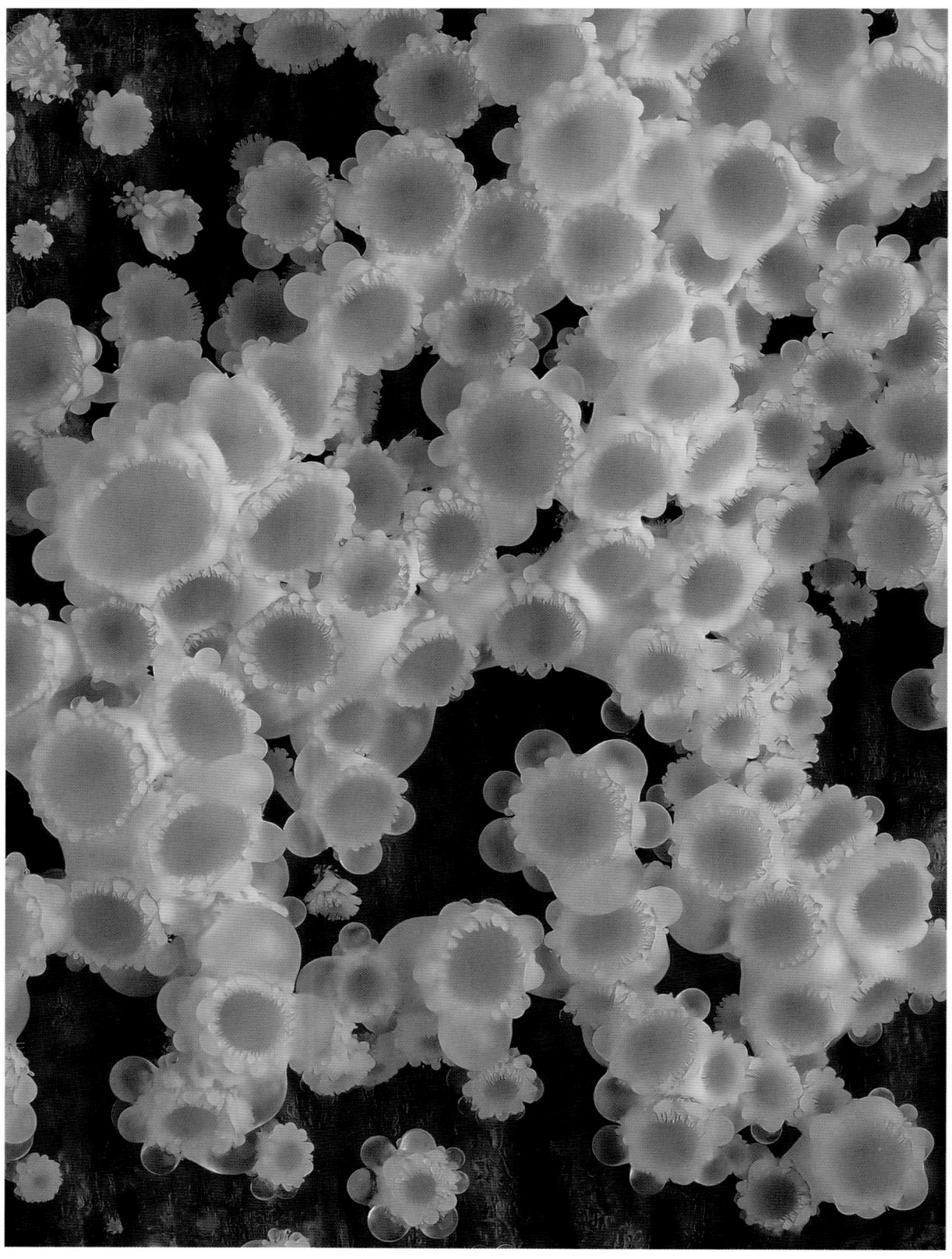

ABOVE: SNOWY DISCO (*LACHNUM VIRGINEUM*). OPPOSITE: PIN MOLD (*MUCORALES* SP.).

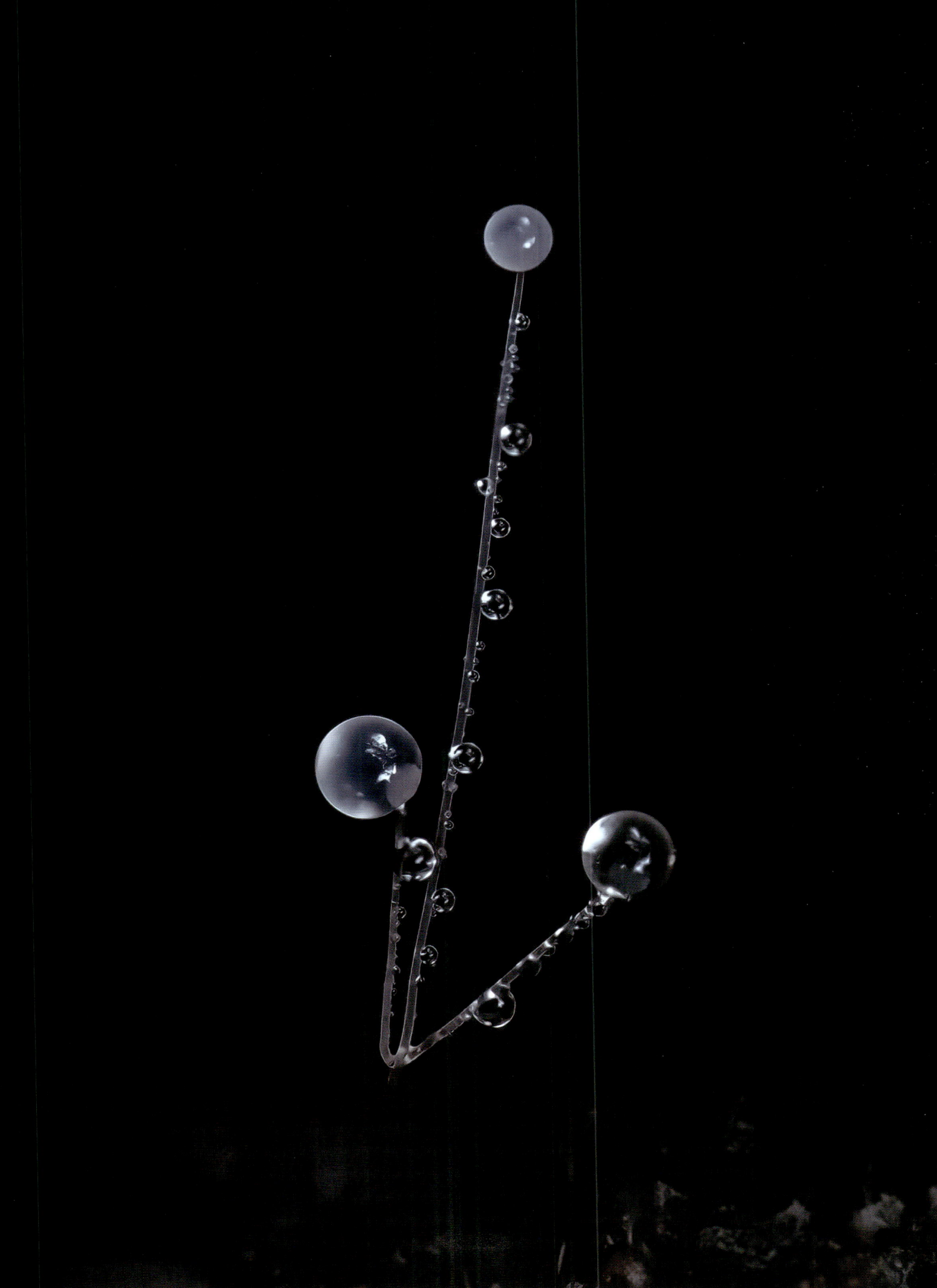

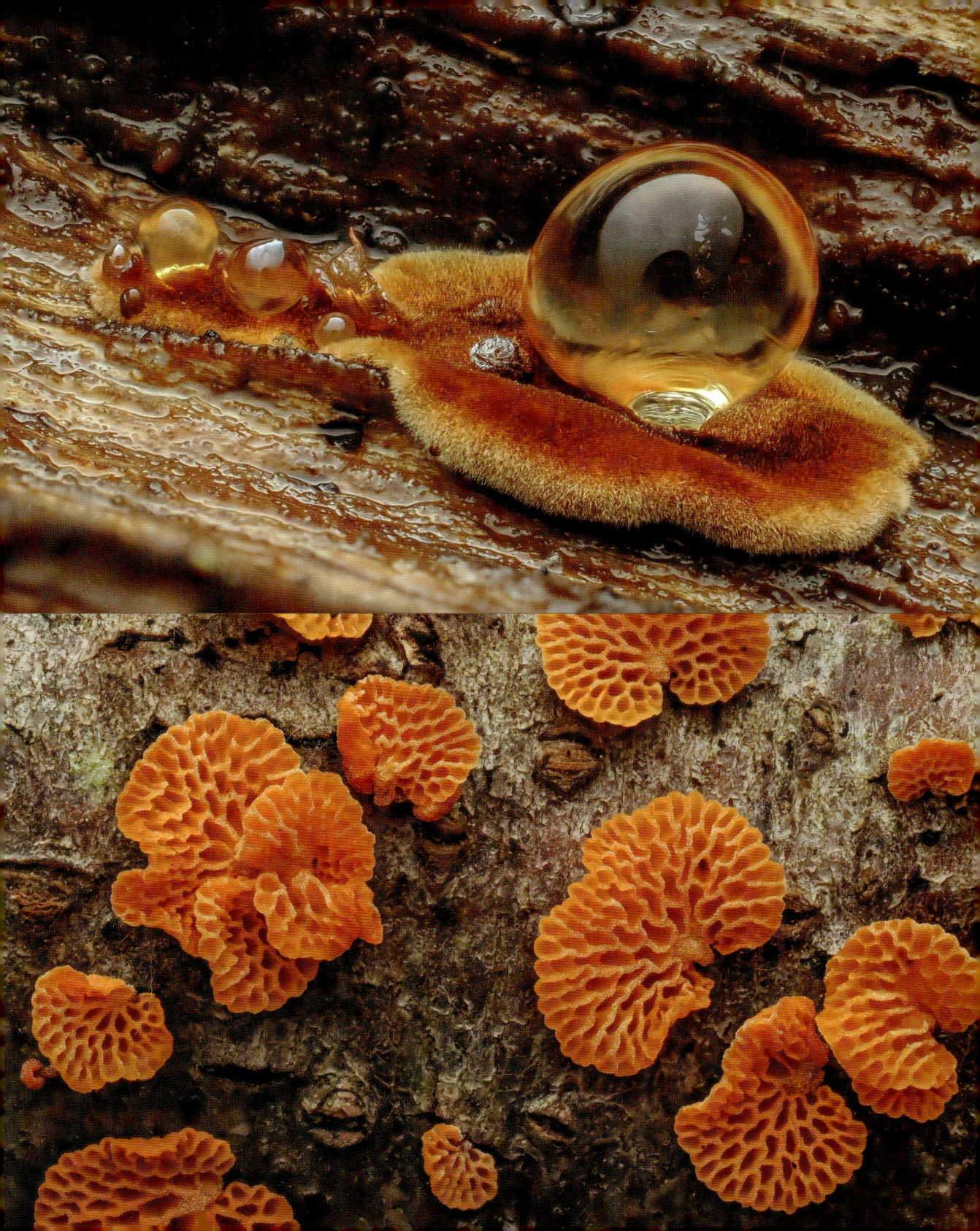

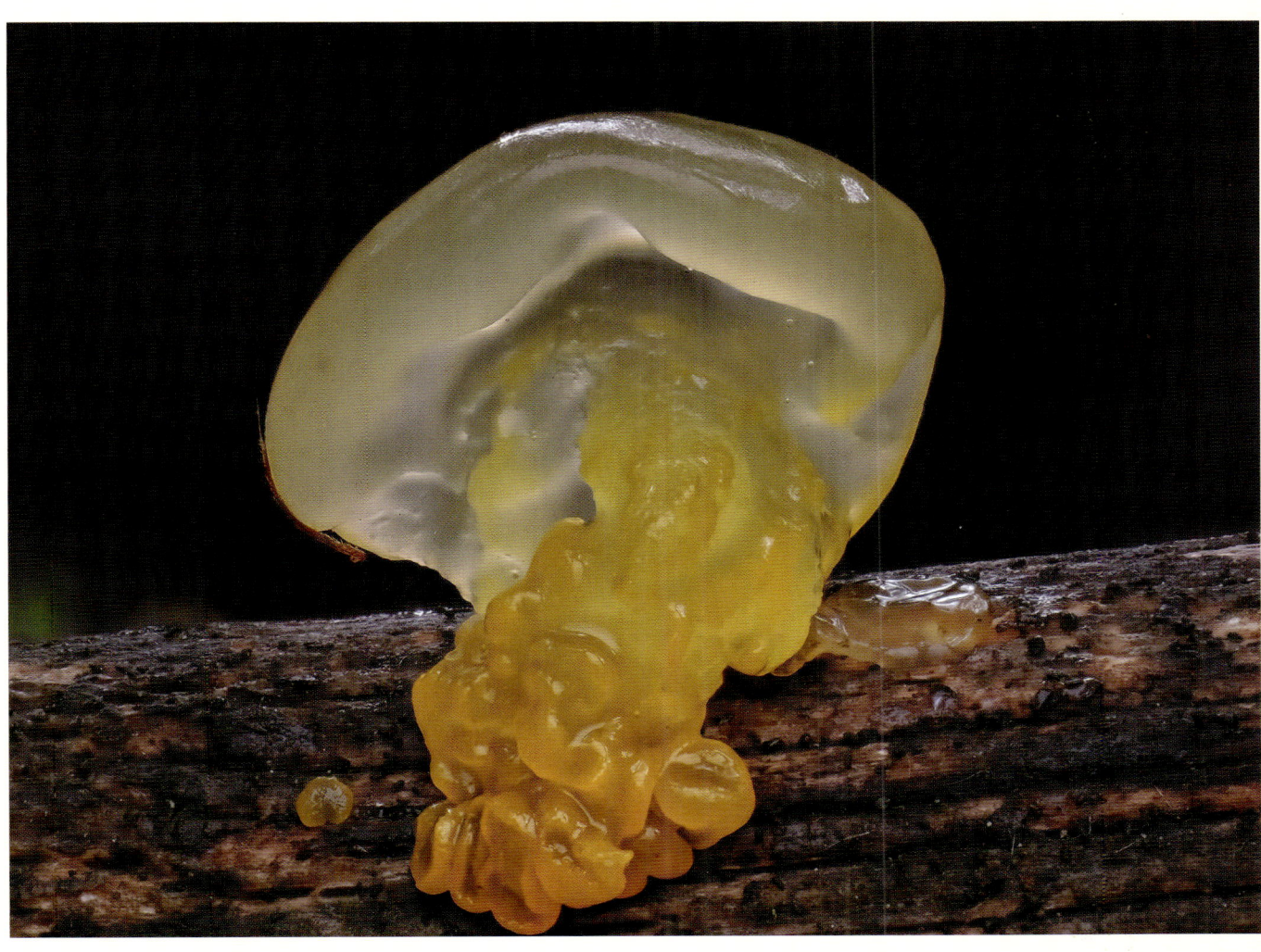

ABOVE: WITCH'S BUTTER (*TREMELLA MESENTERICA*). OPPOSITE ABOVE: CRUST FUNGI SPP. WITH GUTTATION.
OPPOSITE BELOW: ORANGE PING-PONG BAT (*FAVOLASCHIA CALOCERA*).

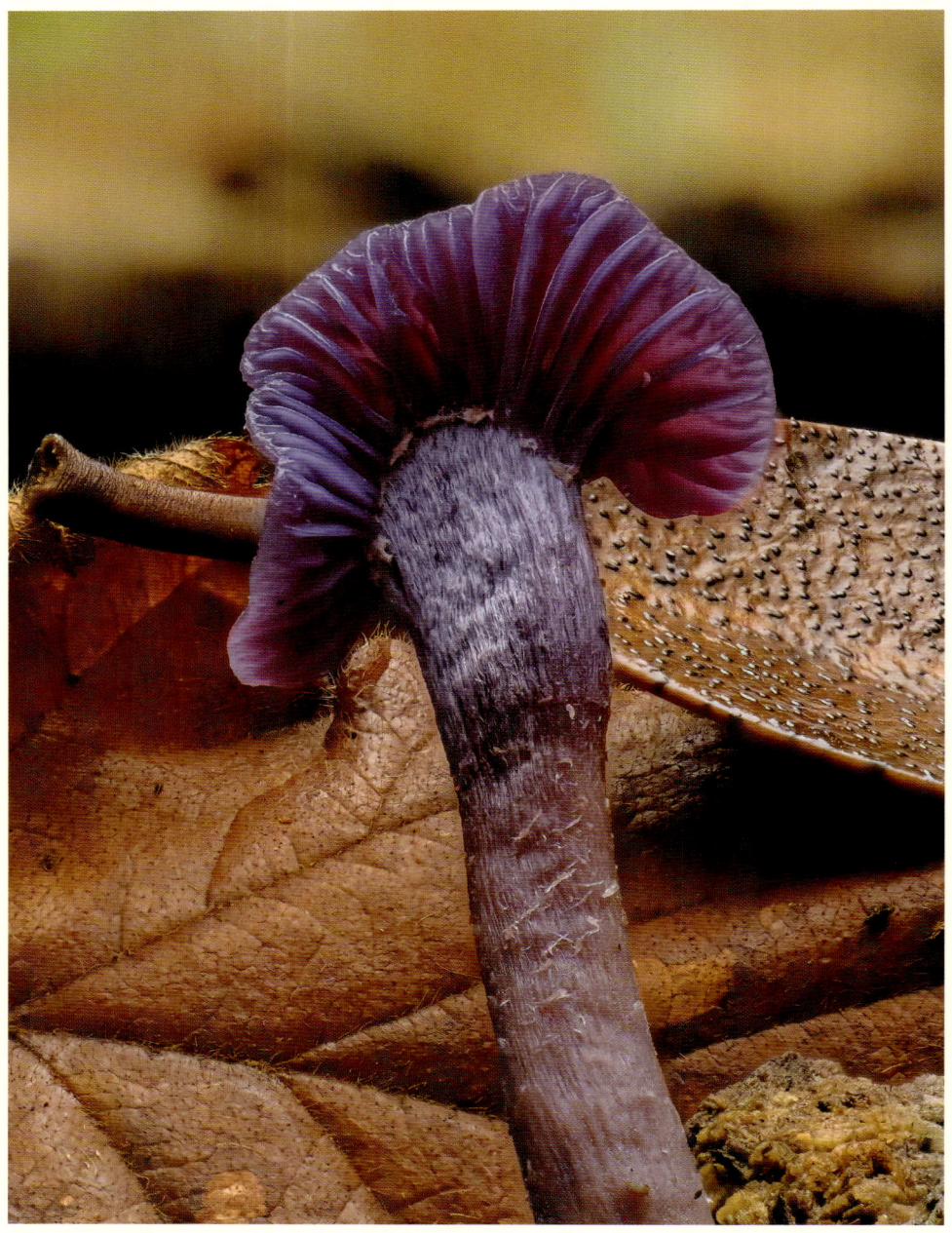

THIS SPREAD AND FOLLOWING SPREAD: AMETHYST DECEIVER (*LACCARIA AMETHYSTINA*).

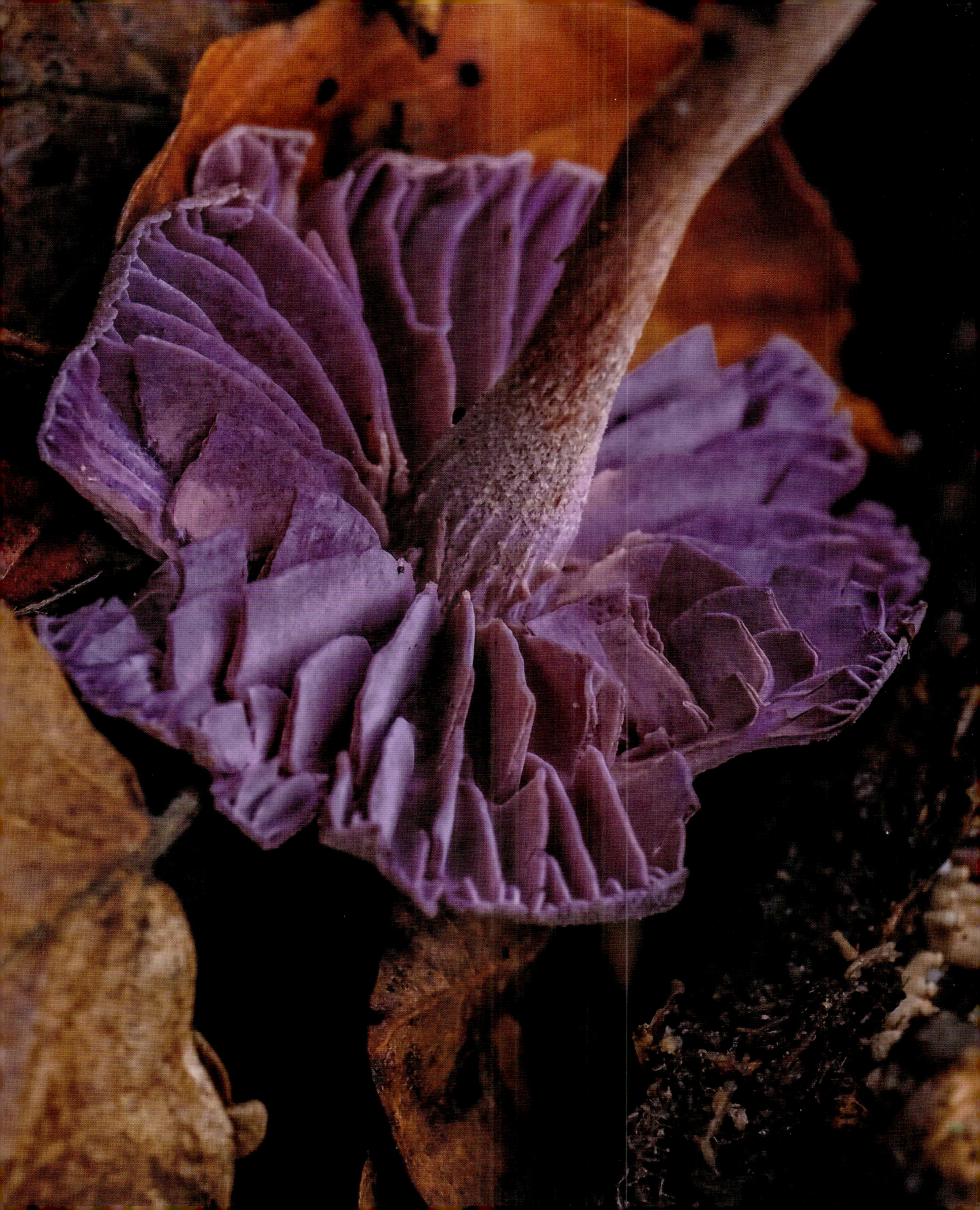

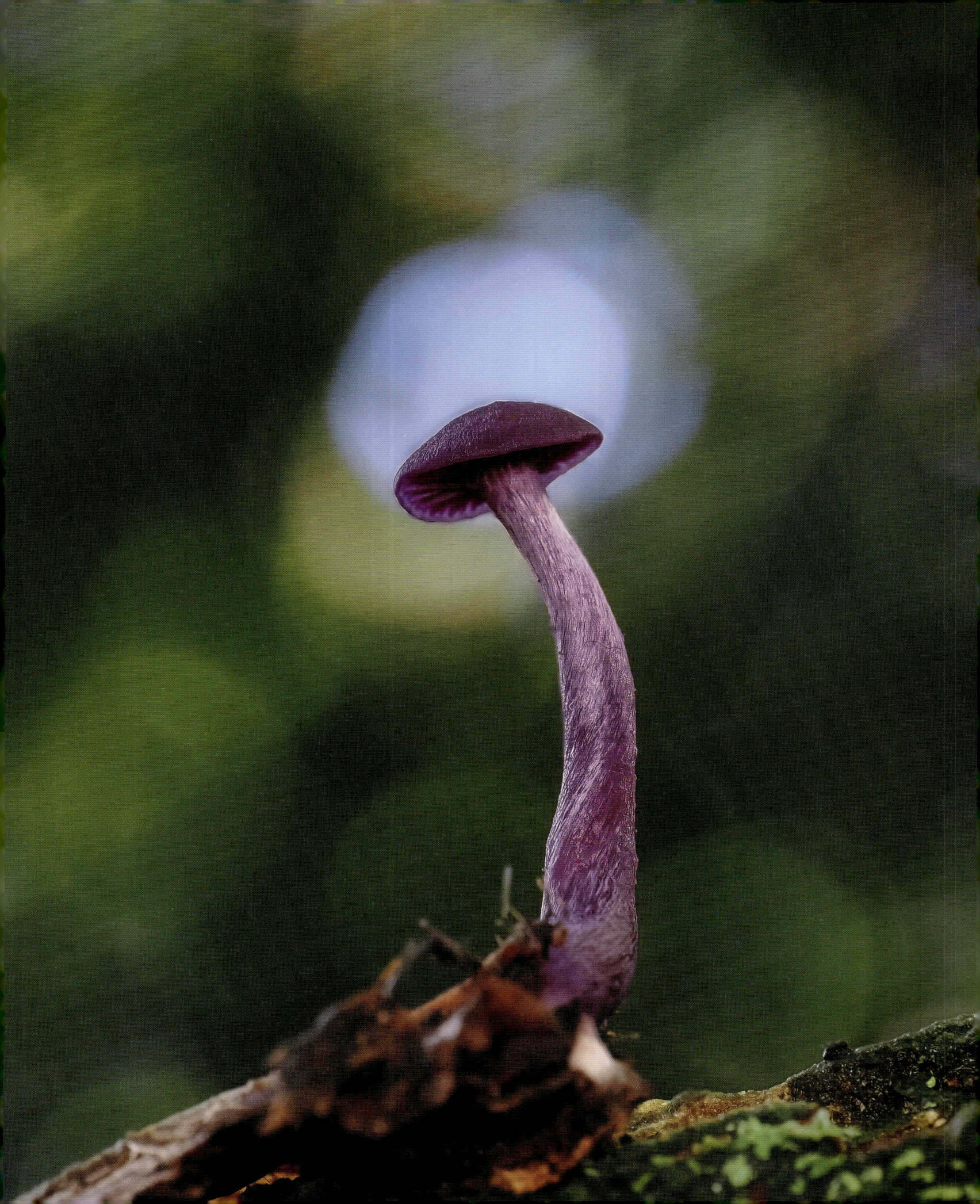

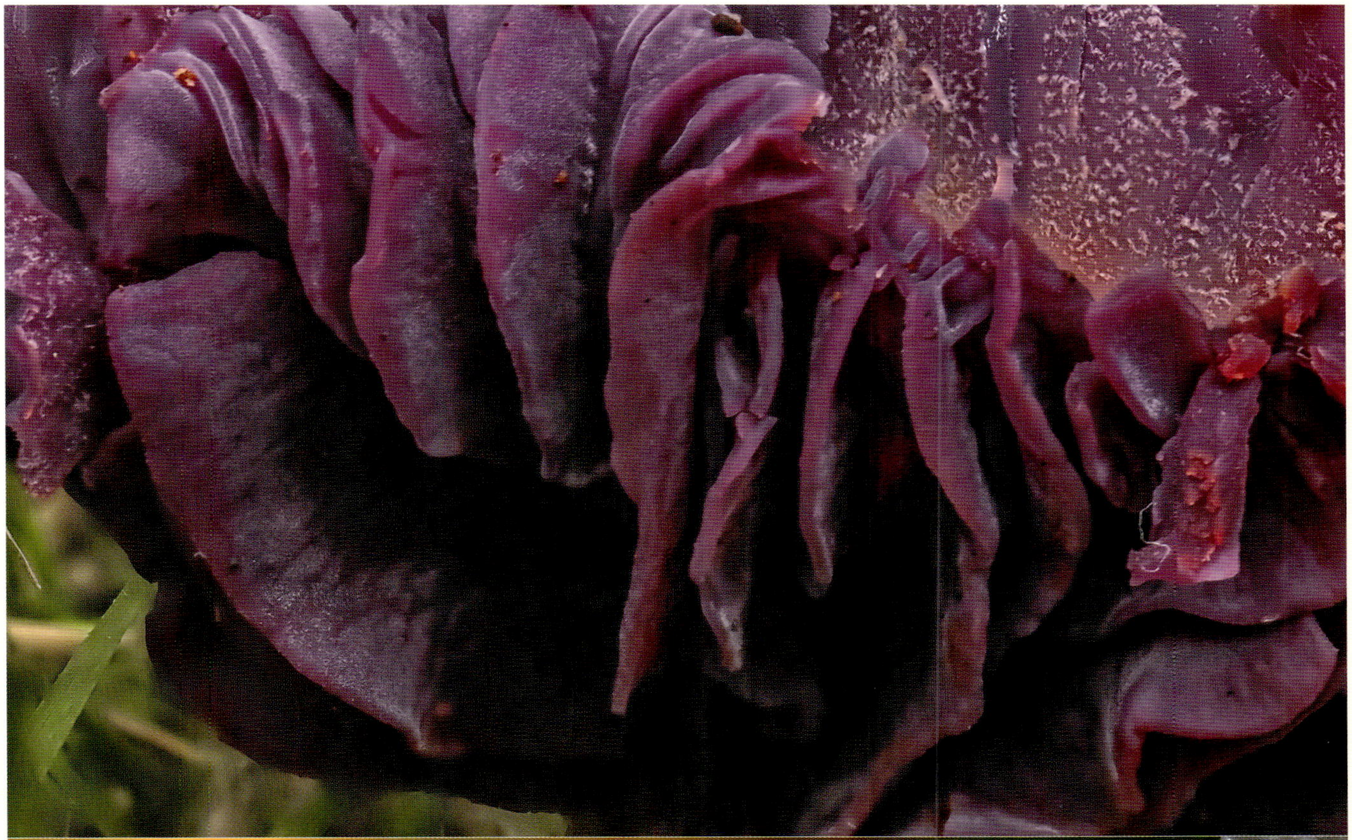

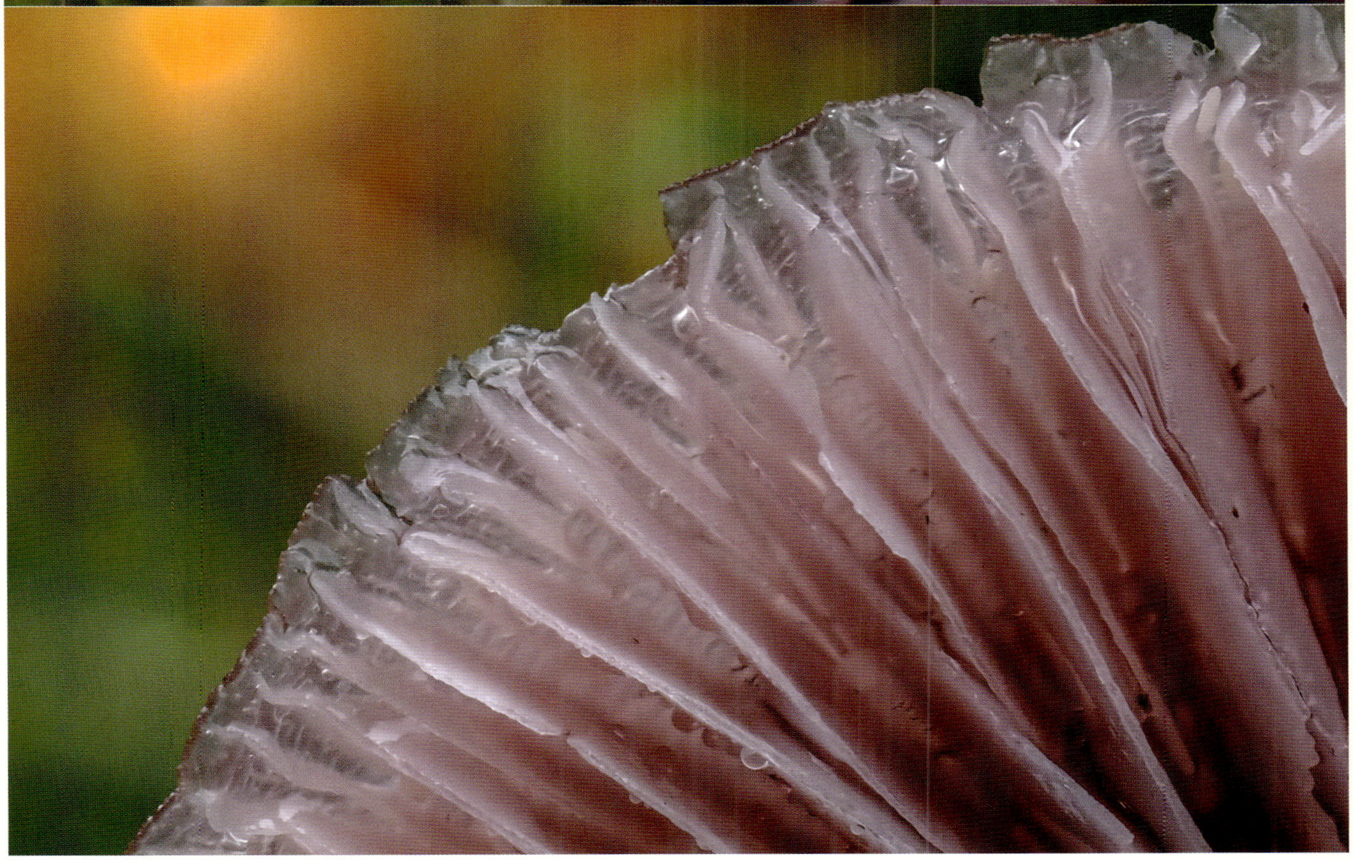

49

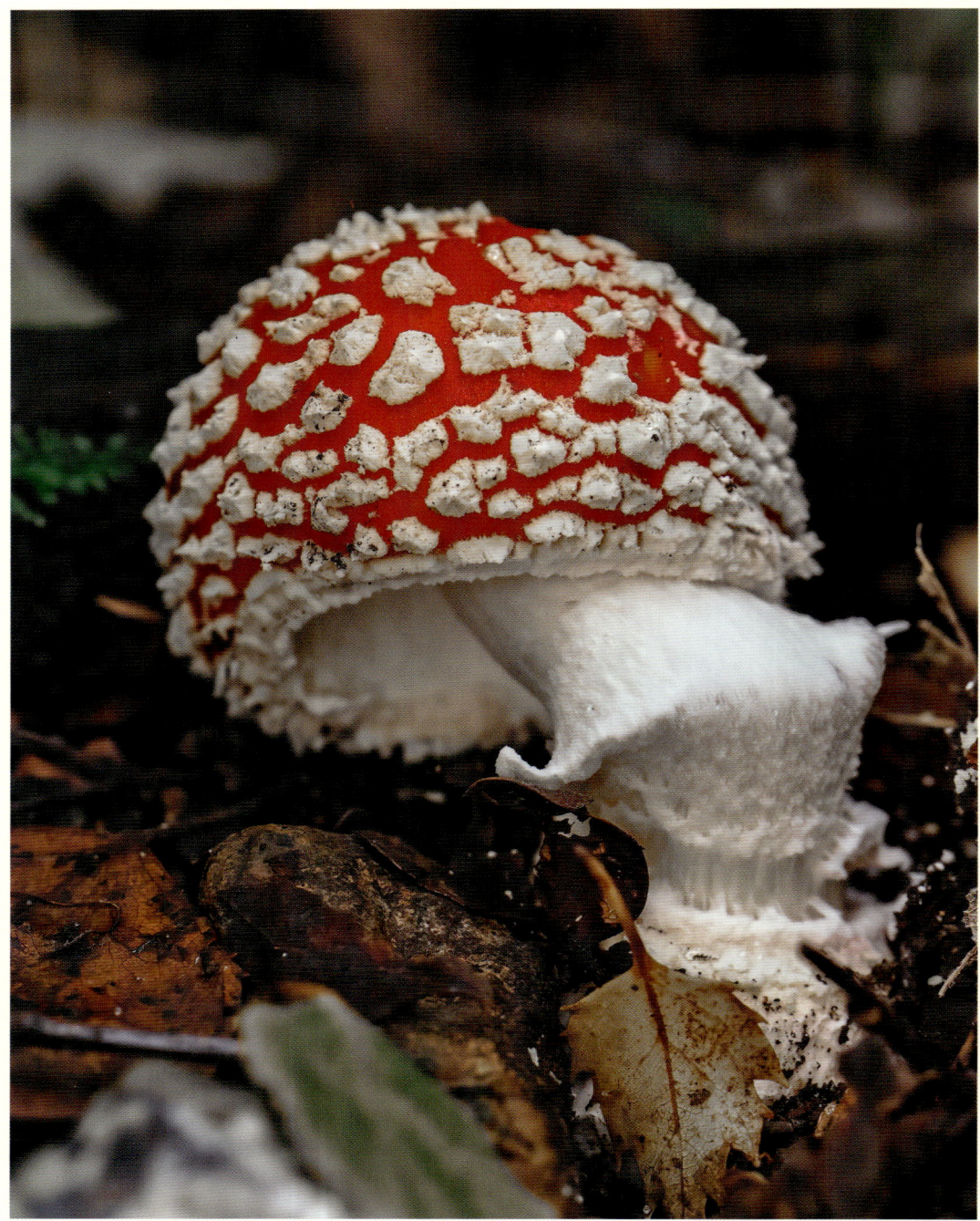

THIS SPREAD AND FOLLOWING SPREAD: FLY AGARIC (*AMANITA MUSCARIA*).

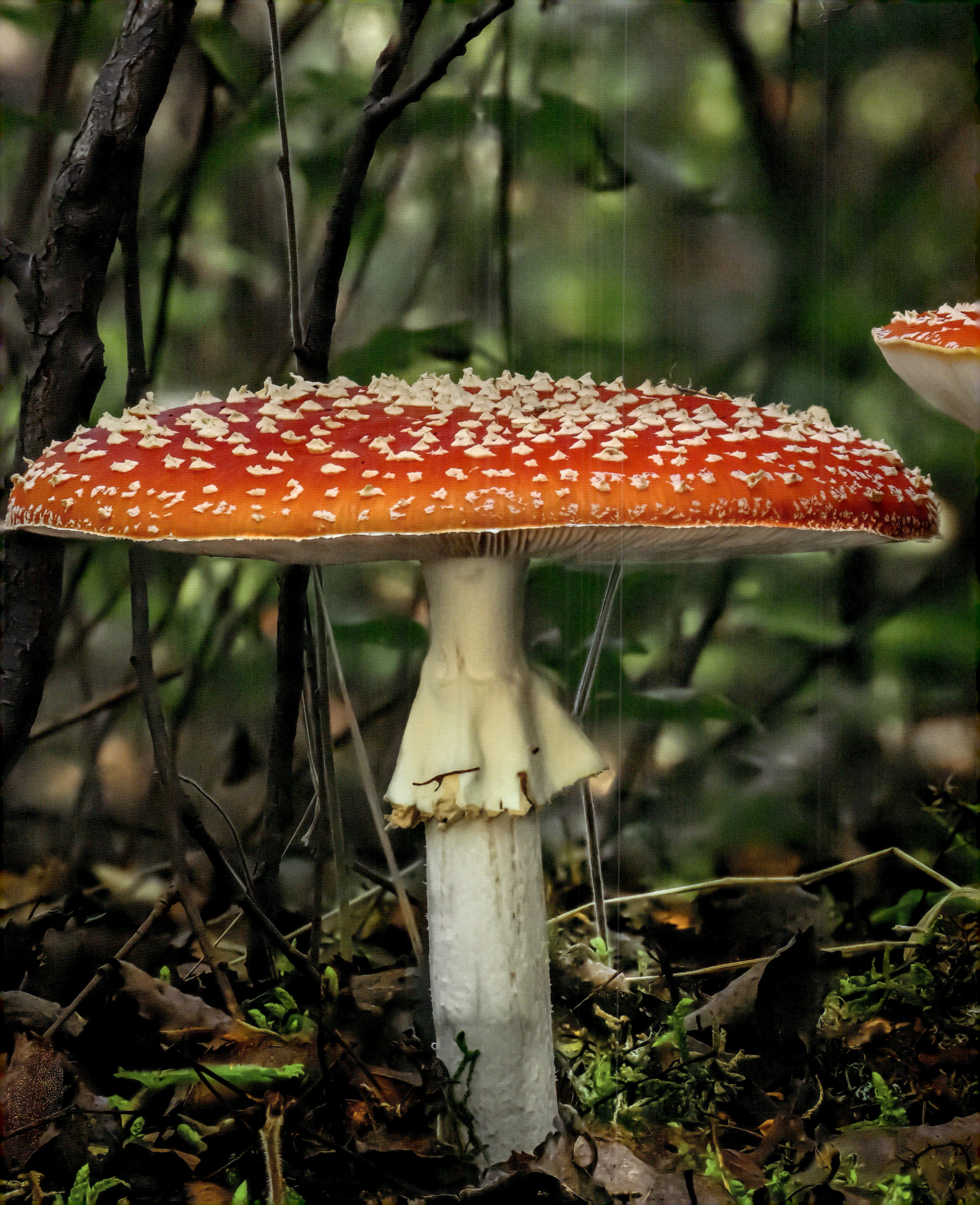

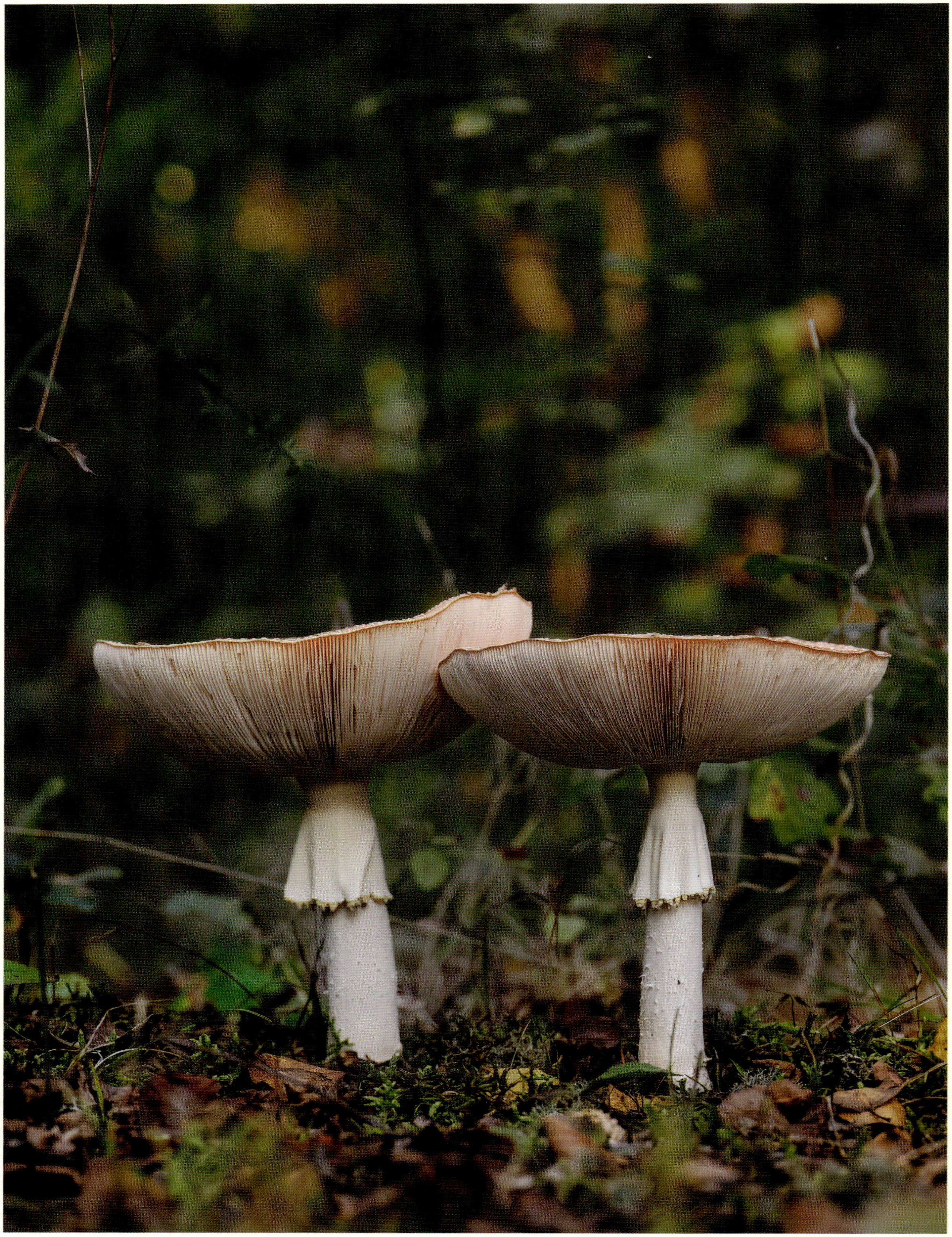

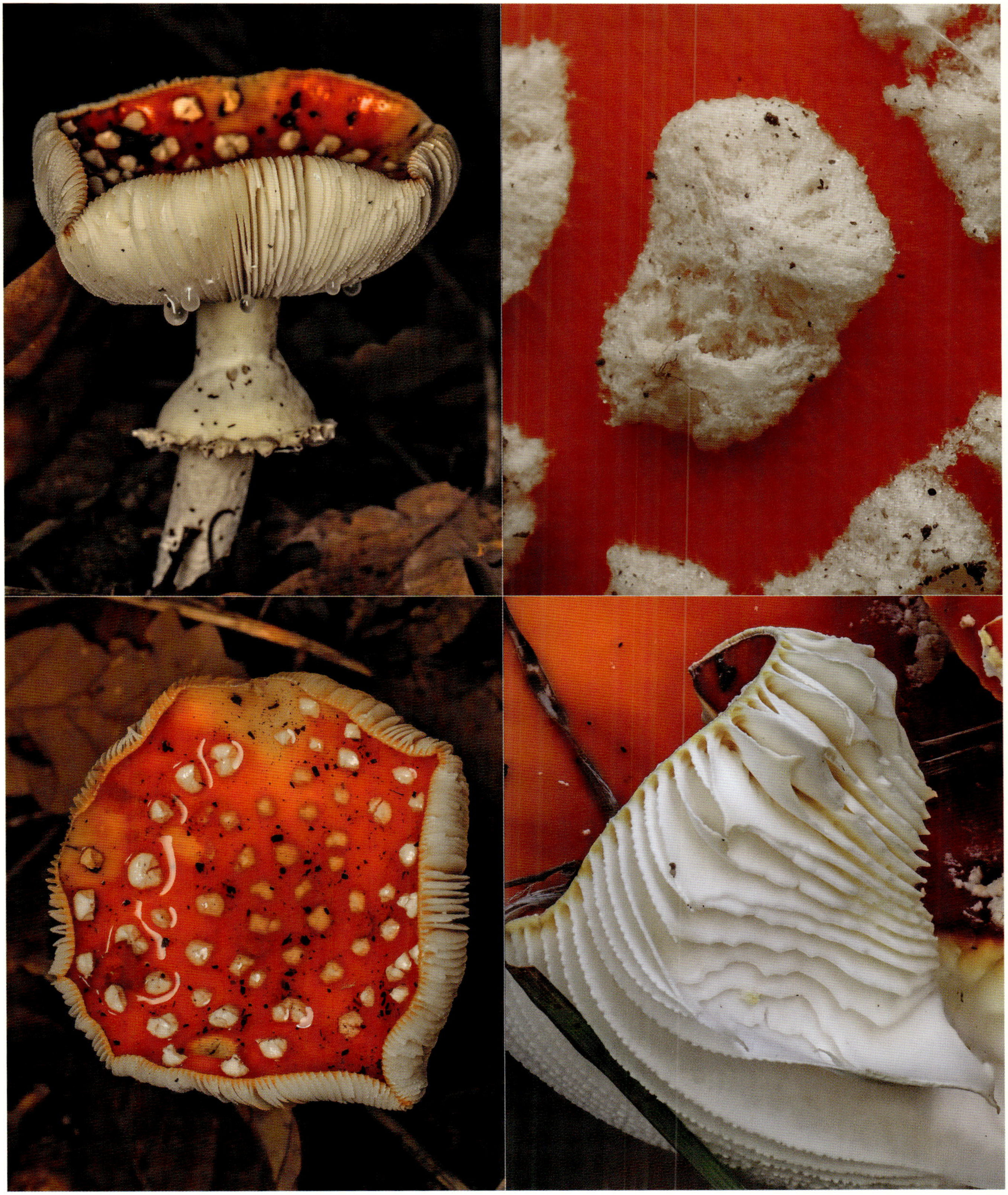

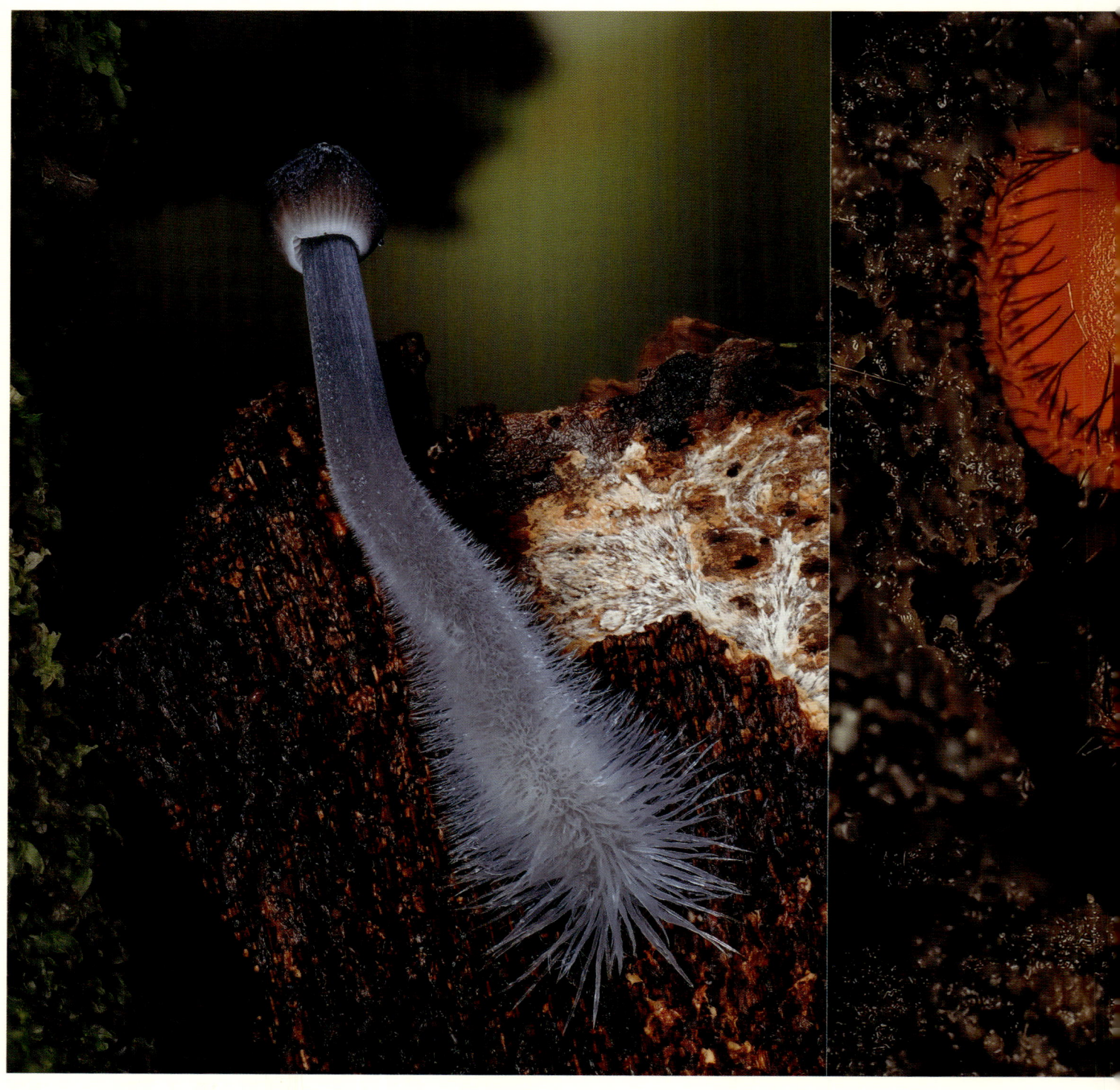

ABOVE: NITROUS BONNET (*MYCENA LEPTOCEPHALA*). OPPOSITE: COMMON EYELASH (*SCUTELLINIA SCUTELLATA*).

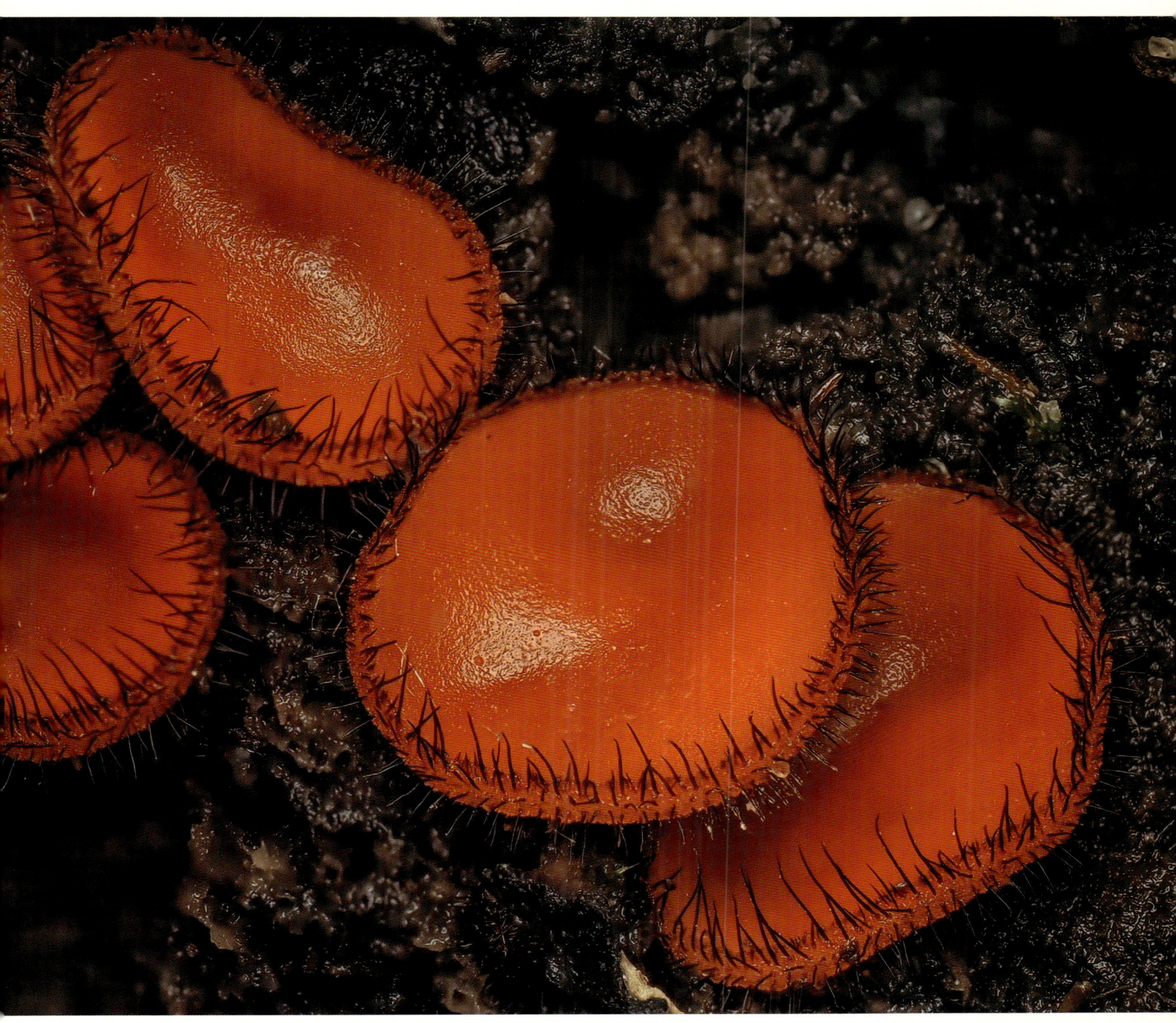

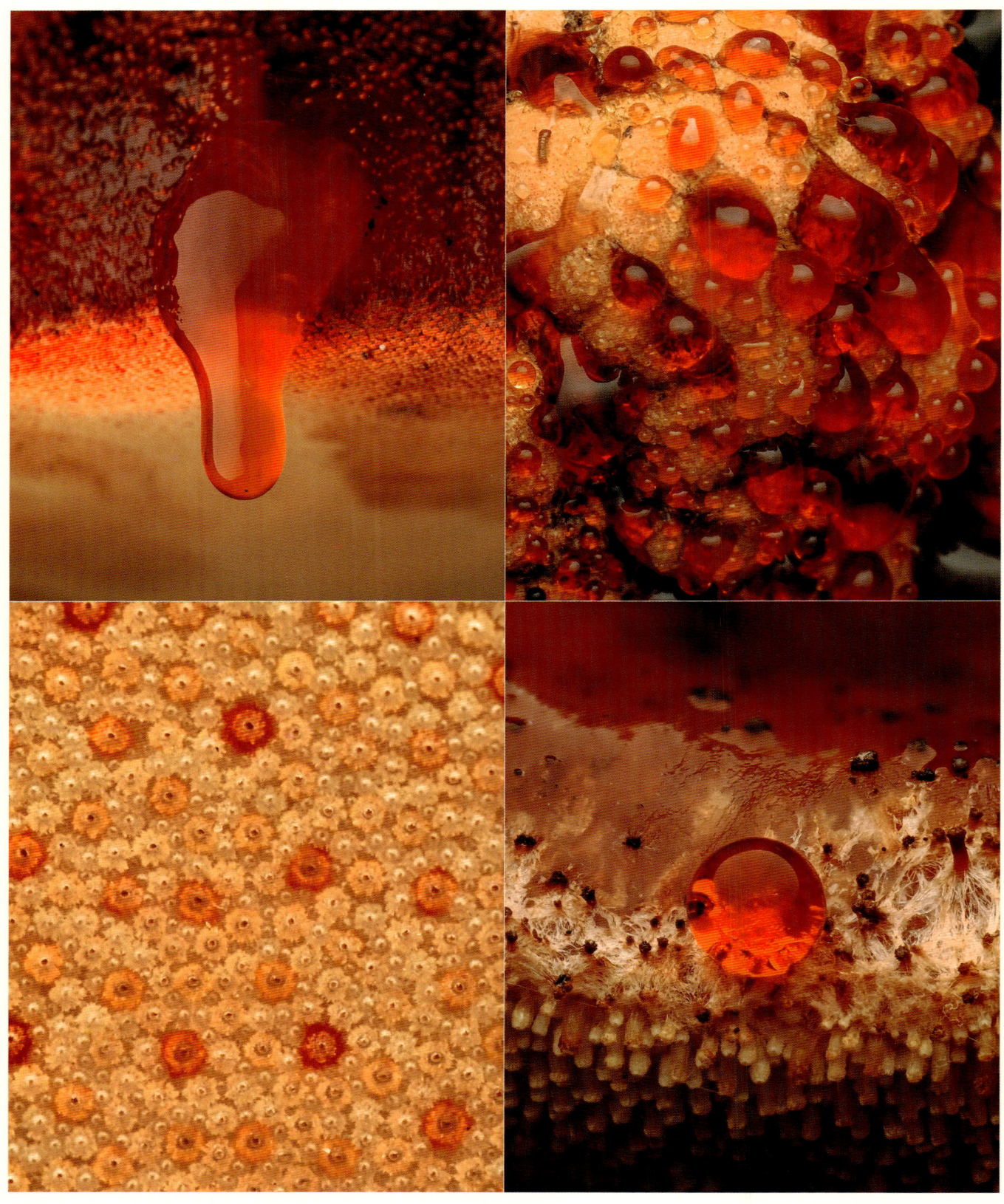

ABOVE AND OPPOSITE: BEEFSTEAK POLYPORE (*FISTULINA HEPATICA*).

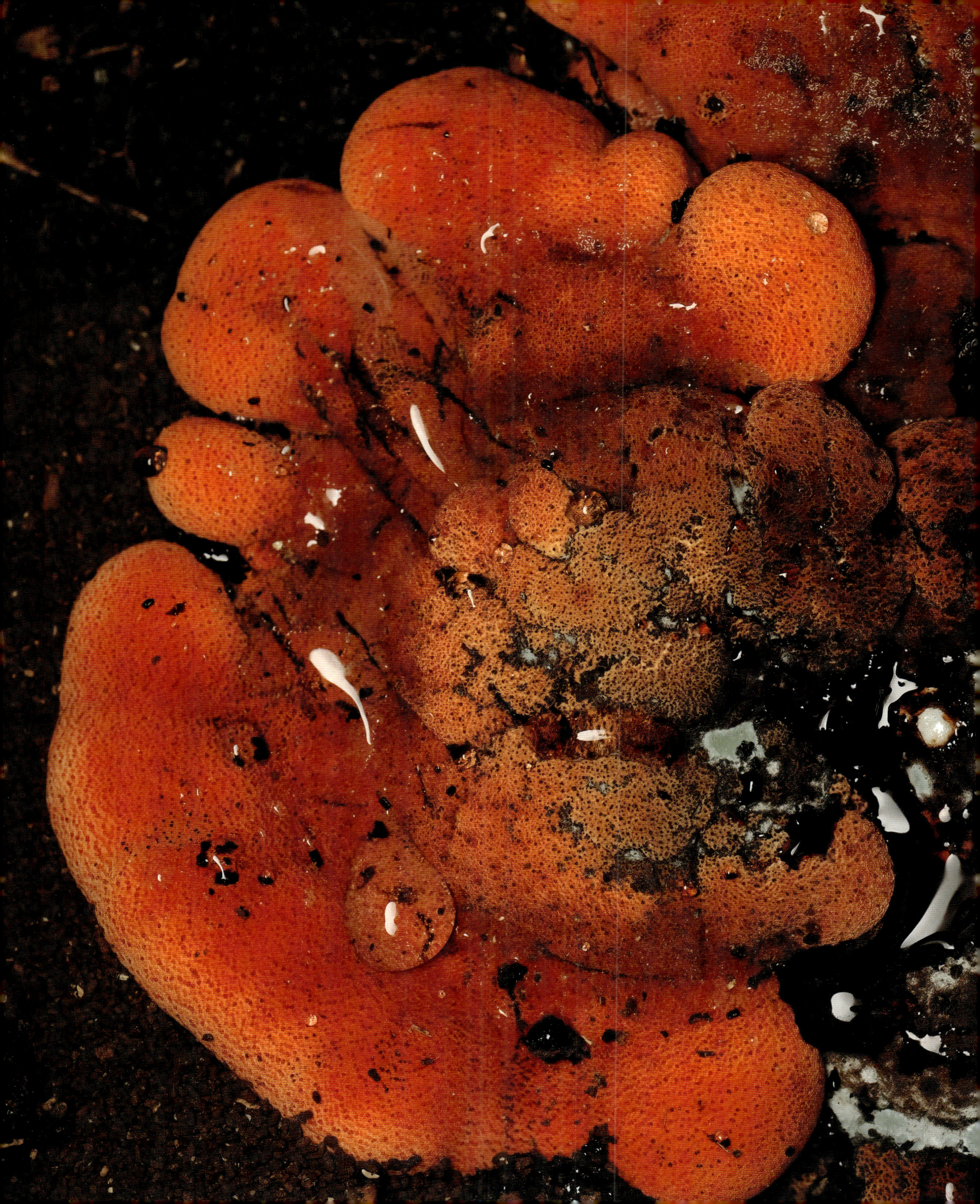

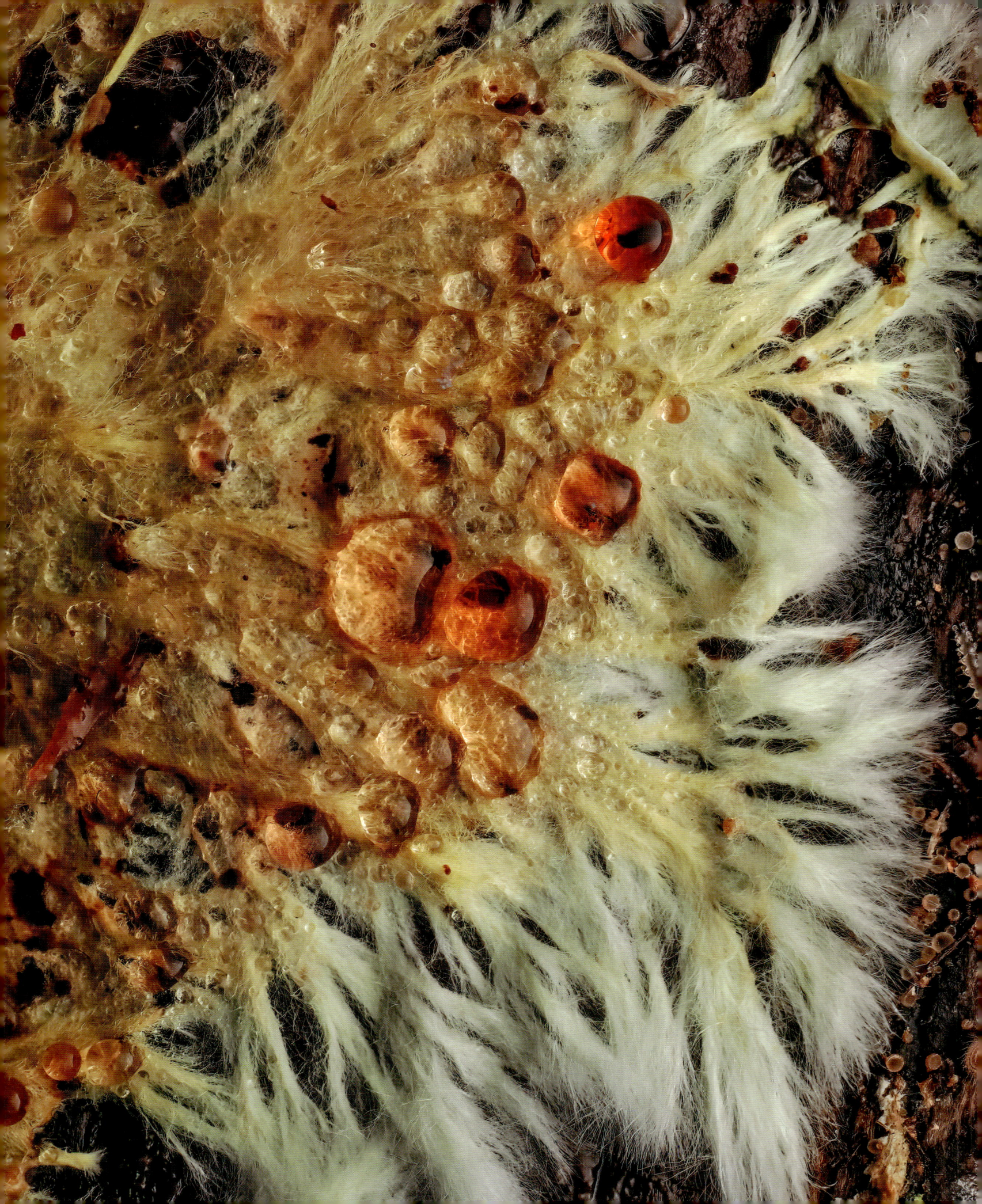

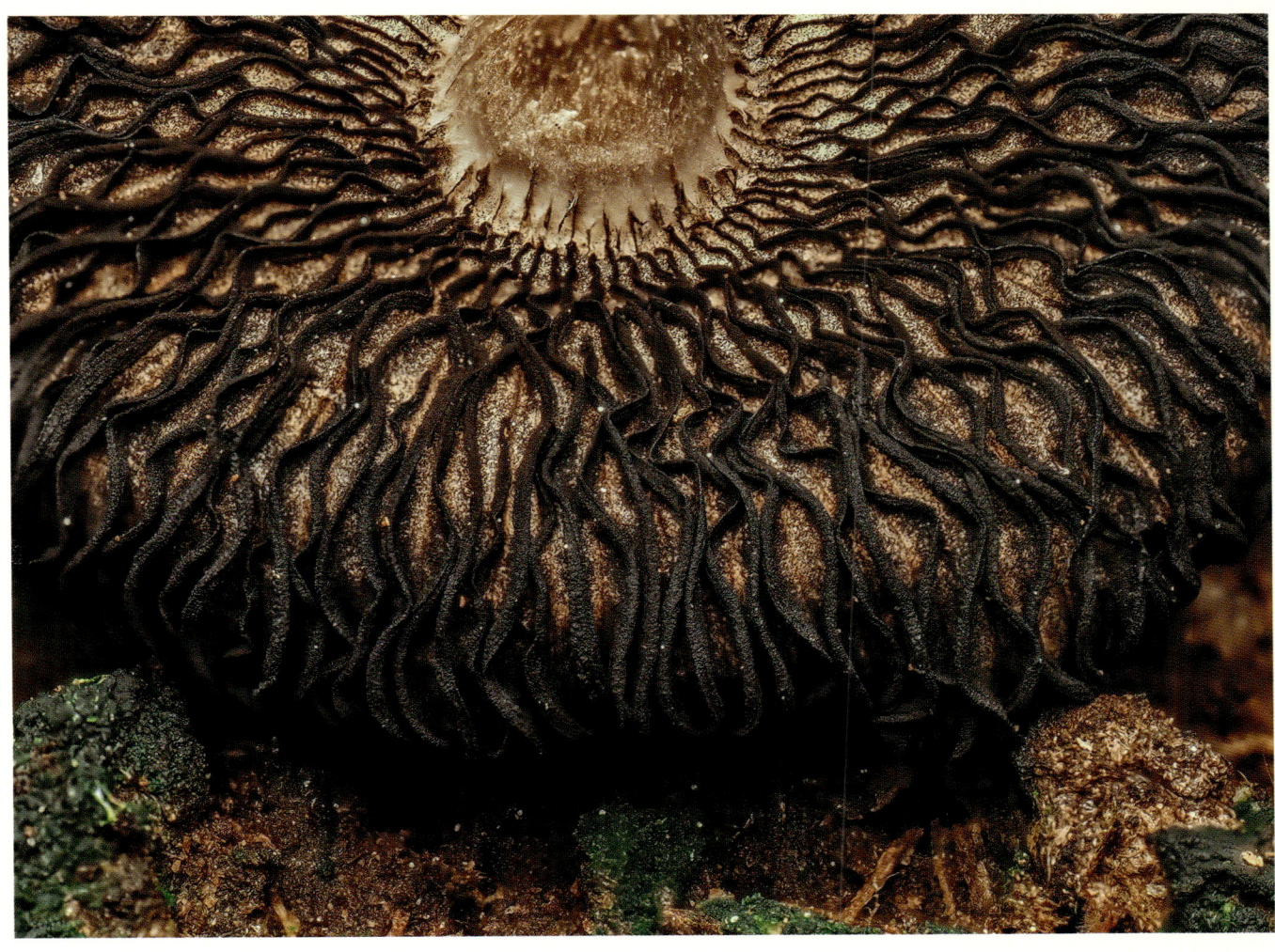

59

ABOVE: *COPRINELLUS* SPP. OPPOSITE: MYCELIUM AND GUTTATION.

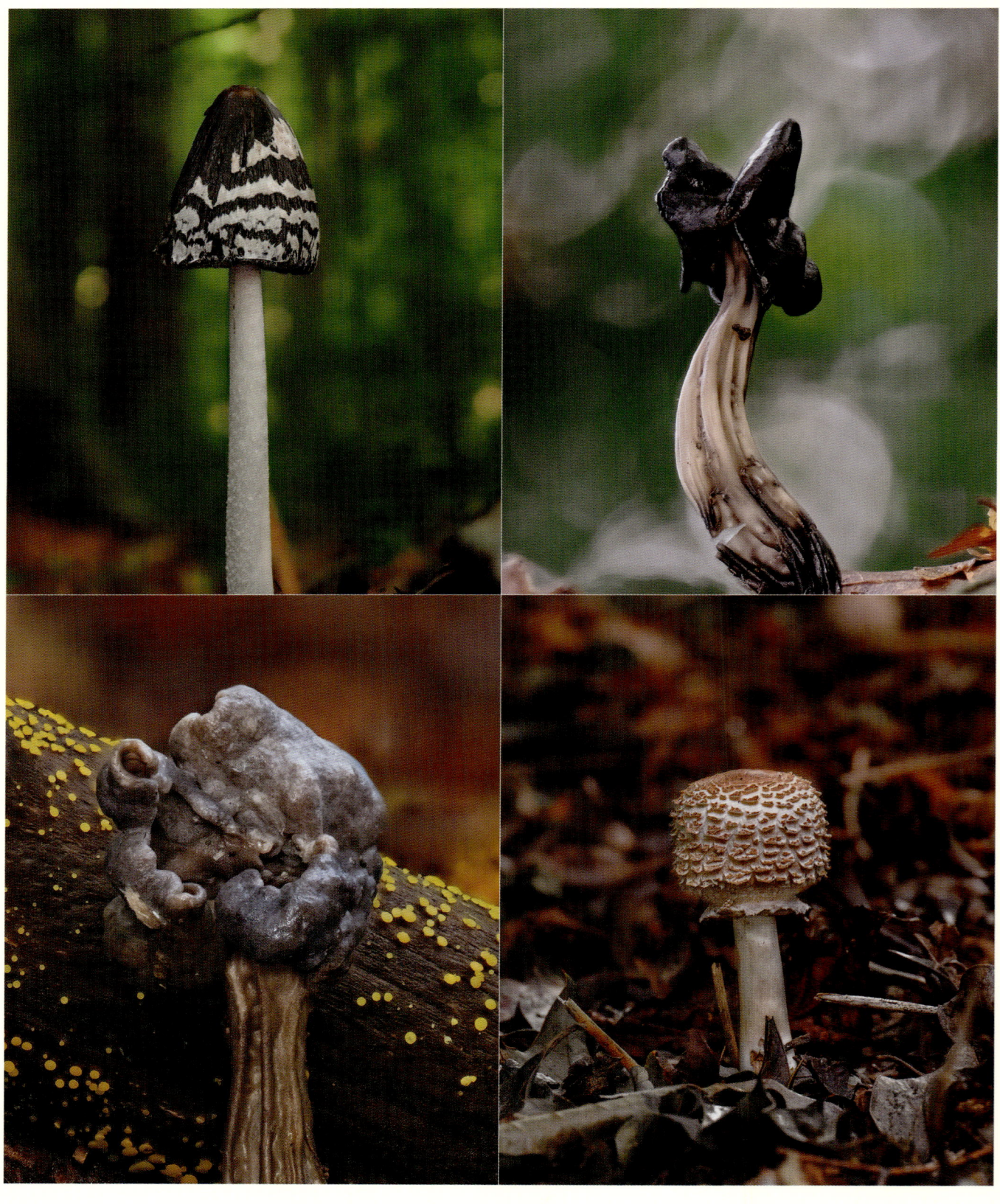

ABOVE LEFT: MAGPIE INKCAP (*COPRINOPSIS PICACEA*). ABOVE RIGHT AND BELOW LEFT: ELFIN SADDLE (*HELVELLA LACUNOSA*). BELOW RIGHT: FALSE PARASOL (*CHLOROPHYLLUM MOLYBDITES*).
OPPOSITE: BLUSHER (*AMANITA RUBESCENS*).

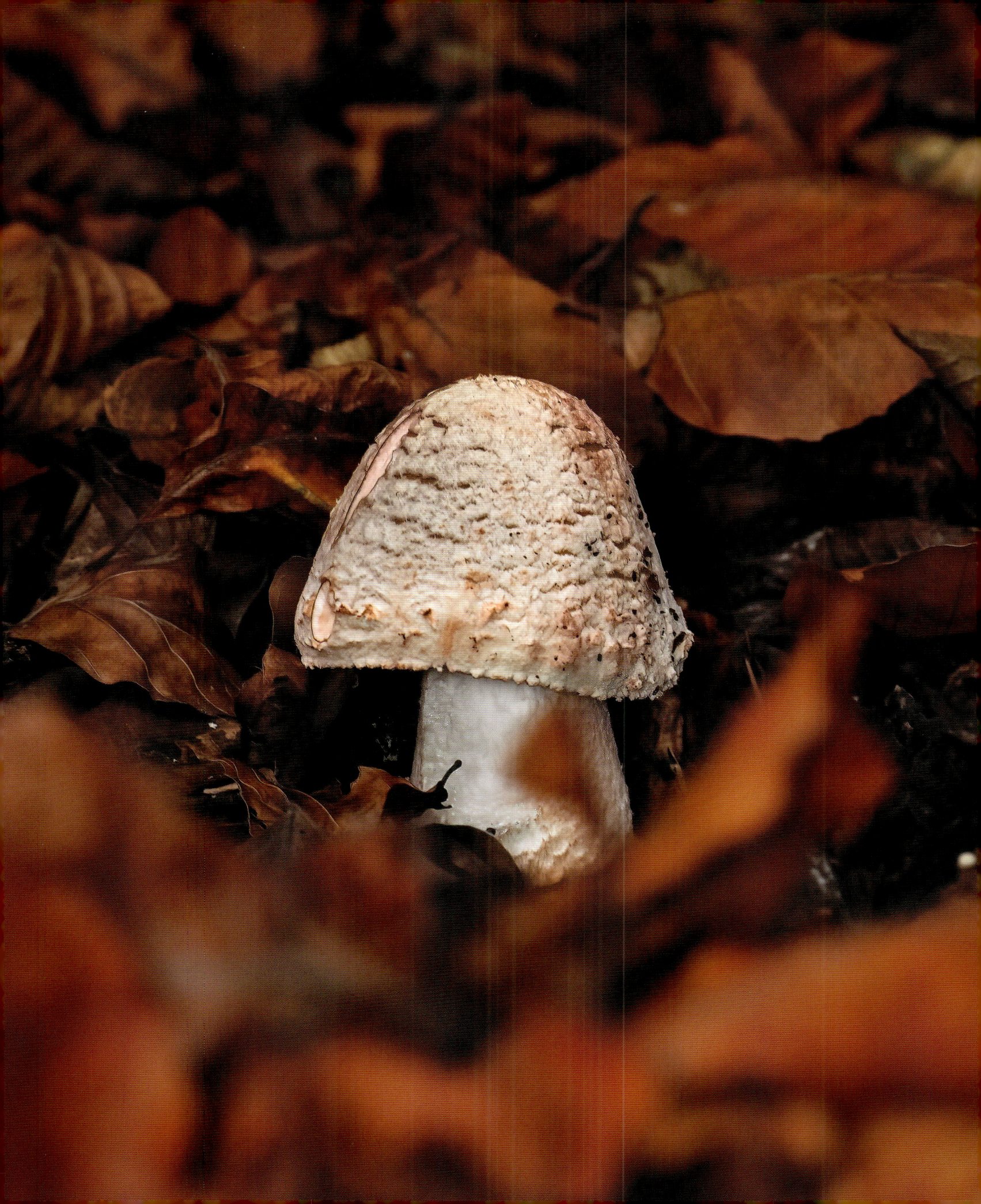

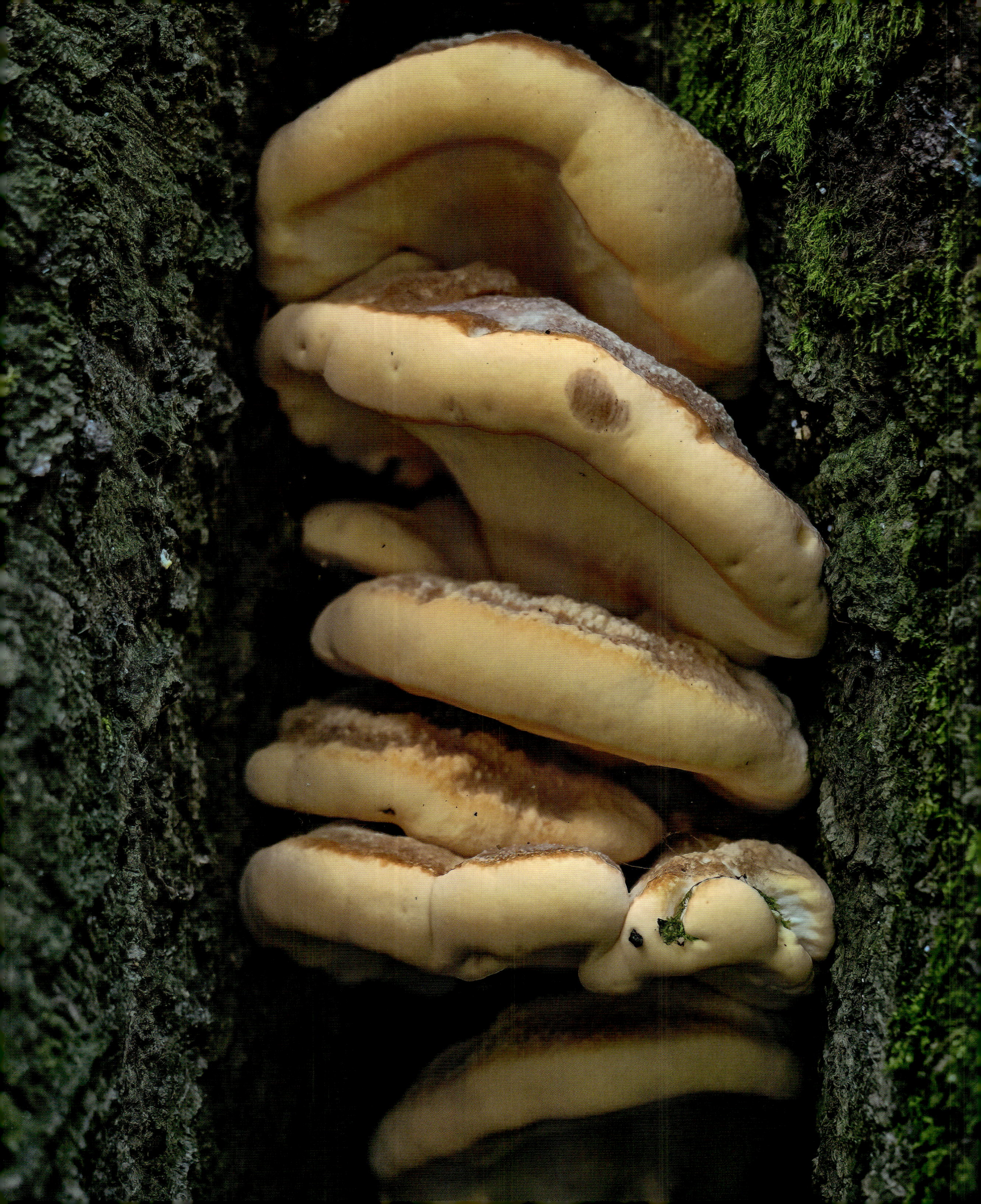

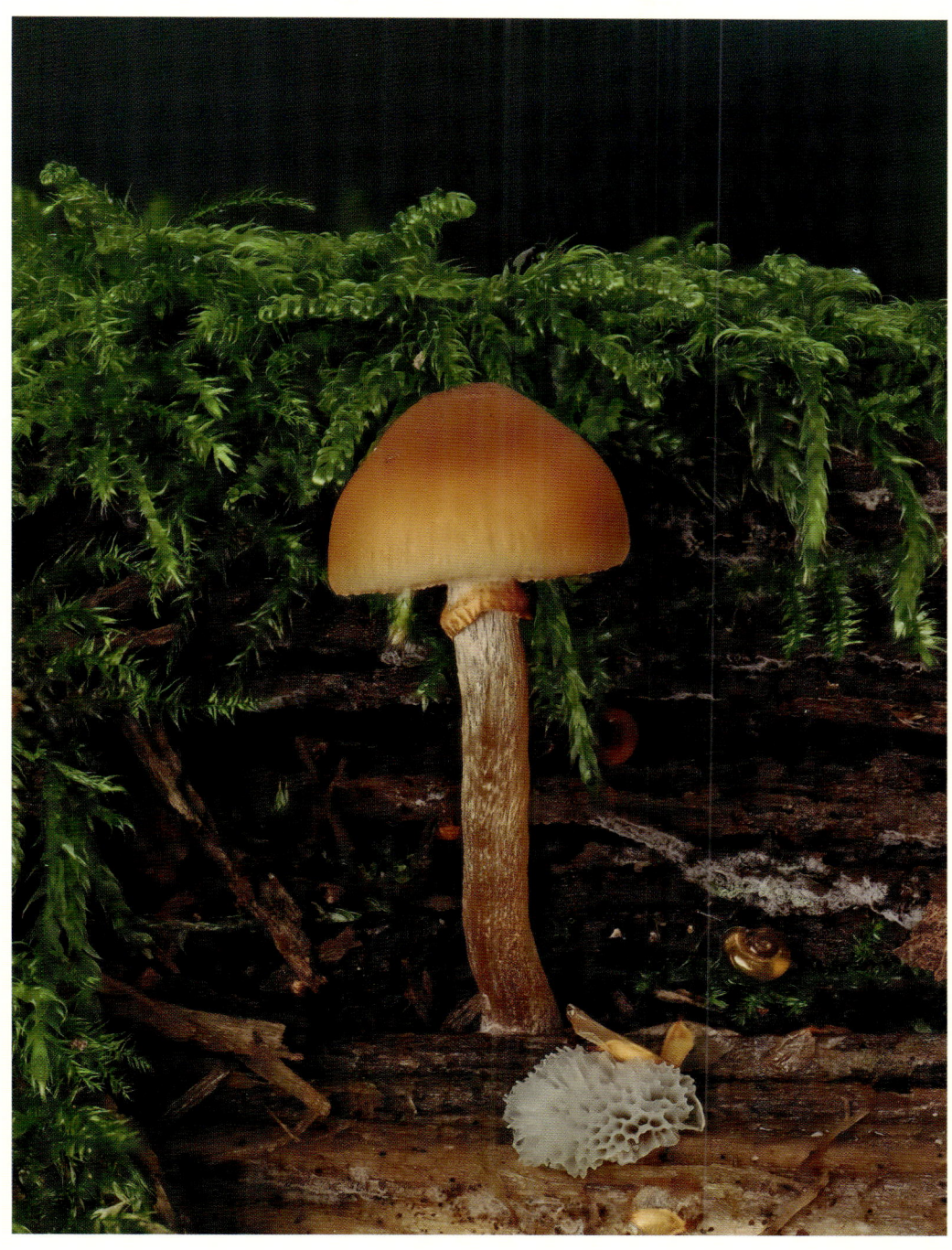

ABOVE: FUNERAL BELL (*GALERINA MARGINATA*). OPPOSITE: GIANT POLYPORE (*MERIPILUS GIGANTEUS*).

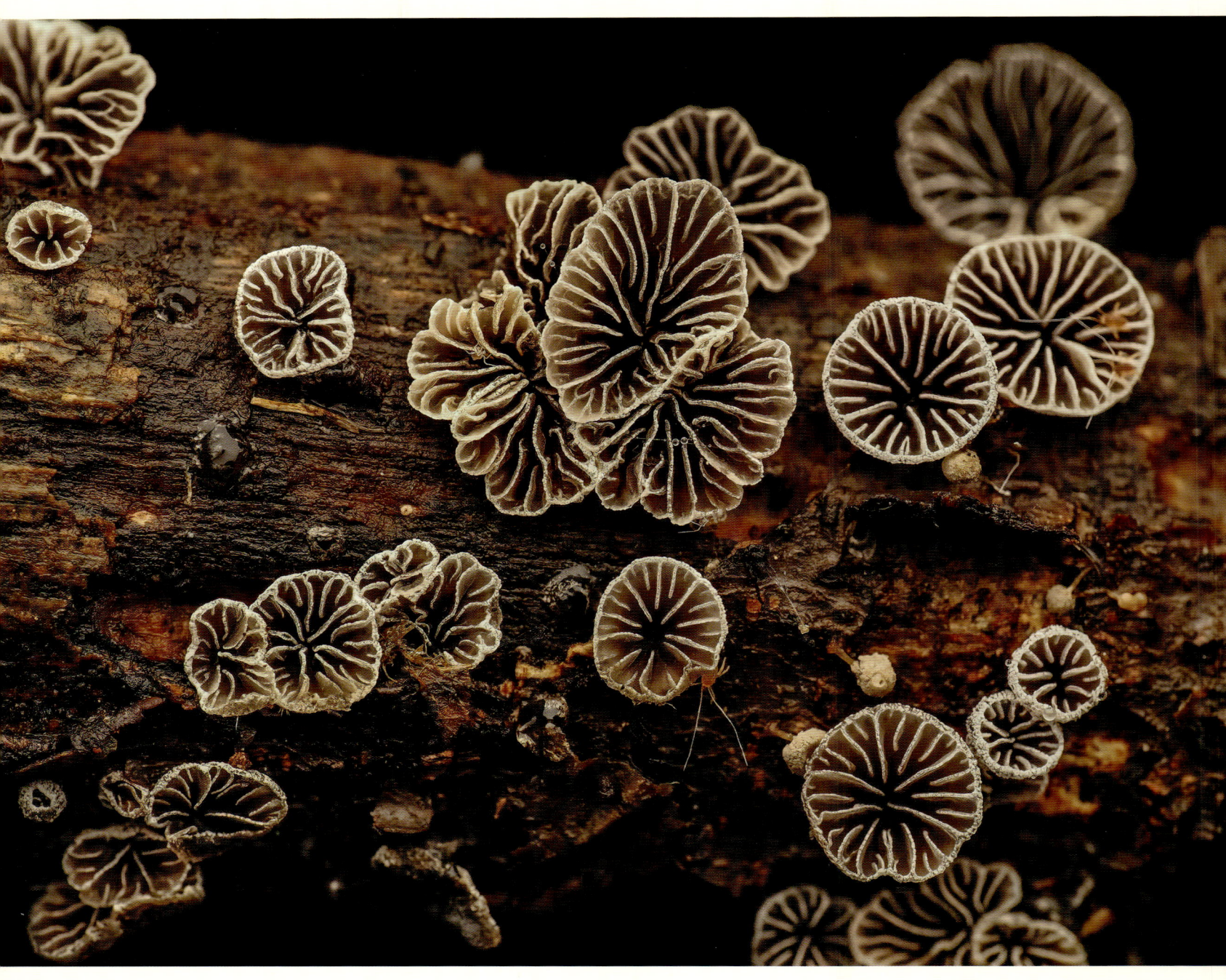

ABOVE: *RESUPINATUS* SPP. OPPOSITE: HARE'S FOOT INKCAP (*COPRINOPSIS LAGOPUS*).

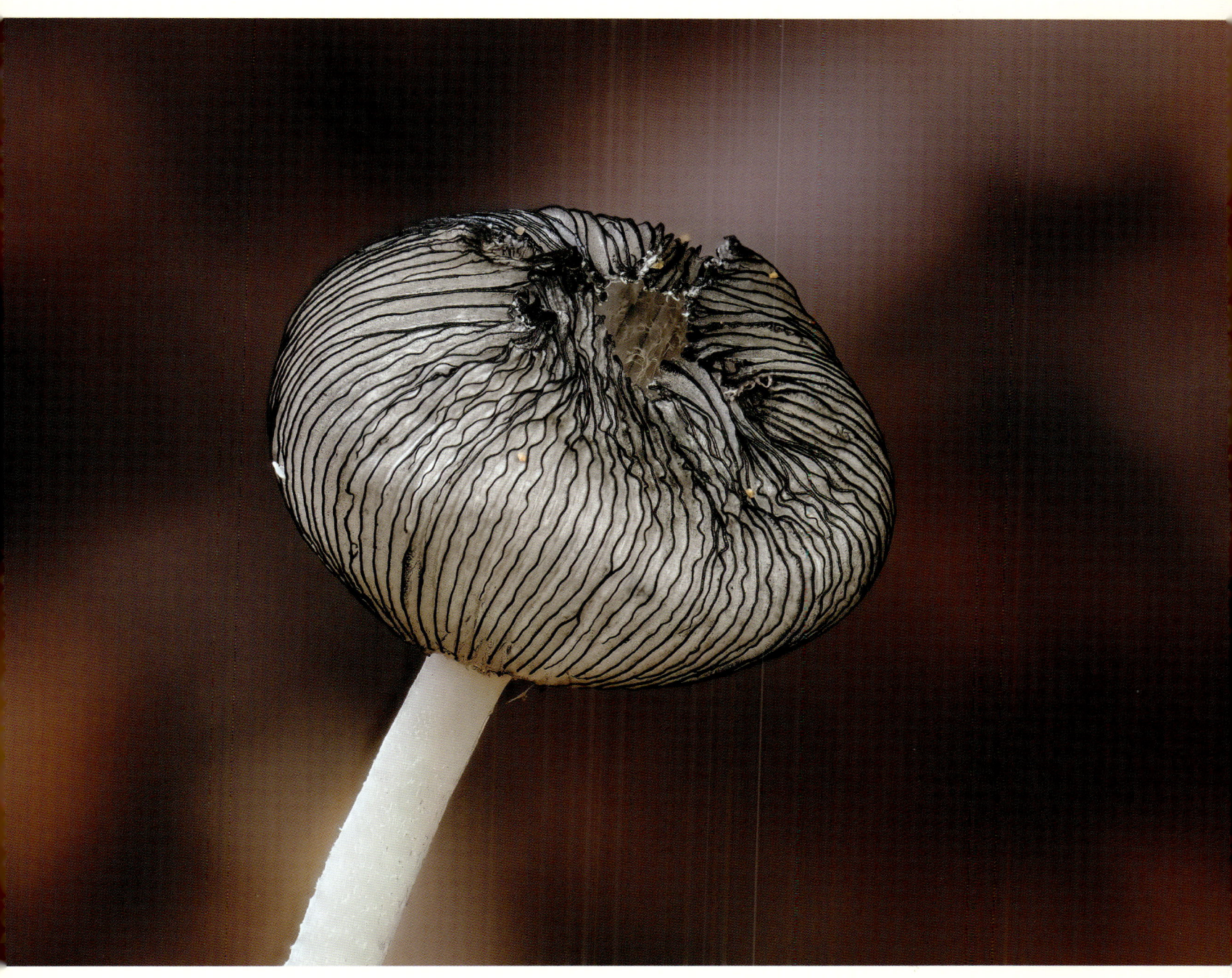

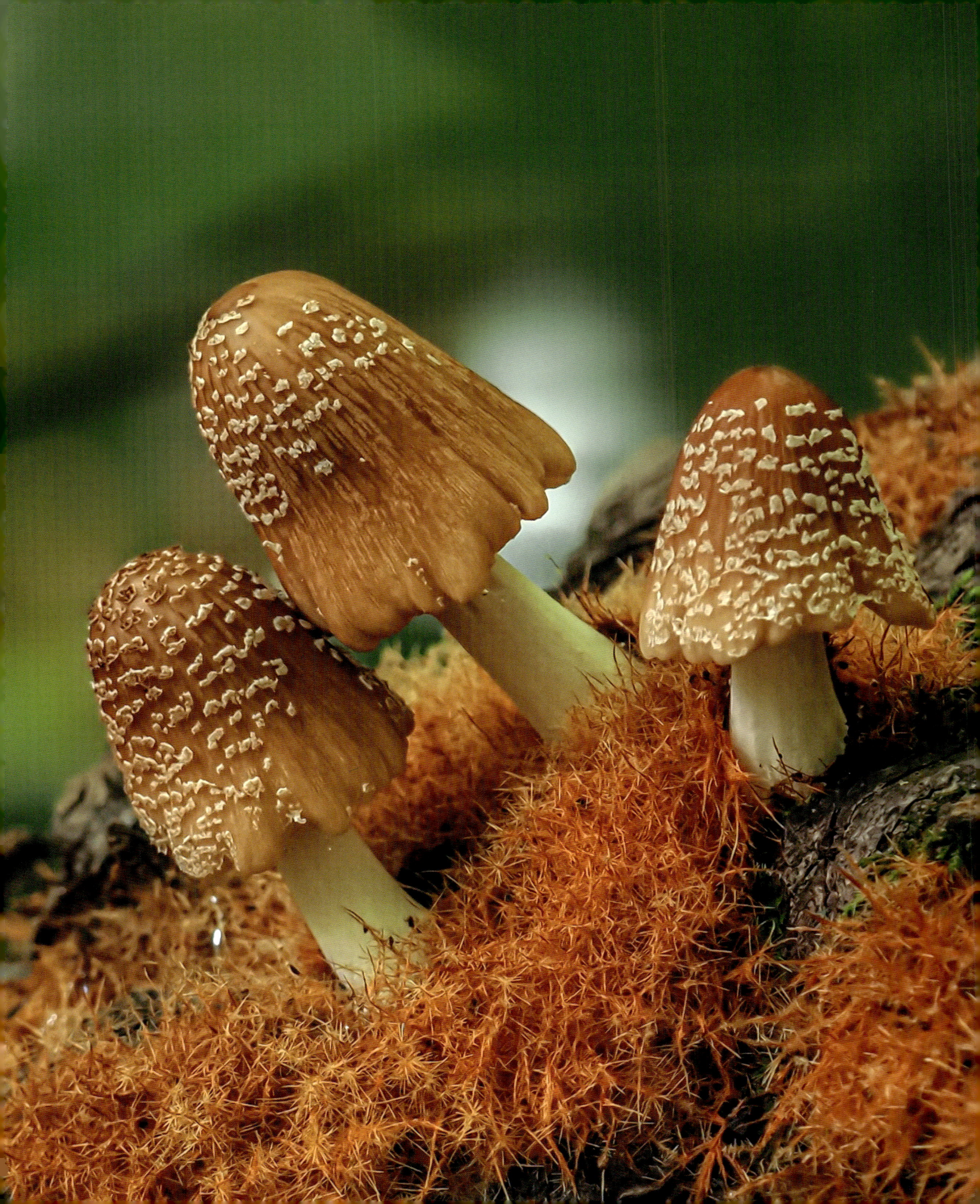

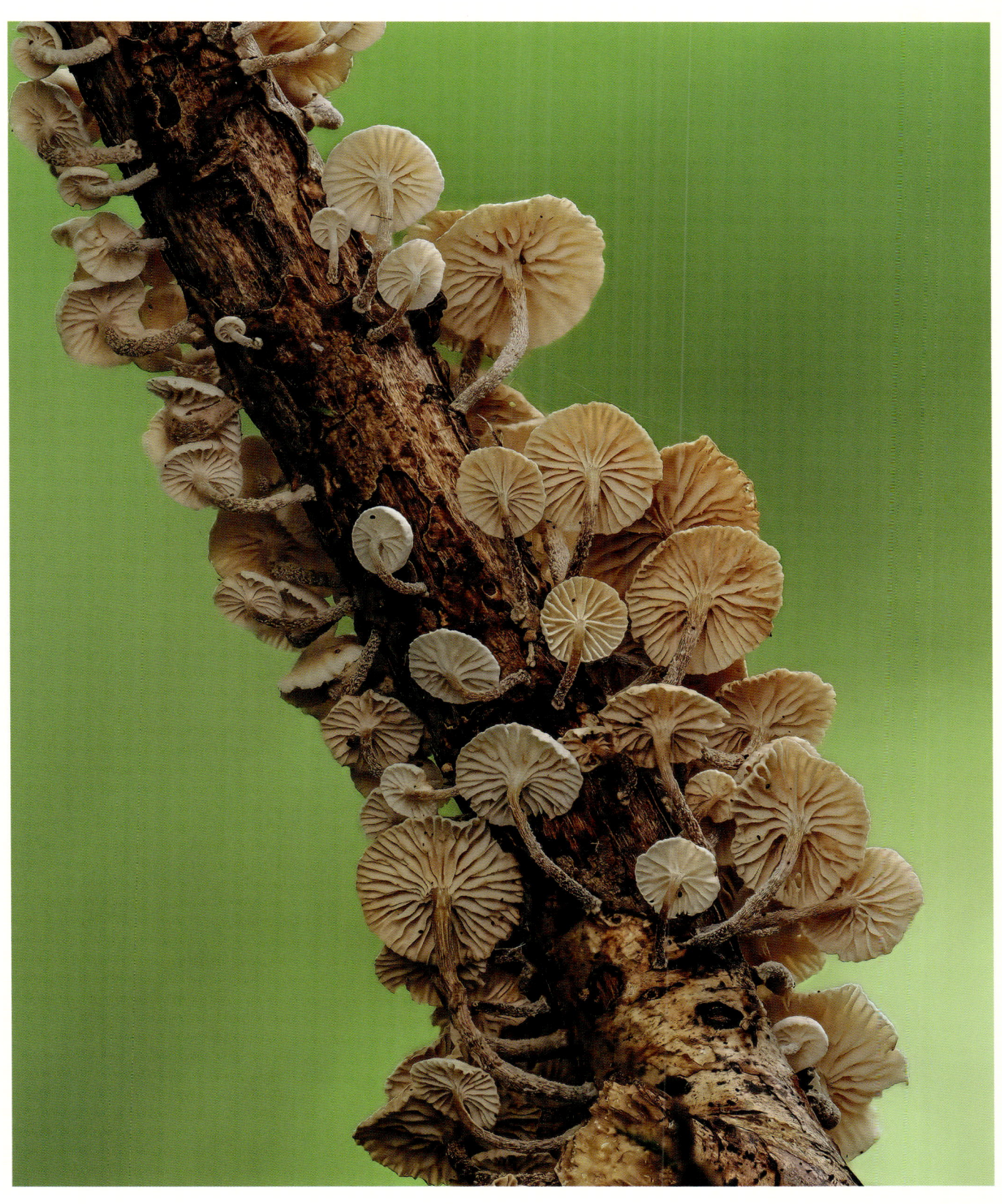

ABOVE: TWIG PARACHUTE (*MARASMIELLUS RAMEALIS*). OPPOSITE: FIRERUG INKCAP (*COPRINELLUS DOMESTICUS*).

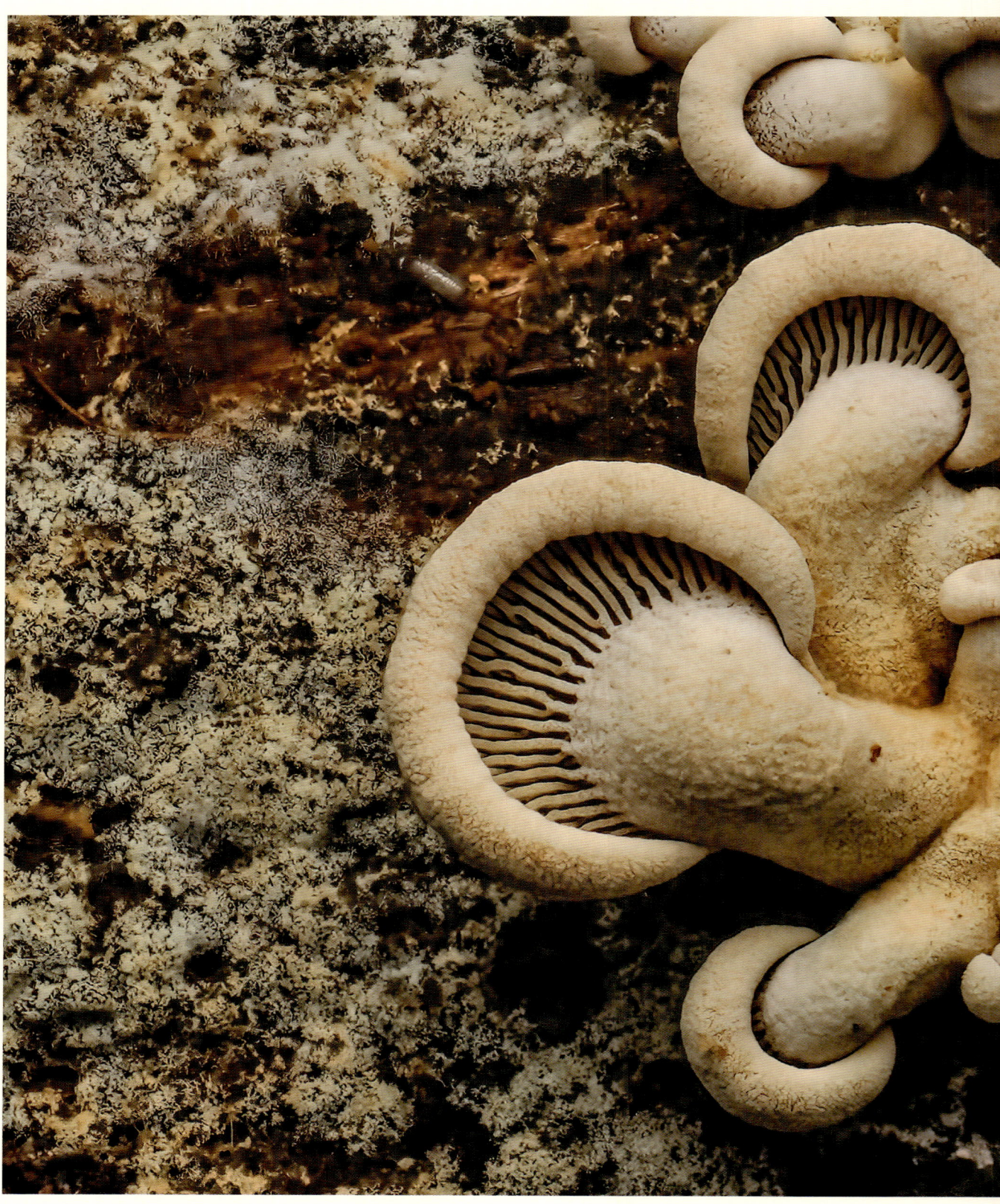

68

ABOVE: BITTER OYSTER (*PANELLUS STIPTICUS*).

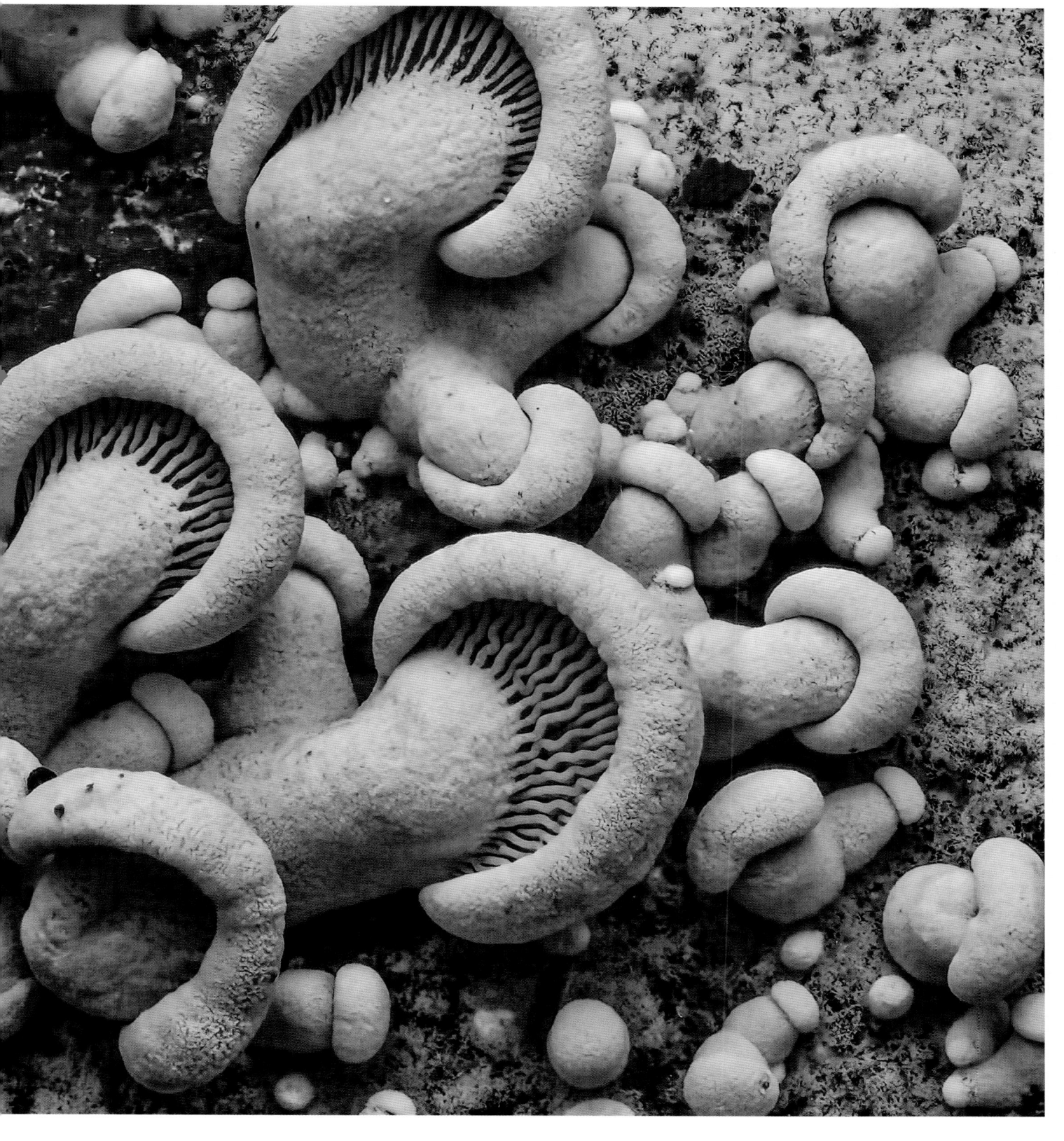

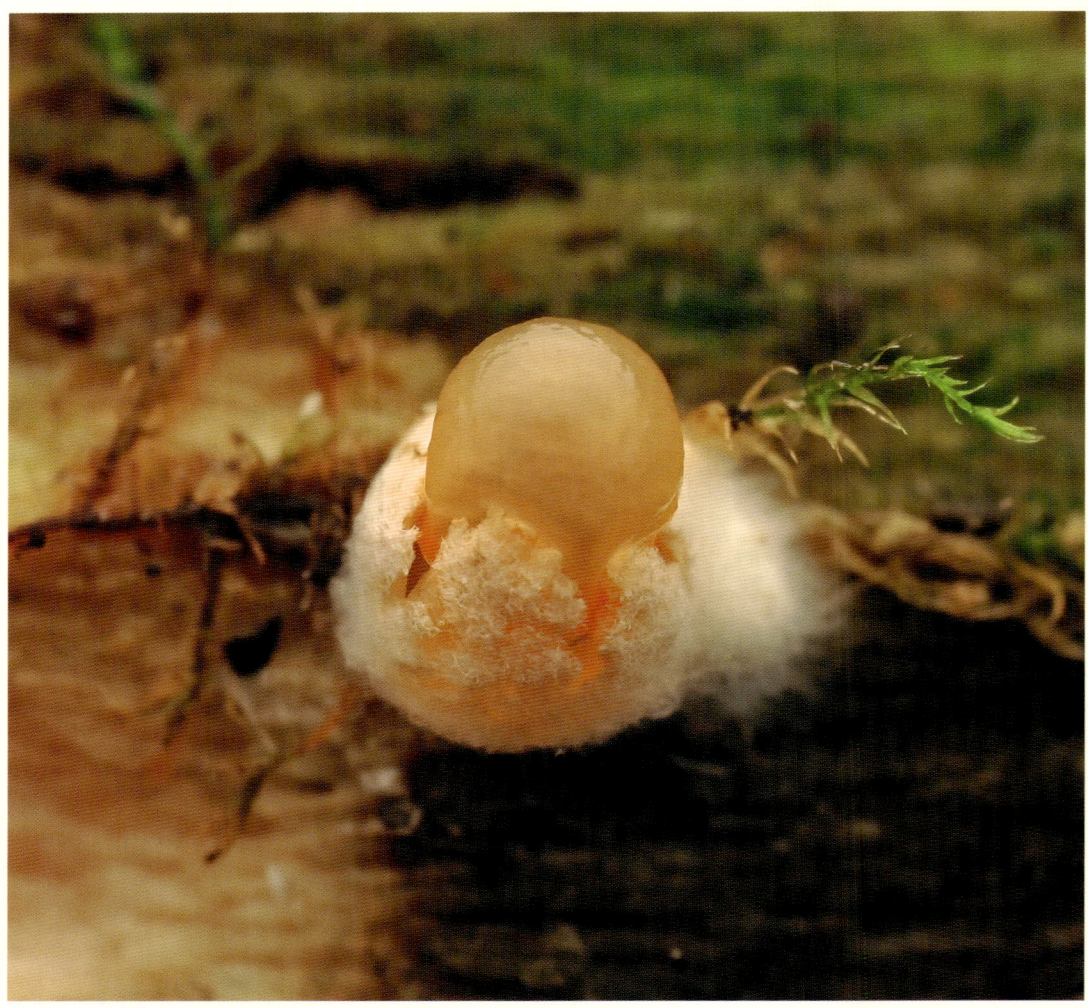

ABOVE AND OPPOSITE: CANNONBALL FUNGUS (*SPHAEROBOLUS STELLATUS*).

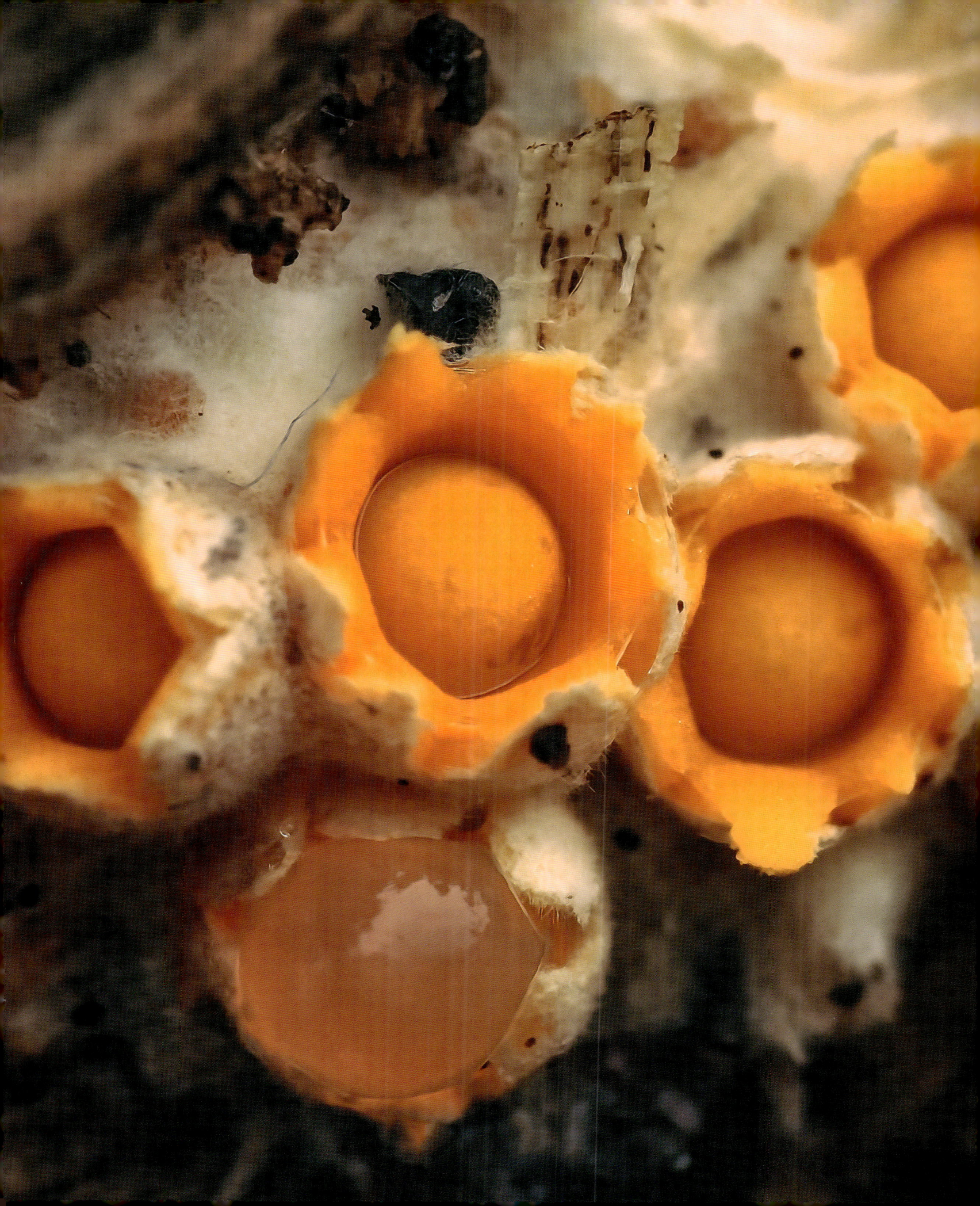

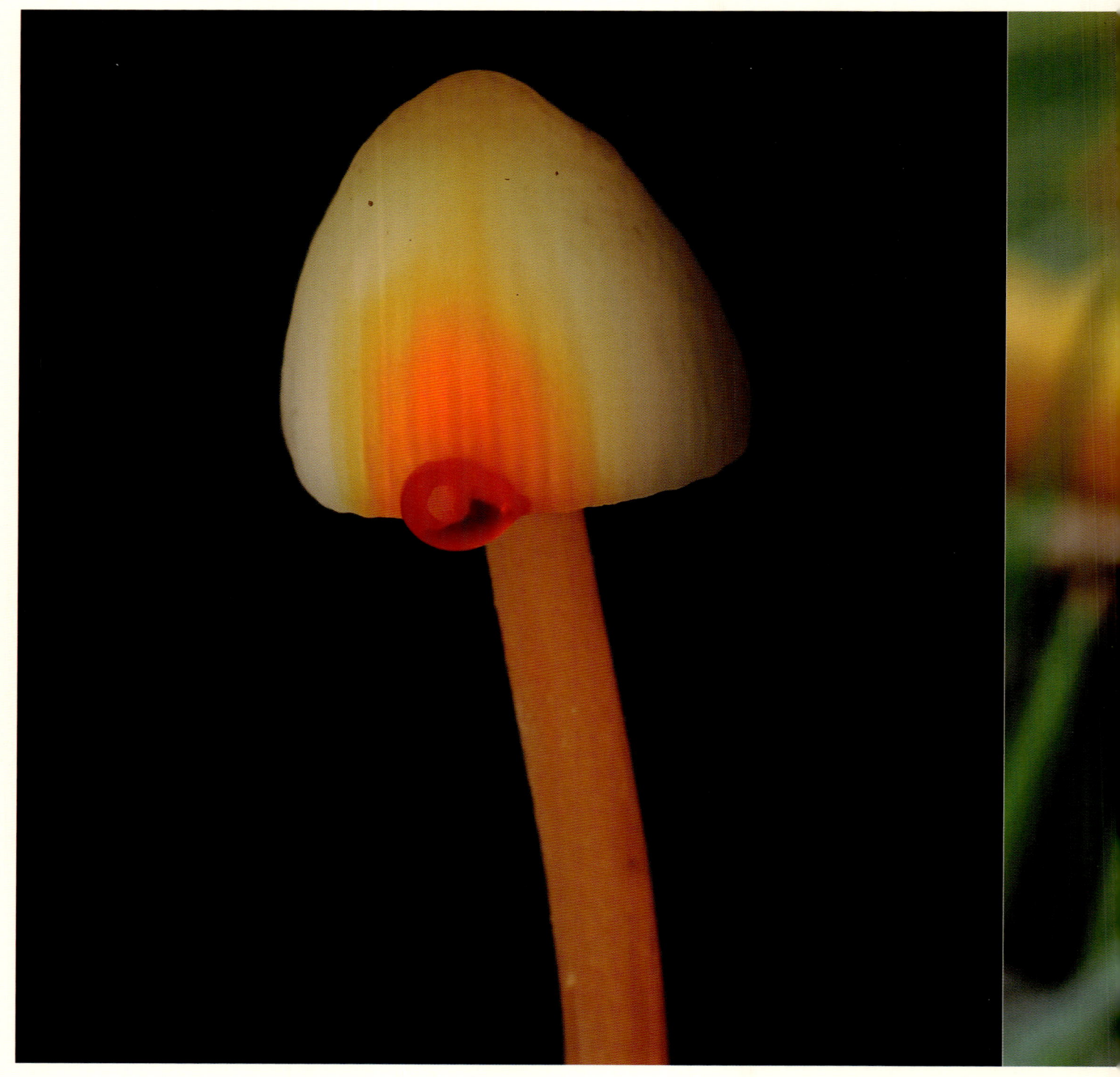

ABOVE AND OPPOSITE: SAFFRONDROP BONNET (*MYCENA CROCATA*).

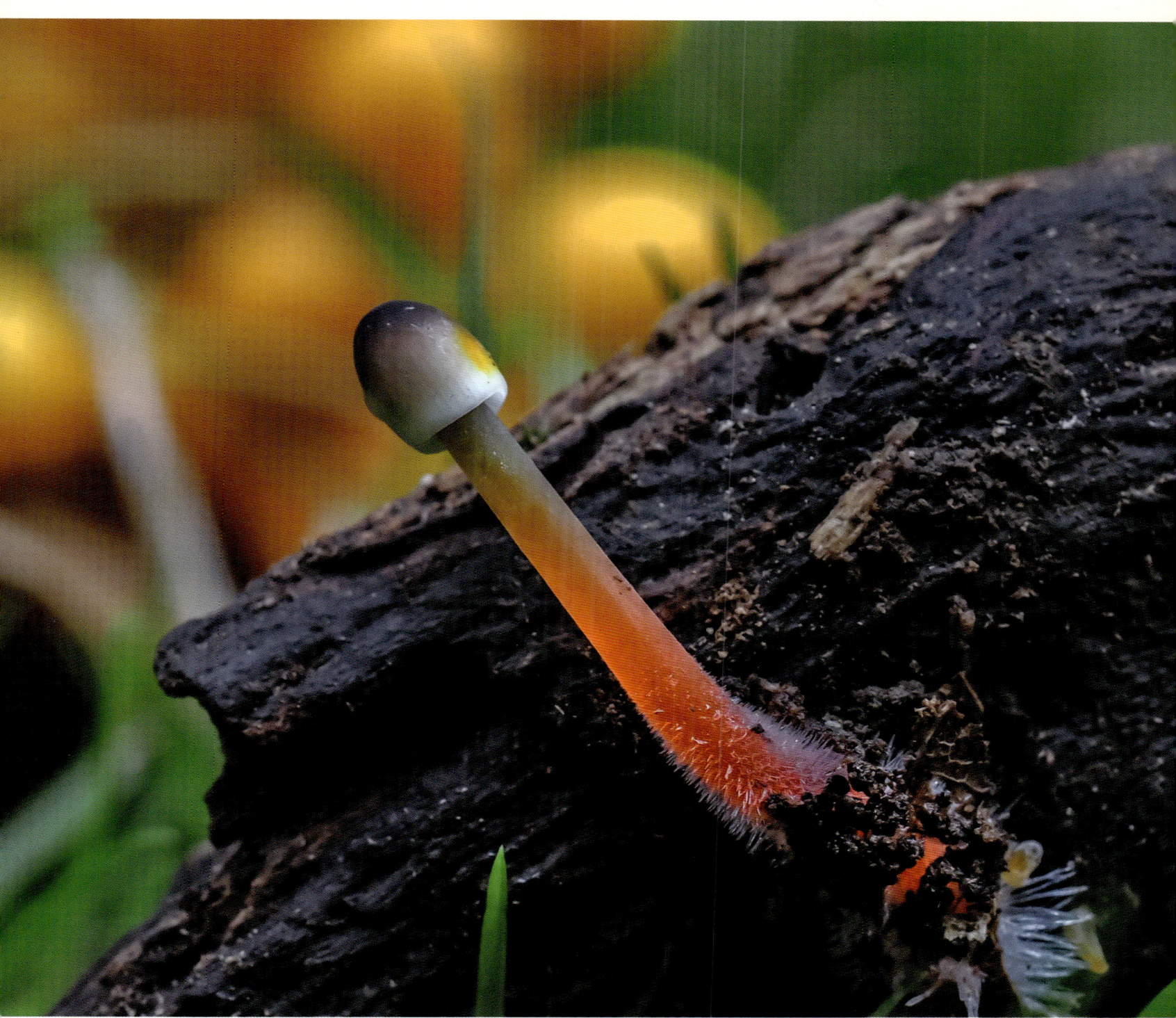

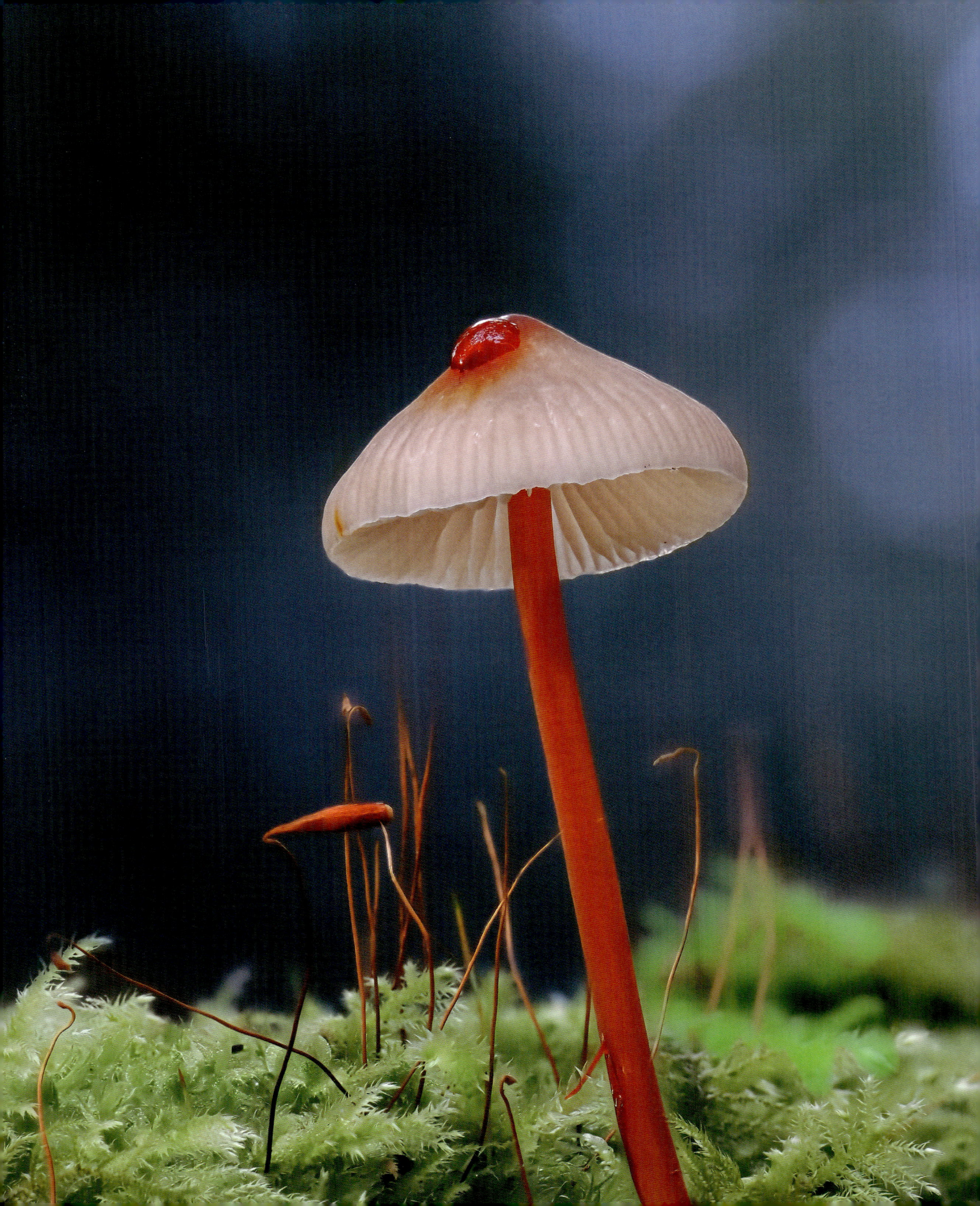

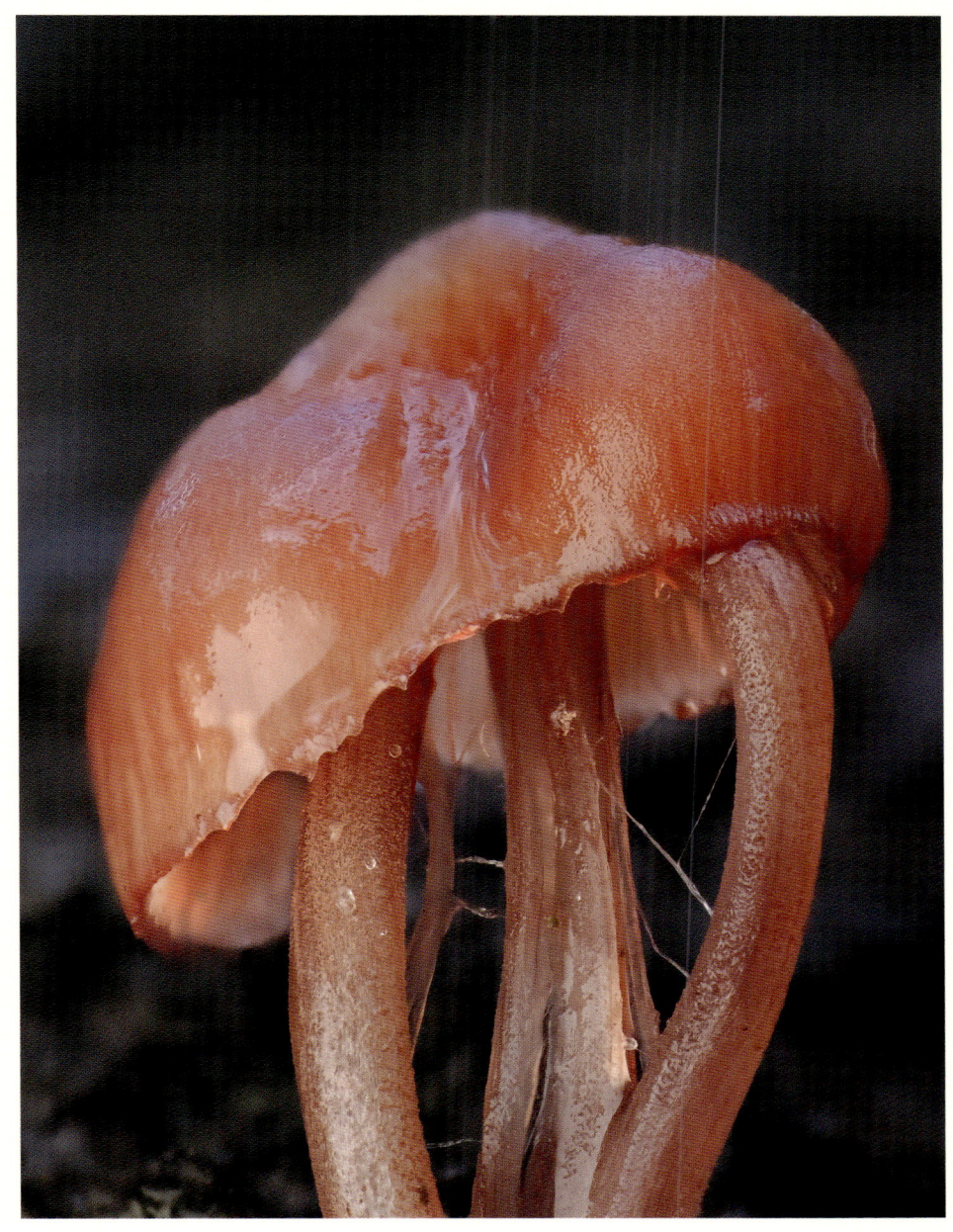

ABOVE: BLEEDING FAIRY HELMET (*MYCENA HAEMATOPUS*). OPPOSITE: SAFFRONDROP BONNET (*MYCENA CROCATA*).

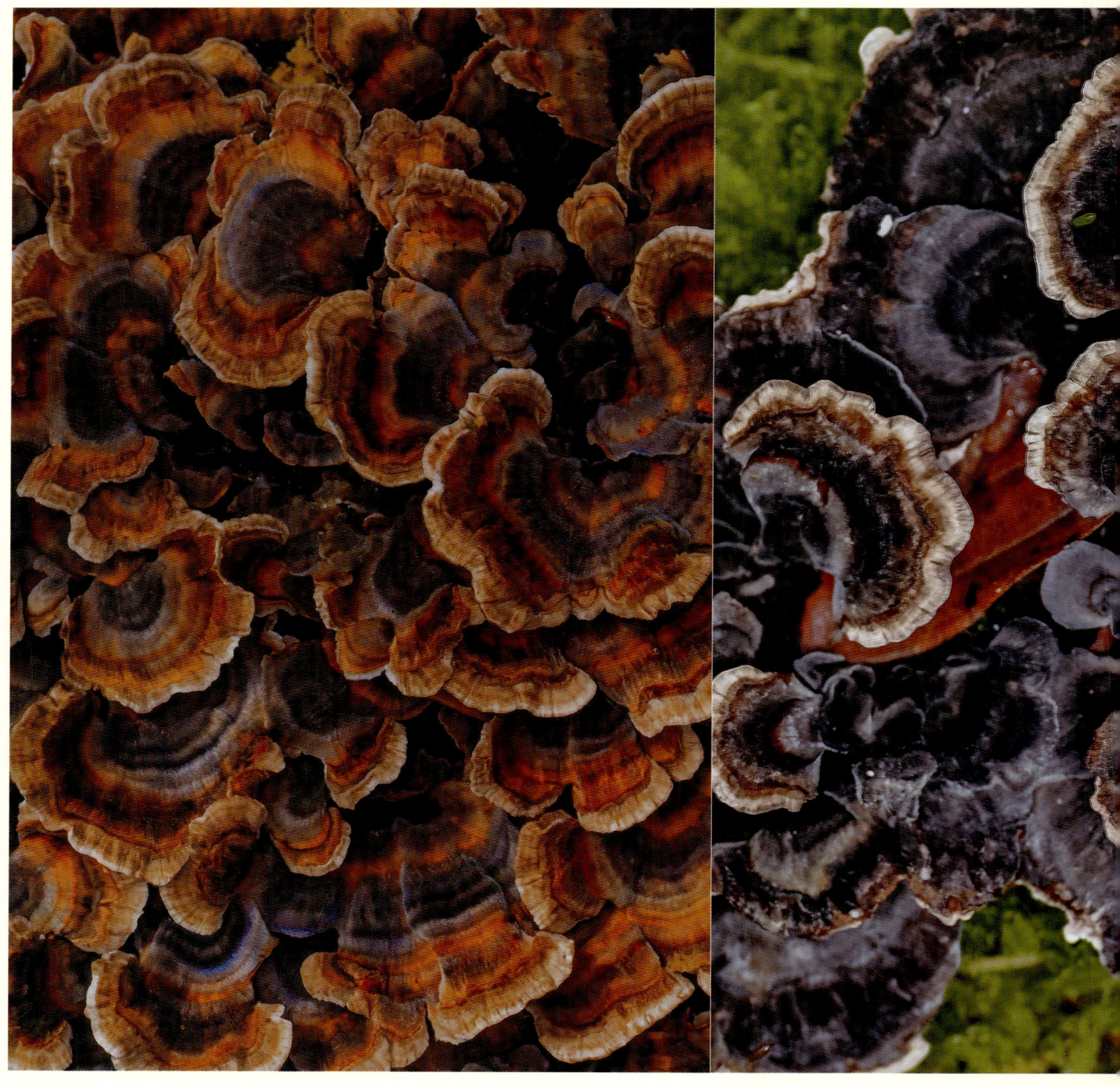

76

ABOVE AND OPPOSITE: TURKEY TAIL (*TRAMETES VERSICOLOR*).

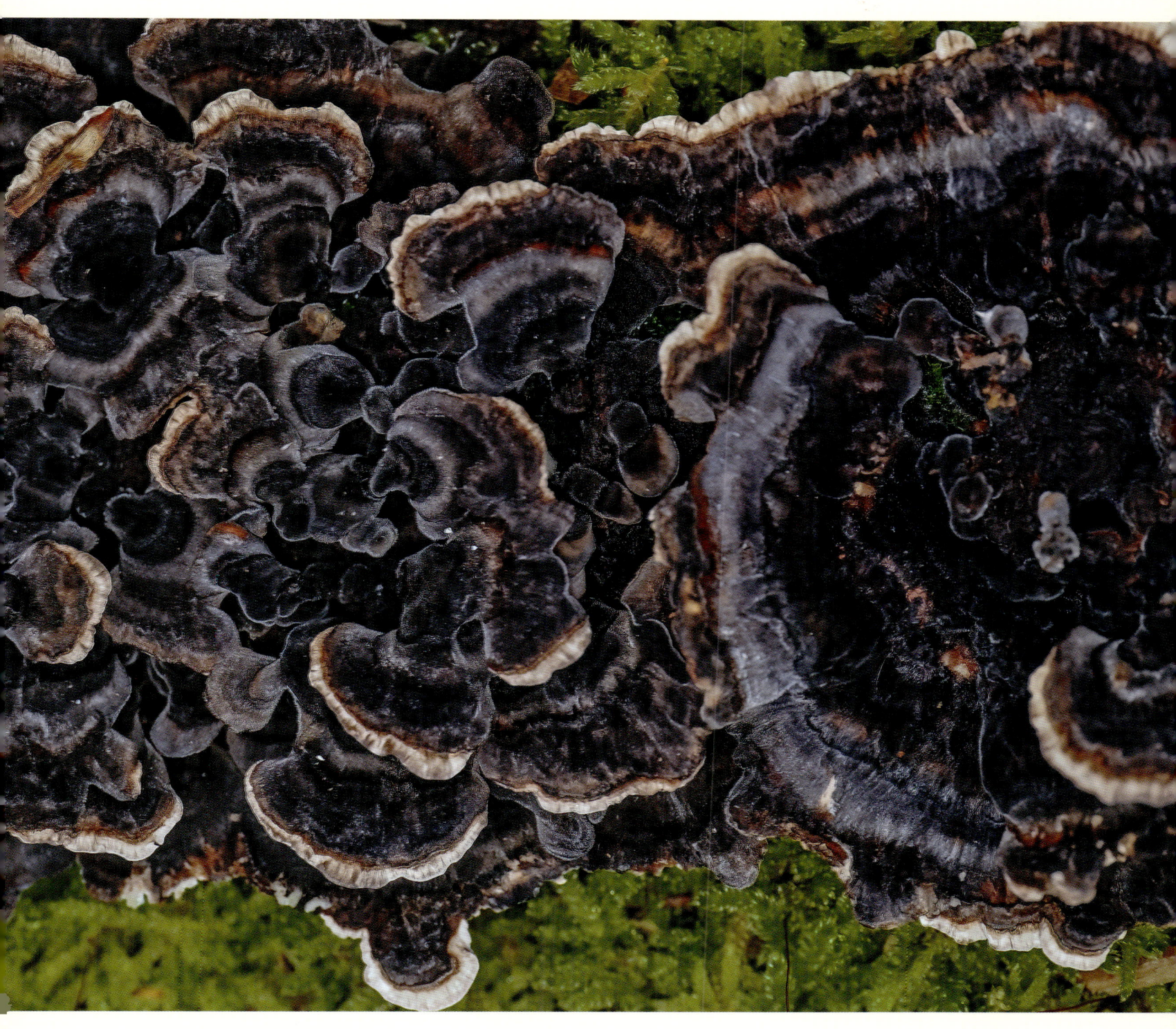

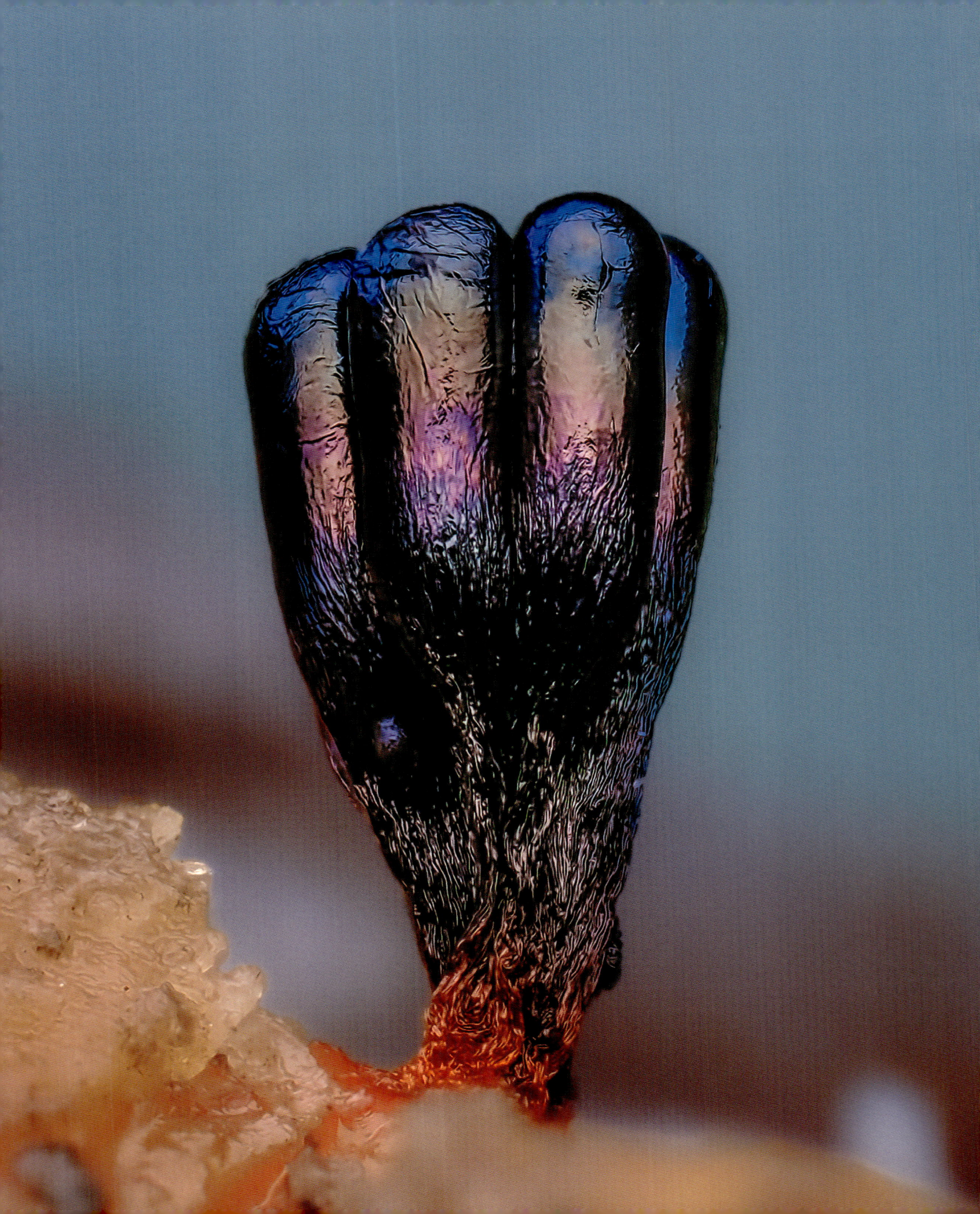

Slime Mold

S lime molds (myxomycetes) are absolutely wild—science-fiction territory awesomeness. I had never heard of slime mold before delving into macrophotography. Most people are probably unaware of the weird and wonderful existence of slime mold, and I am extremely excited to introduce it through photography.

In my extremely biased opinion, slime molds are among the most intriguing and enigmatic organisms on earth, straddling the boundary between the world of fungi and animals through their unique life cycle and behavior. These organisms are not true fungi, though they often resemble them in their spore-producing phase. They are also not animals, though their feeding stage exhibits behavior reminiscent of single-celled amoebae. There is still much debate on how slime molds should be classified for complex ultra-scientific reasons that I am not going to go into, so for now, they are classified in the kingdom Protozoa.

OPPOSITE: WASP'S NEST SLIME MOLD (*METATRICHIA VESPARIA*).

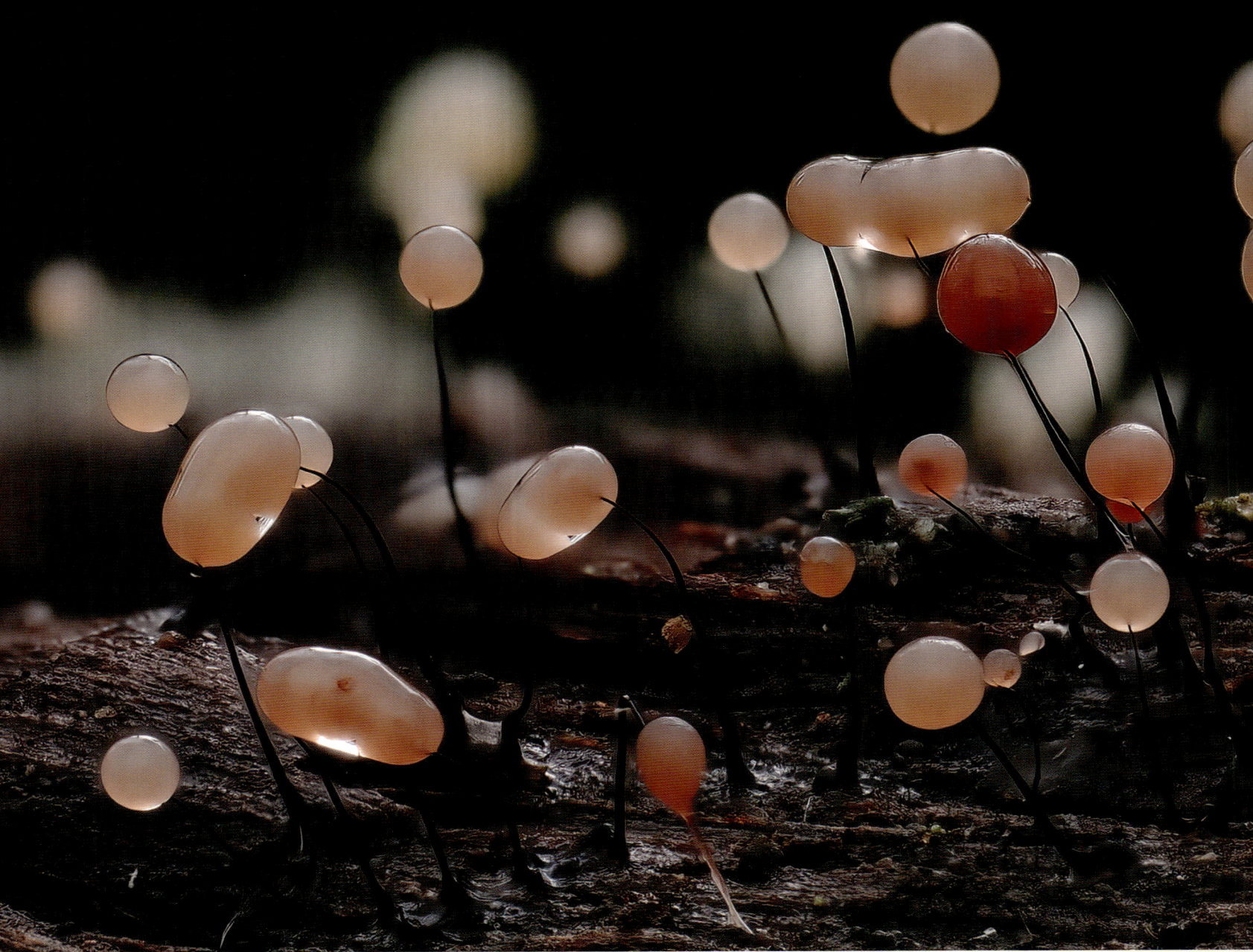

The life cycle of slime molds is a fascinating journey from a single cell to collective intelligence. In their active stage, they exist as amoeba-like cells, known as plasmodial slime, or as swarm cells in cellular types. These individual cells roam the forest floor, consuming bacteria, fungal spores, and other microscopic particles in a similar way to amoebae.

When food is abundant, these cells thrive independently, but scarcity triggers the most alien and incredible transformation. In response to adverse conditions, many individual cells congregate to form a single, larger organism called a plasmodium, which can move and behave as a single entity. This plasmodium can crawl, undulating across dead leaves, wood, and soil with a slow, persistent movement that challenges our usual understanding of what single-celled organisms can do. I've taken a few timelapse videos and can confirm it. It's a breathtakingly wicked thing to witness!

The plasmodium navigates the forest floor, intelligently searching for food and optimal

ABOVE AND FOLLOWING SPREAD: *COMATRICHA* SPP.

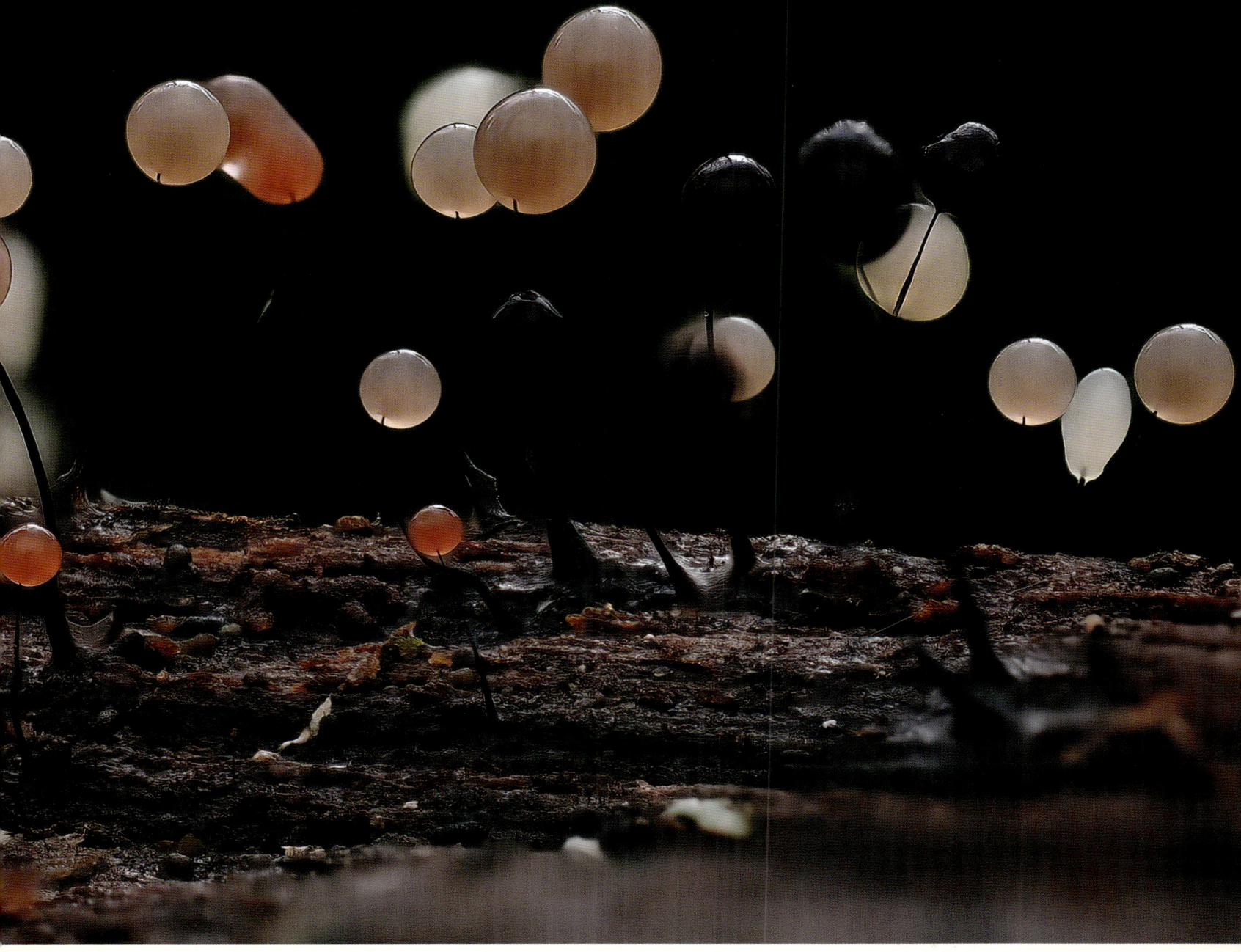

environmental conditions. When the environment becomes suboptimal, or when the plasmodium has exhausted its food sources, slime molds initiate a wondrous phase of their life cycle: reproduction. The gooey plasmodium transforms into a fruiting body, which can take various forms—stalks, tiny droplet-studded towers, or intricate fragile sculptures—all designed to disperse spores into the air, much like fungi. These fruiting bodies are often super-fun colors, displaying neon hues of yellow, orange, red, or even iridescent blues,

purples, and greens. Like fungi, slime molds are forest decomposers. By consuming bacteria and other decaying materials, they help to clean up the forest floor and recycle vital nutrients back into the ecosystem. Interestingly, some fungi compete with slime molds for resources, parasitizing them and impacting their growth and reproduction. Protozoa and other microorganisms prey on the spores of slime molds, maintaining a balance in their populations and contributing to the microbial diversity of the ecosystem. These parasitized slime

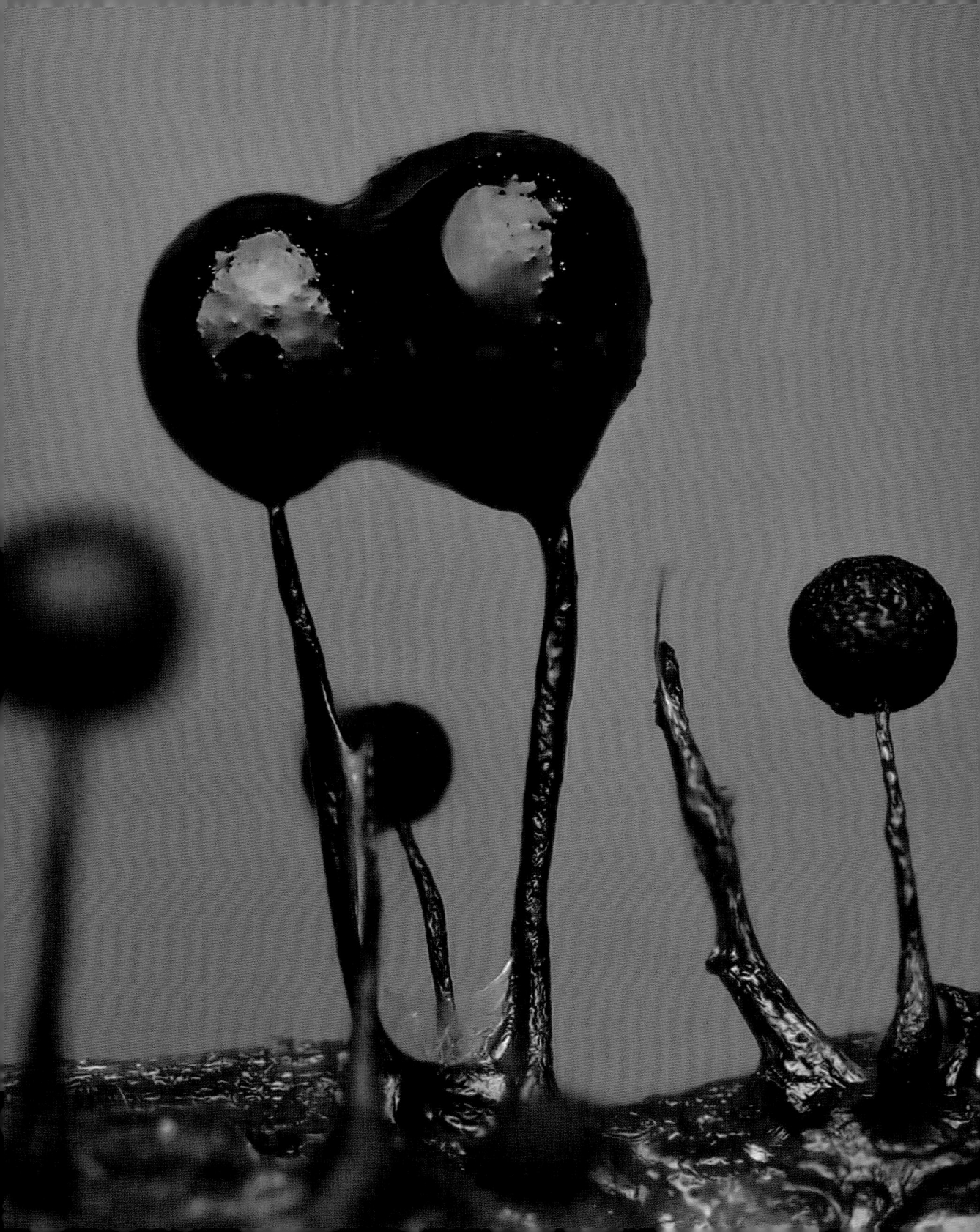

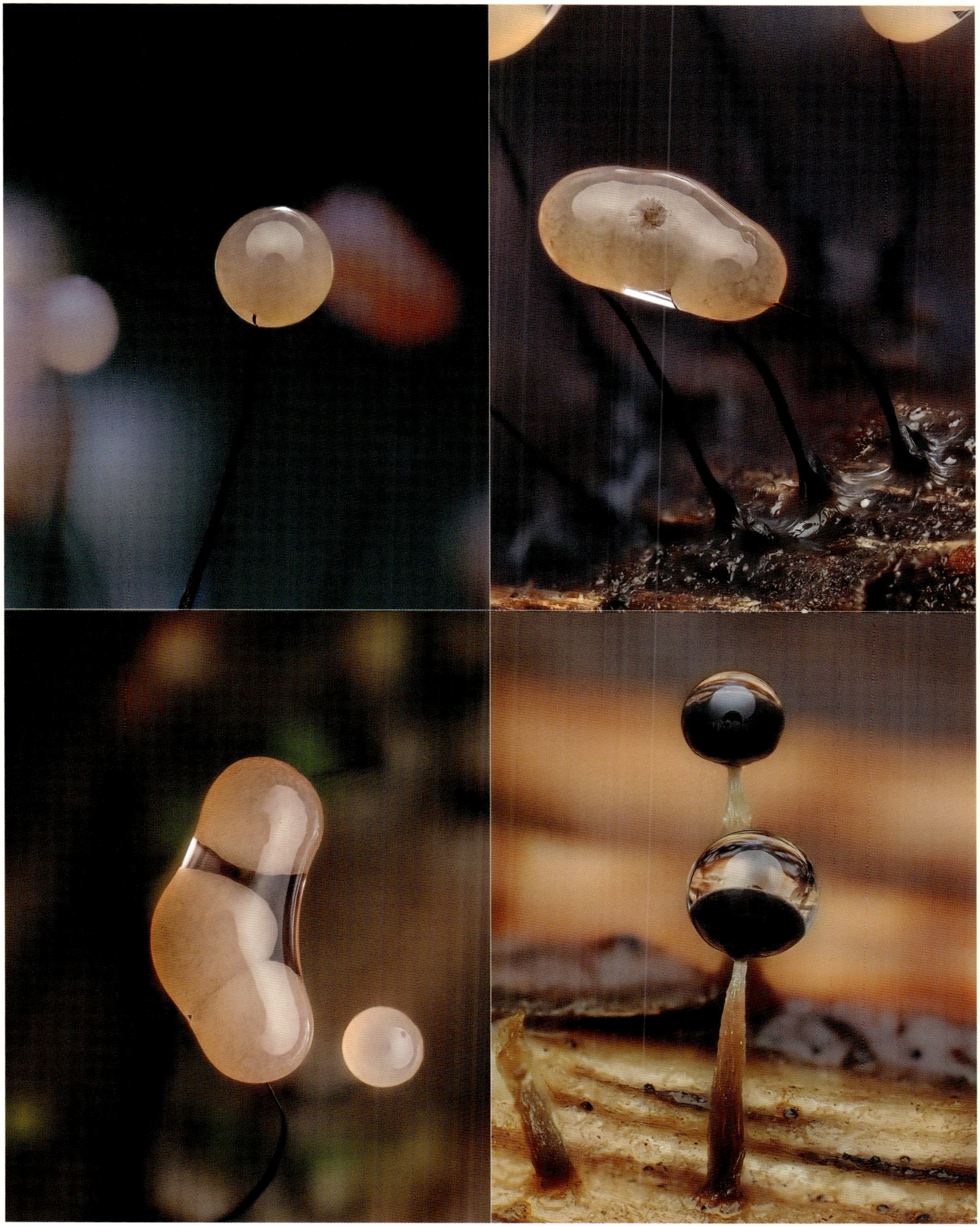

subjects are rare for me to find but make for some of my most abstract and unbelievable shots.

I don't know if it's just my experience, but slime molds are either nowhere to be seen or can be found in startling abundance—feast or famine. Slime molds seem to grow pretty much anywhere in the forest and can be found on leaf litter, dead wood, and anything close to the ground. As long as there is adequate moisture, and temperatures don't get too extreme, slime molds can thrive.

I live in an area that is loved by slime molds. I have had incredible luck finding it year-round and in abundance. Through an overwhelming amount of dedicated urban-forest excursions, I have figured out where and when I'm most likely to come across slime molds. I'm no expert, but sometimes, when the conditions seem all wrong, there it is!

The sweet spot for me seems to be temperate days after a week of rain, followed by a day or two of sunshine. London has a microclimate and generally has mild winters with the odd heatwave in summer or a brief uneventful snowfall every few years. The temperature averages around 52°F (11°C) much of the year, which, I've found, is perfect for slime molds.

Photographing slime molds can be tricky. Some start out as white blobs, clinging together in a big mass, and are either highly reflective or located in a difficult-to-reach place on the underside of a fallen tree or beneath a cluster of thorny stinging plants. Certain slime molds are also impossibly small and can require a lot of patience and extremely steady hands or the assistance of a tripod to get the shot. Slime molds such as *Comatricha* and *Hemitrichia* form thin structures known as sporangia during their reproductive phase. These sporangia usually measure about 0.5 to 2 millimeters in height and reliably force me to work extra hard to achieve a decent photograph.

Slime molds can also be difficult to identify properly in their developing stages. When I first started photographing them, I made many identification assumptions and mistakes; I'm far more cautious now with identification. Sometimes I crowdsource for IDs (thanks, Rosie!), but I've become more confident in the last few years since I tend to find the same species in the same areas year after year.

One of the joys of photographing slime molds is that once inspecting it under a macro lens, it is often inhabited by an array of tiny forest creatures. Much like fungi, slime molds play a critical yet understated role, providing nutrition to a diverse array of organisms. Slugs and snails annoyingly eat a lot of slime molds. They feed on both the plasmodial stage of slime molds and the fruiting bodies, or the reproductive structures. I really love gastropods, but they often beat me to the best species before I can get a shot. Springtails and beetles are another key group of slime mold consumers. It is very rare for me to photograph slime molds without a springtail photobomber or two.

My favorite slime mold to photograph is quite common: *Ceratiomyxa*, or coral slime mold. My tendency with macrophotography is to lean into abstract composition, and coral slime never lets me down in that regard. It comes in a few different varieties and the shapes it creates are quite varied. It can be bright white, peachy pink, or neon yellow. It can take the form of glossy and translucent toothlike nubs, fuzzy finger-like tentacles, a fluffy and vast fur-like shroud completely wrapping a log, a shockingly

84

OPPOSITE: PUSH PIN SLIME MOLD (*HEMITRICHIA CALYCULATA*).

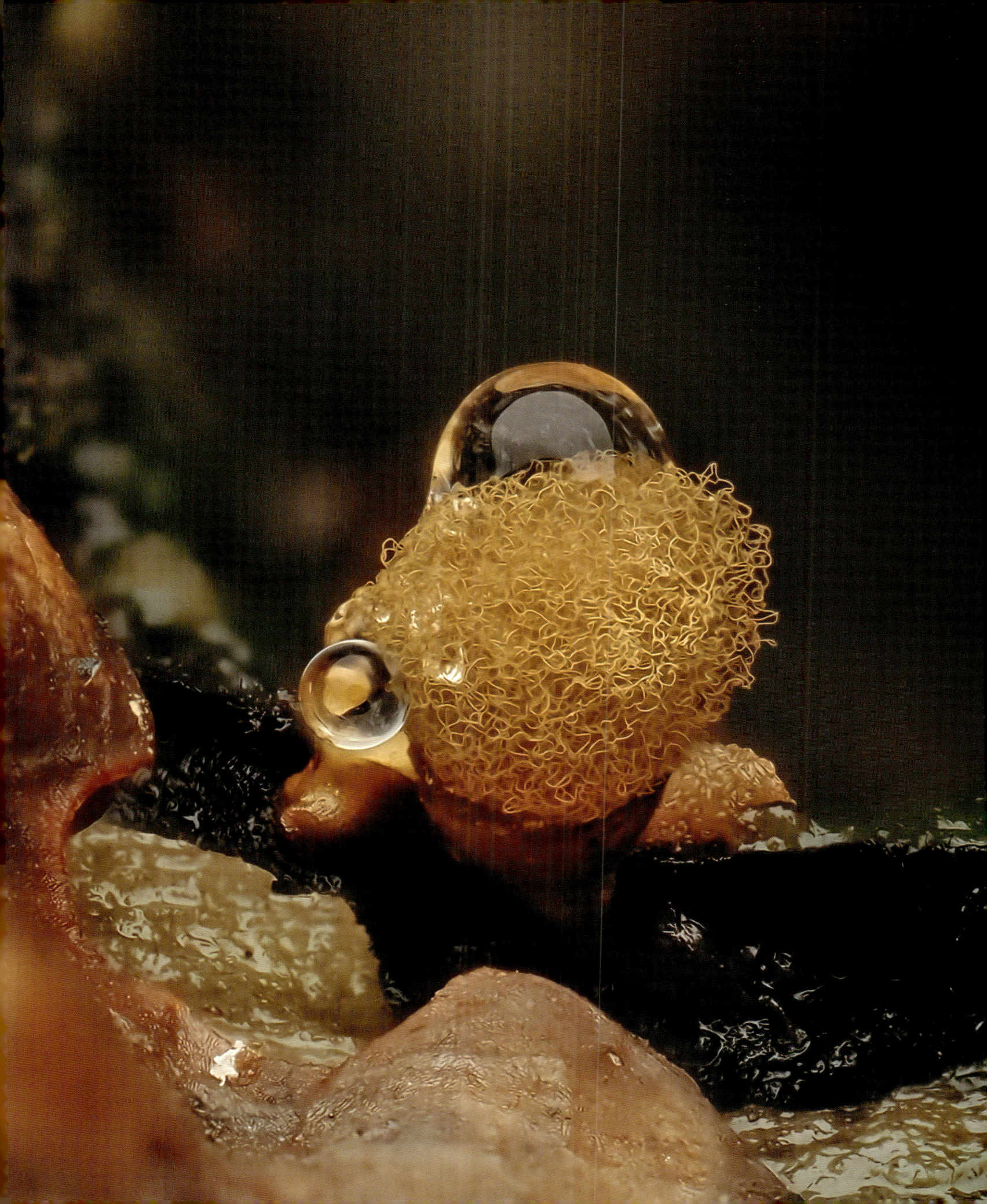

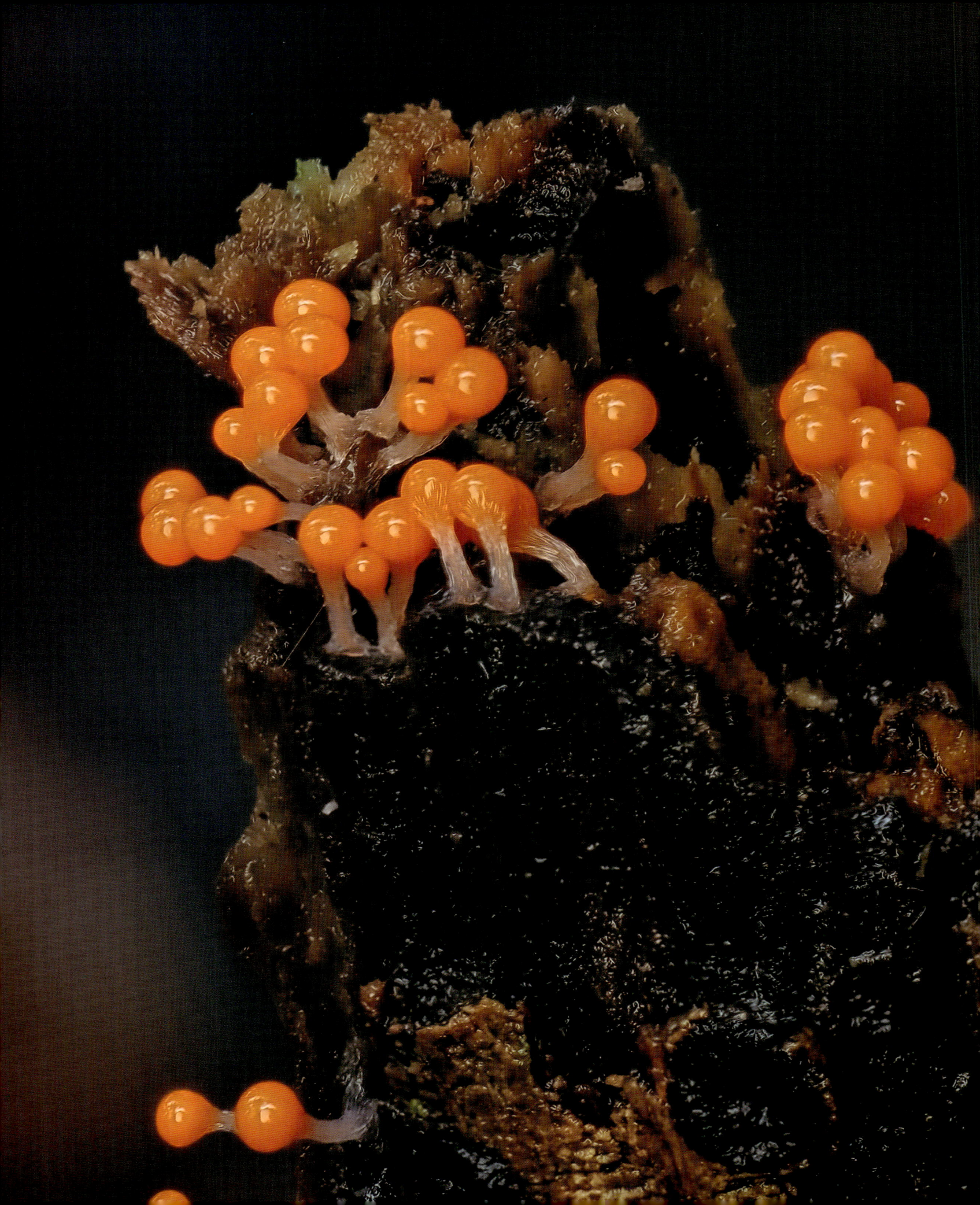

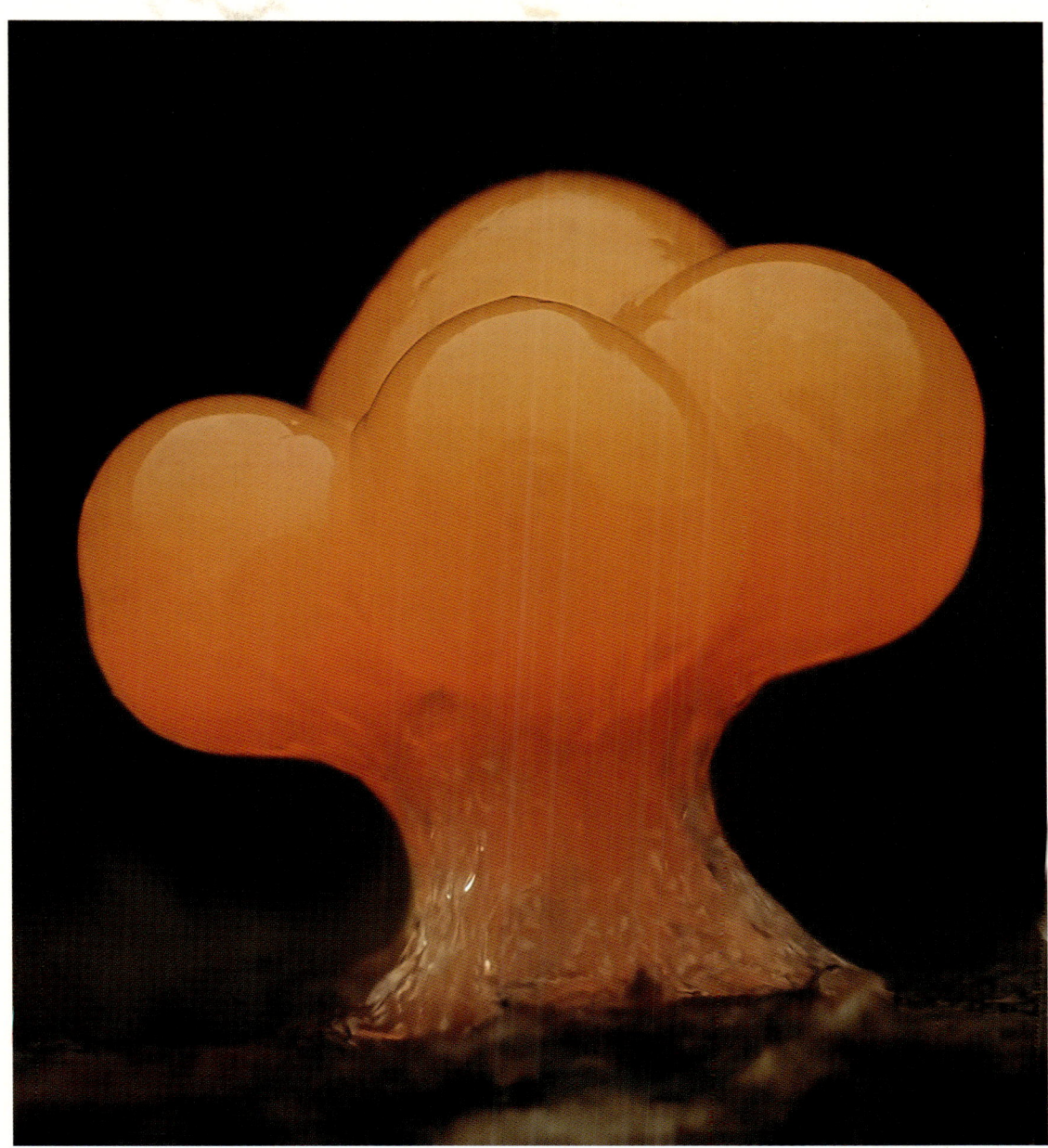

yellow complex honeycomb, or a perfectly formed blanket of spikes. It is so fascinating, and I never tire of it.

There are countless species of slime mold, and I've only seen most of them online. It's my greatest macro-hope to one day travel somewhere like Tasmania, (where slime mold seems to rule the day) and have the opportunity to check some more species off my bucket list. There is so much to see and learn. I have barely scratched the surface of slime mold information here. I strongly encourage you to seek out additional literature written by actual scientists dedicated to myxomycetes magic.

ABOVE AND OPPOSITE: SALMON-EGGS (*HEMITRICHIA DECIPIENS*).

88

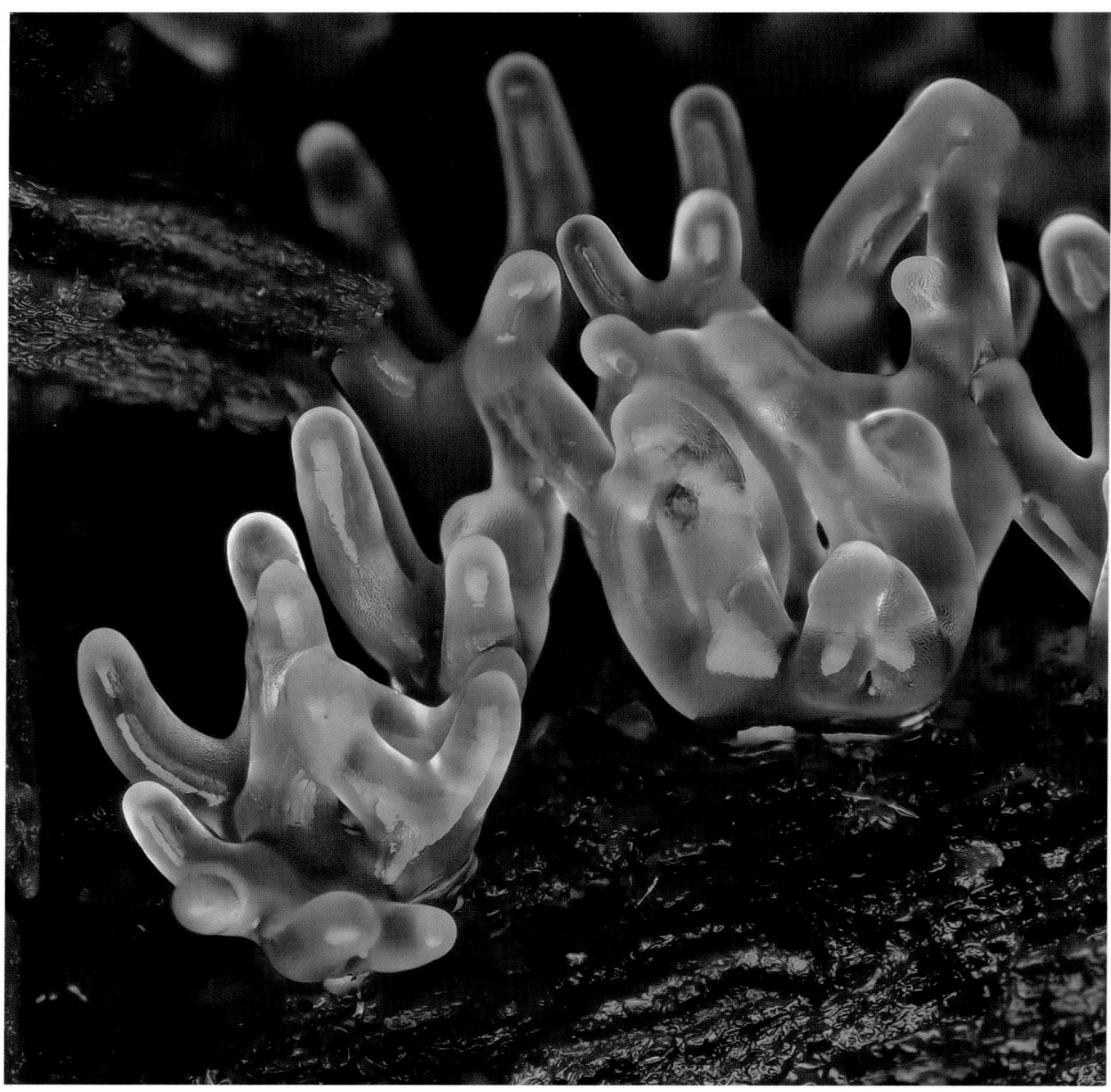

ABOVE: COMMON CORAL SLIME (*CERATIOMYXA FRUTICULOSA*). OPPOSITE: *CERATIOMYXA FRUTICULOSA F. FLAVA.*

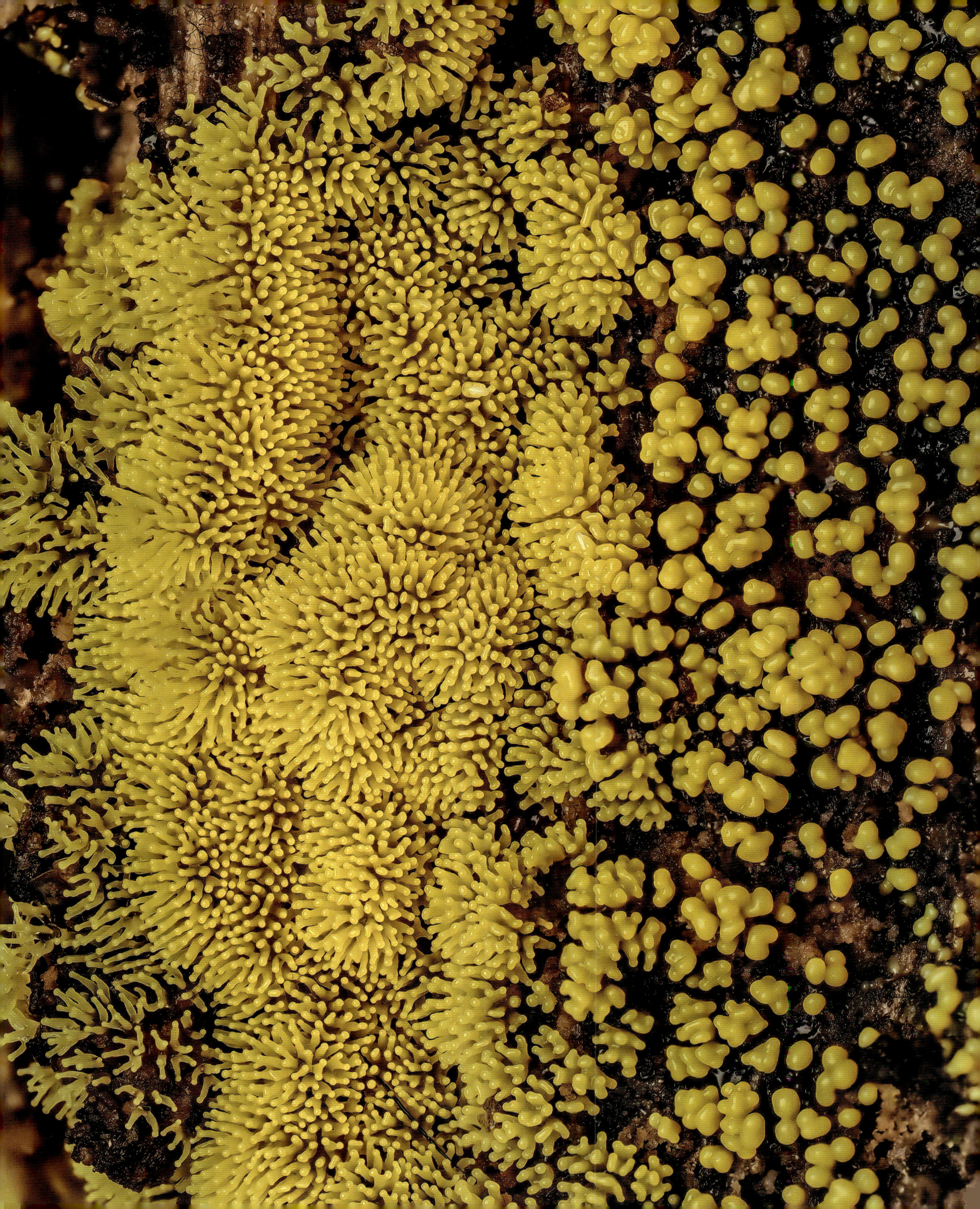

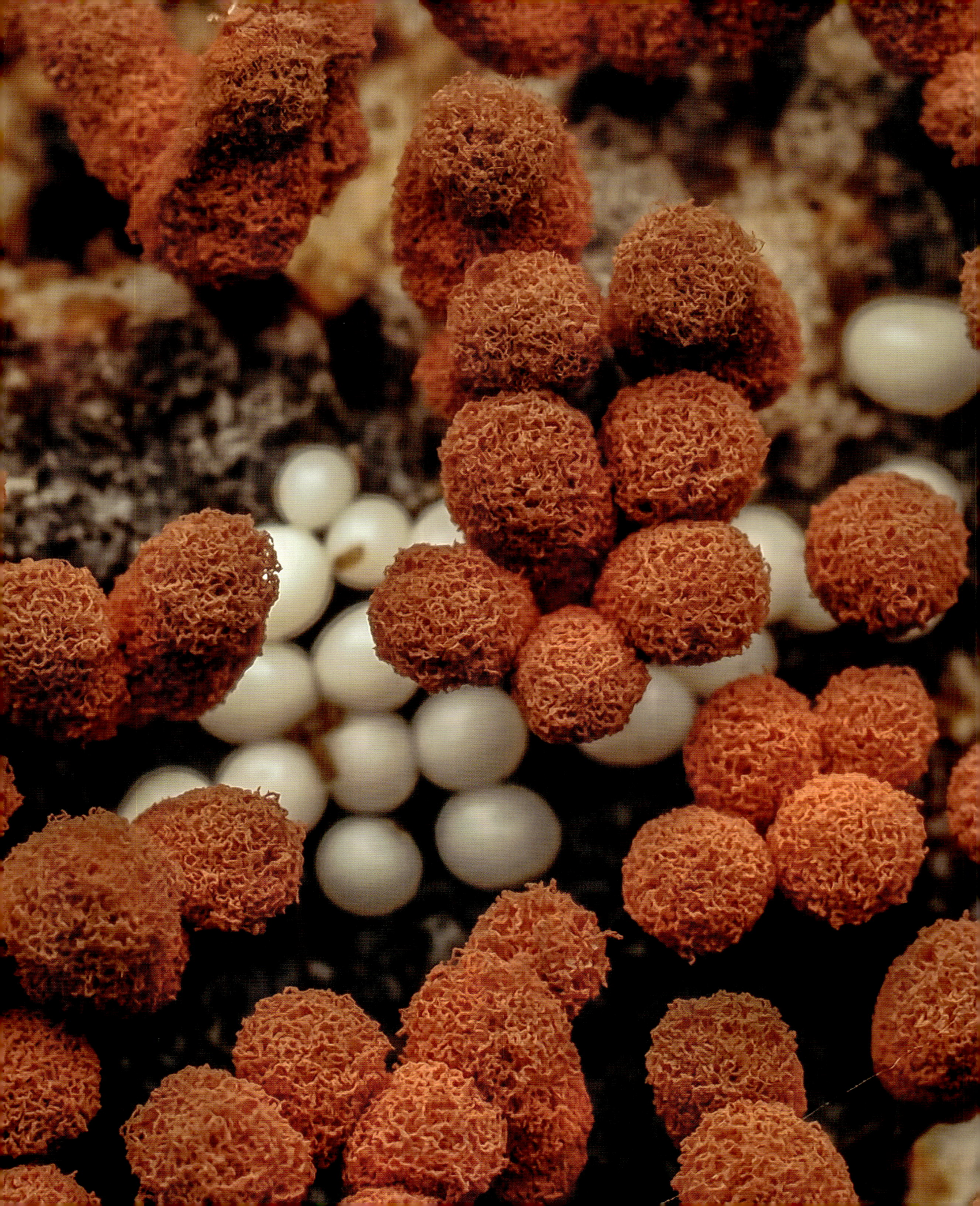

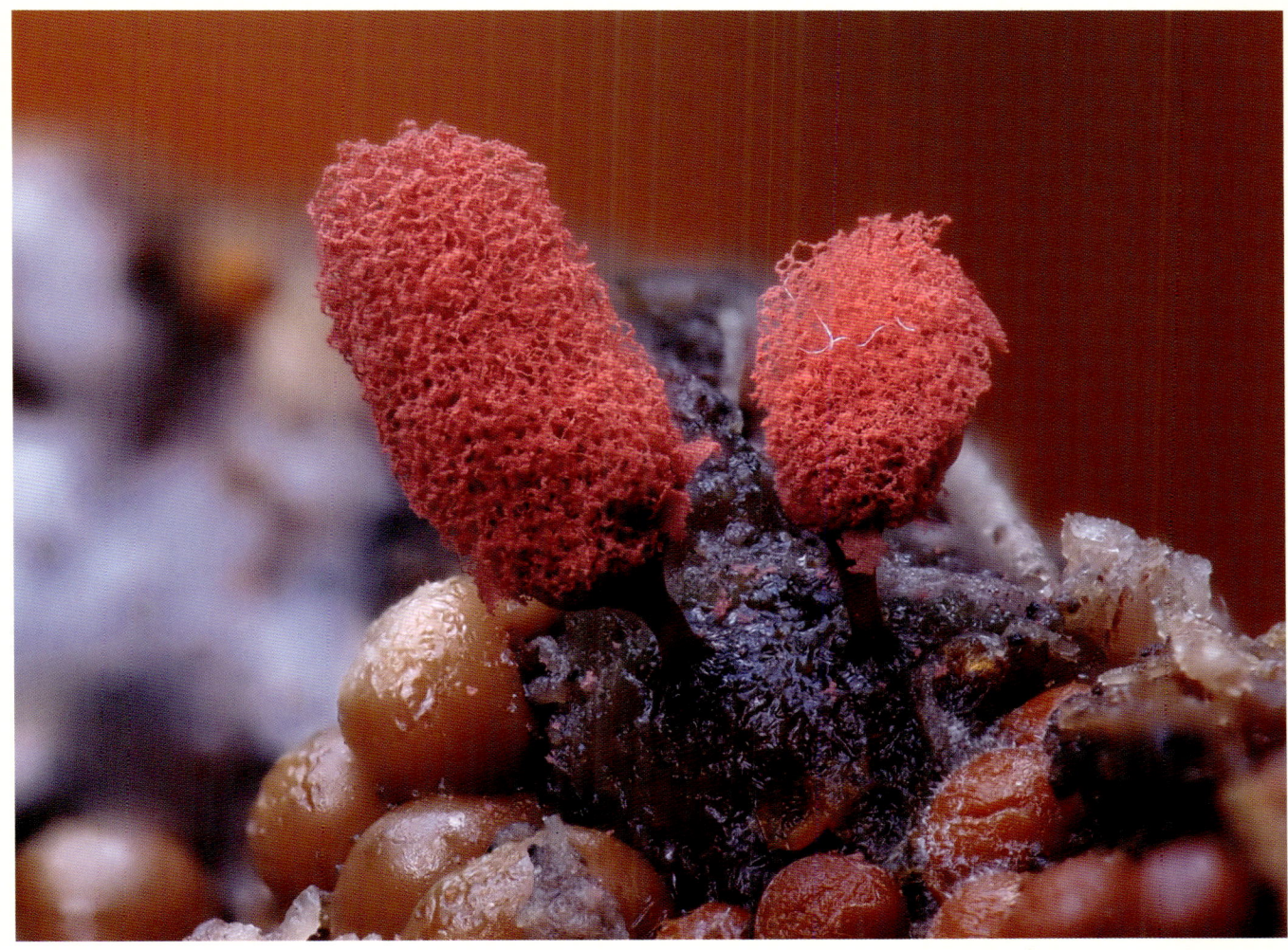

ABOVE: *ARCYRIA* SPP. FOUND IN LONDON, UK. OPPOSITE: *ARCYRIA* SPP. FOUND IN NEW FOREST NATIONAL PARK, UK.

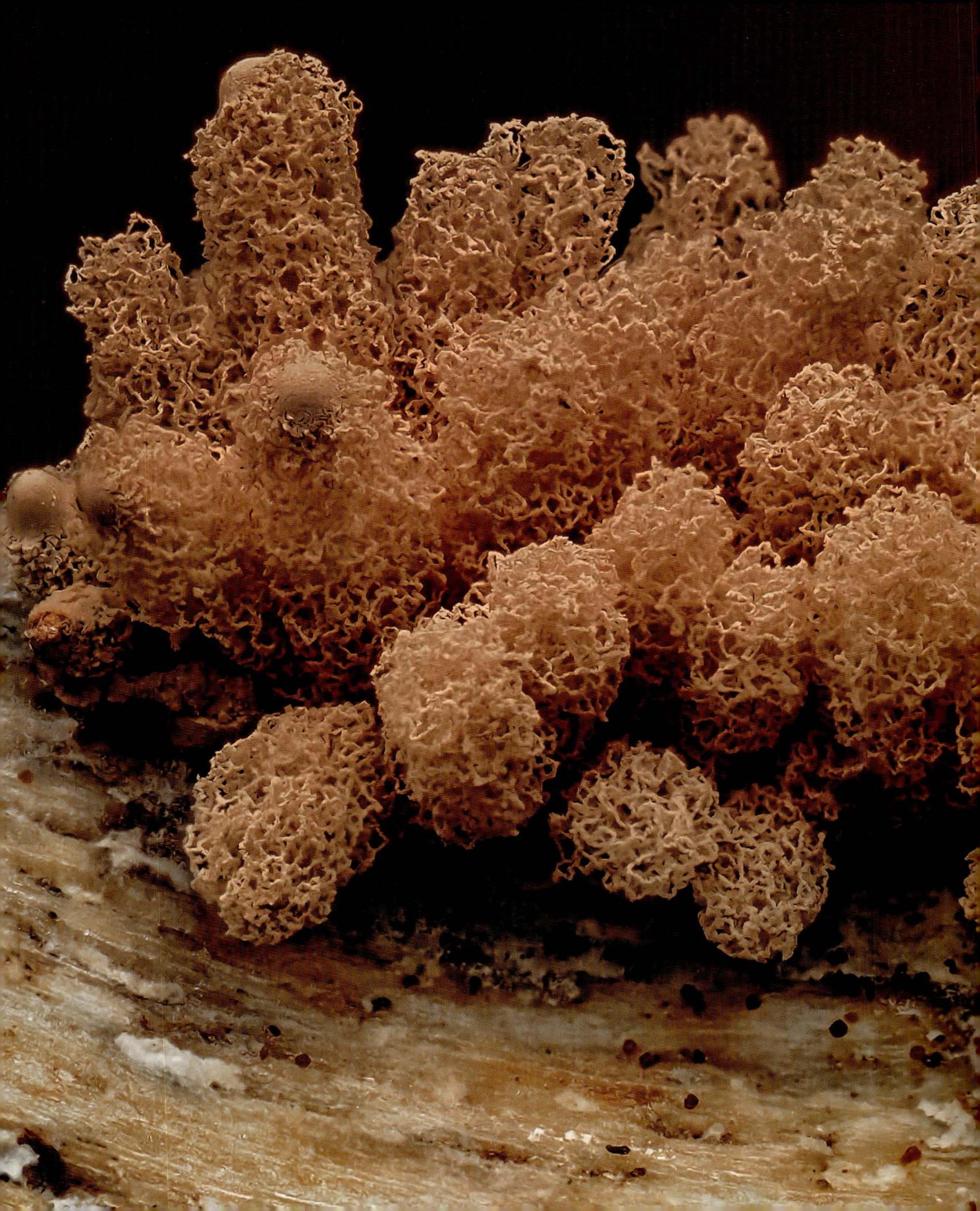

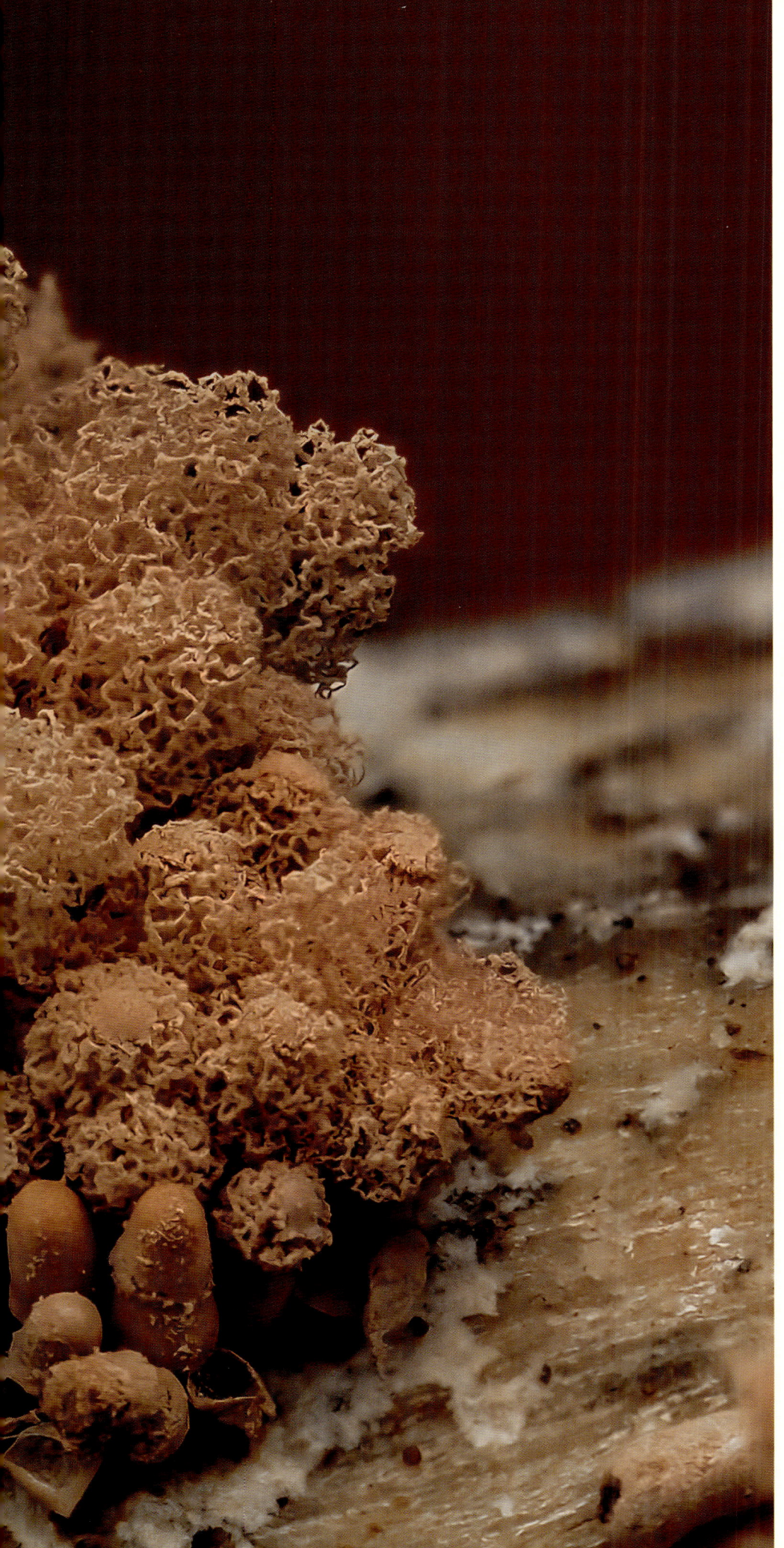

93

LEFT: *ARCYRIA* SPP.

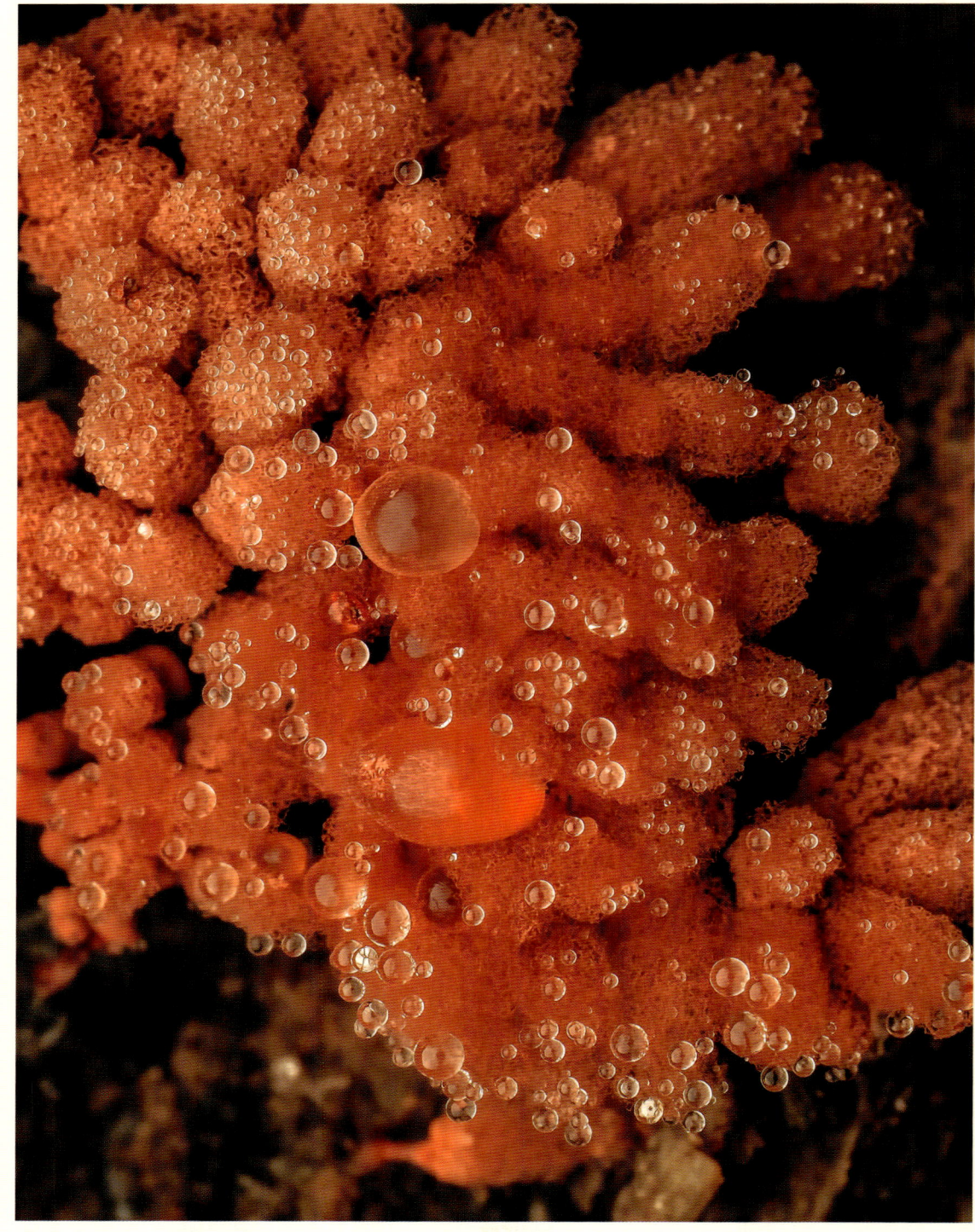

ABOVE: *ARCYRIA* SPP. WITH WATER DROPLETS. OPPOSITE: FROZEN *ARCYRIA* SPP.

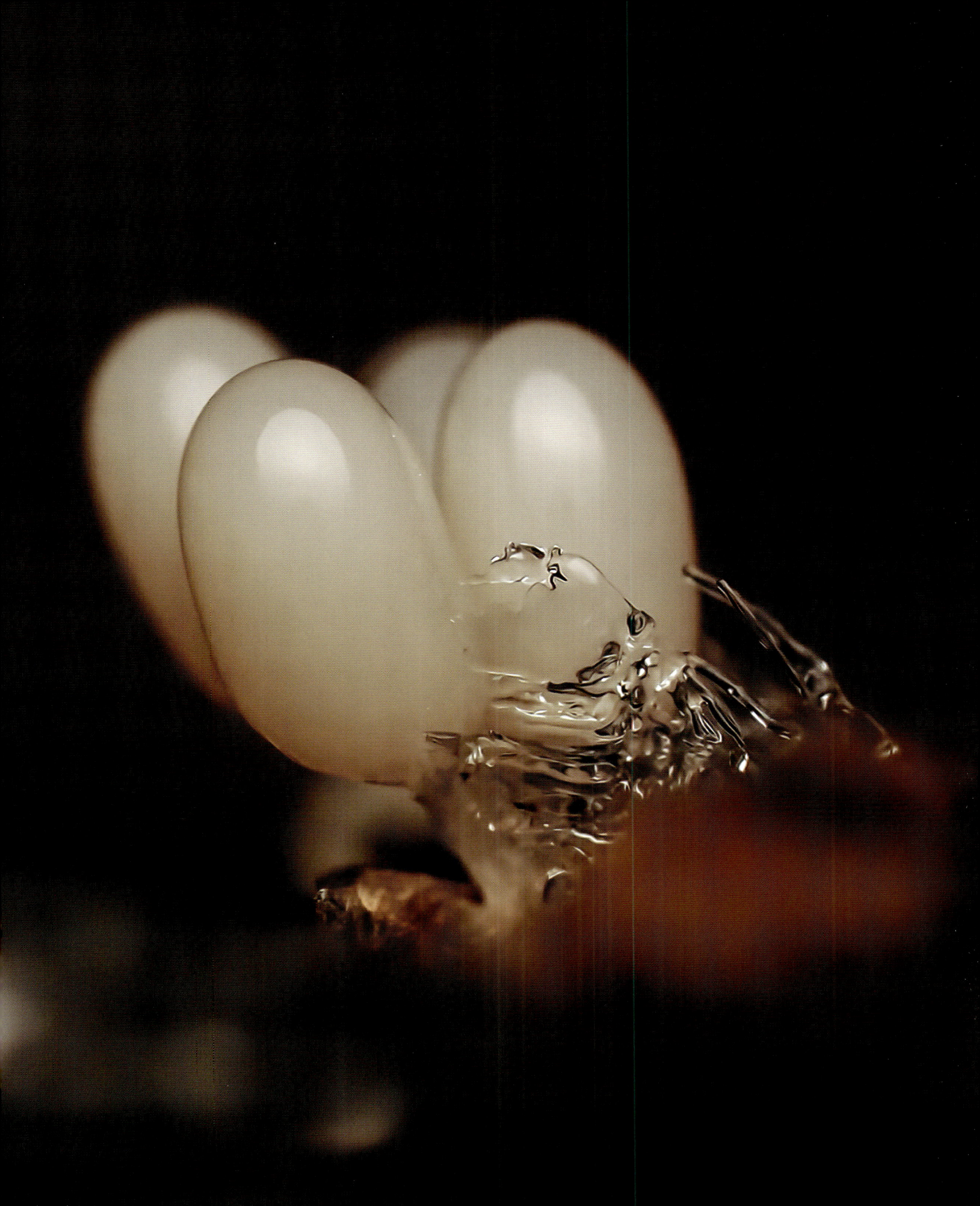

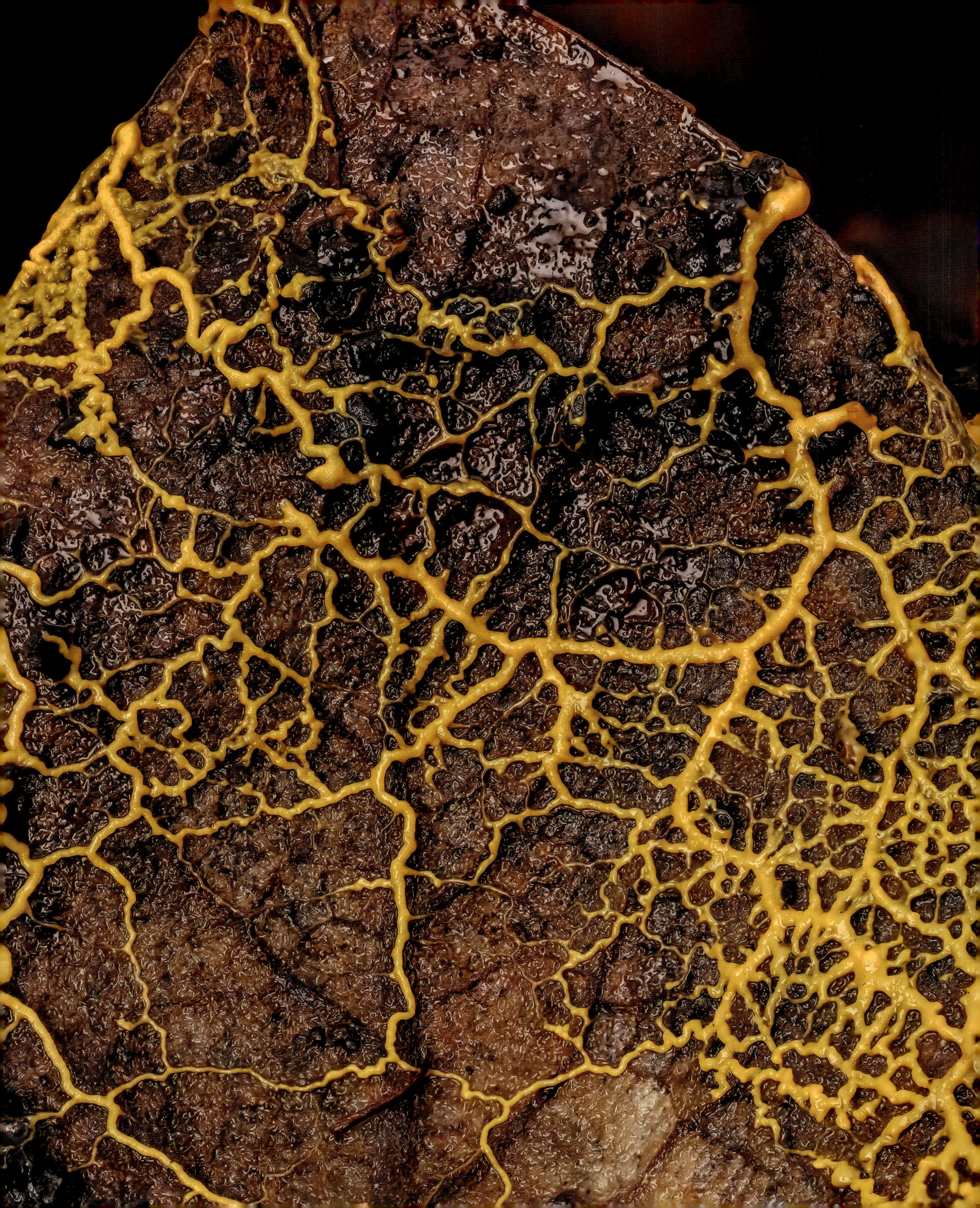

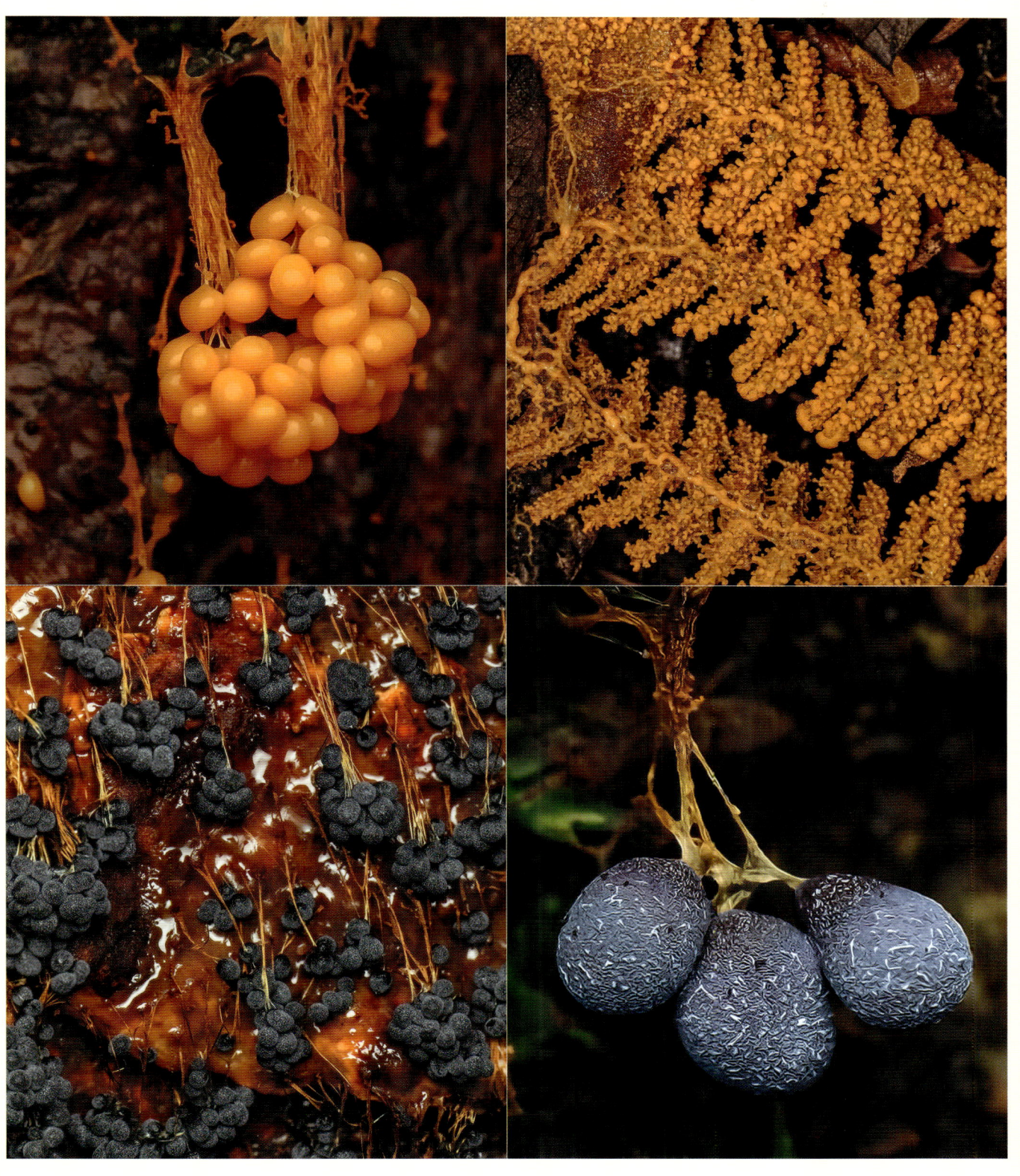

ABOVE LEFT, BELOW LEFT, AND BELOW RIGHT: HANGING SLIME MOLD (*BADHAMIA UTRICULARIS*). ABOVE RIGHT: HANGING SLIME MOLD (*B. UTRICULARIS*) ON A FERN. OPPOSITE: HANGING SLIME MOLD (*B. UTRICULARIS*) ON A LEAF.

98

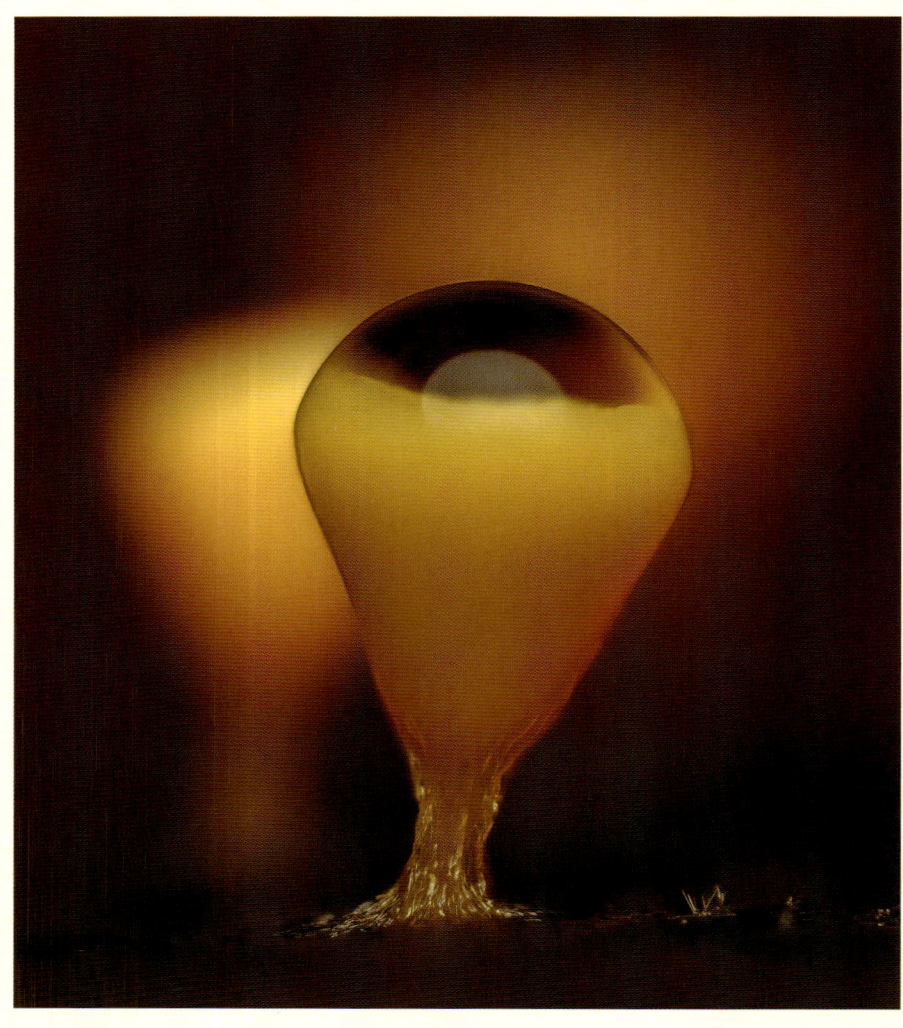

ABOVE: JUVENILE *CRATERIUM* SP. OPPOSITE: JUVENILE *CRATERIUM* SP.

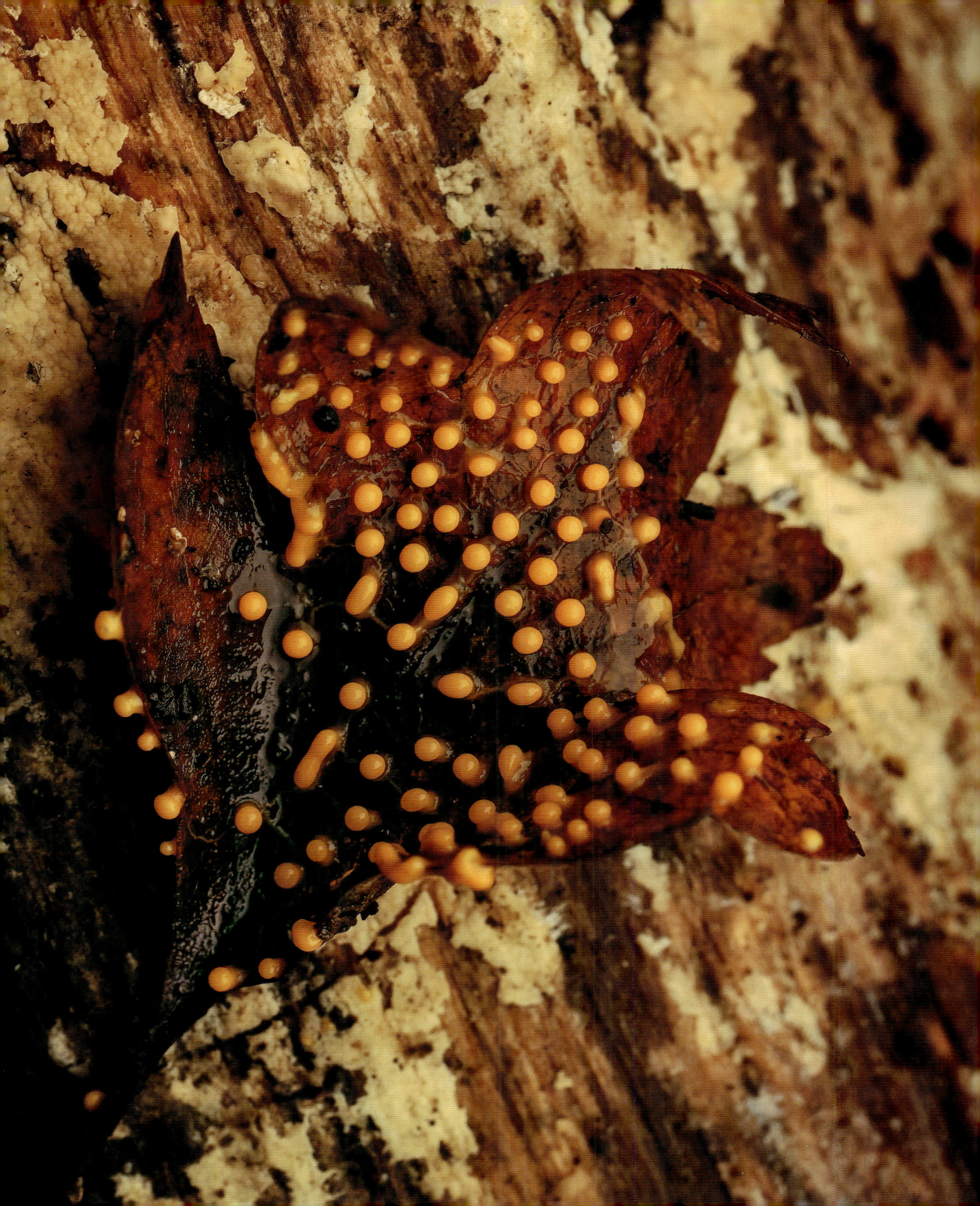

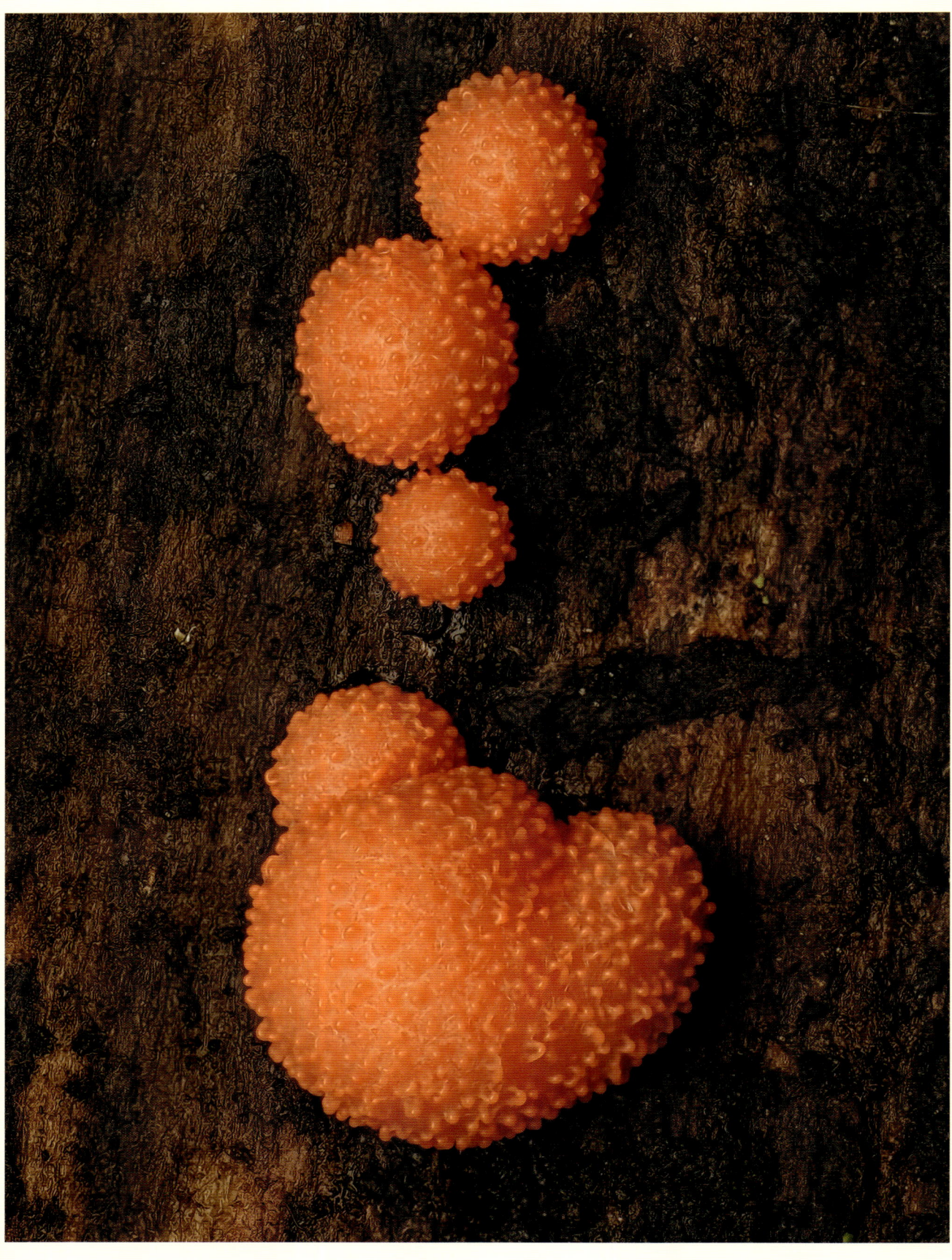

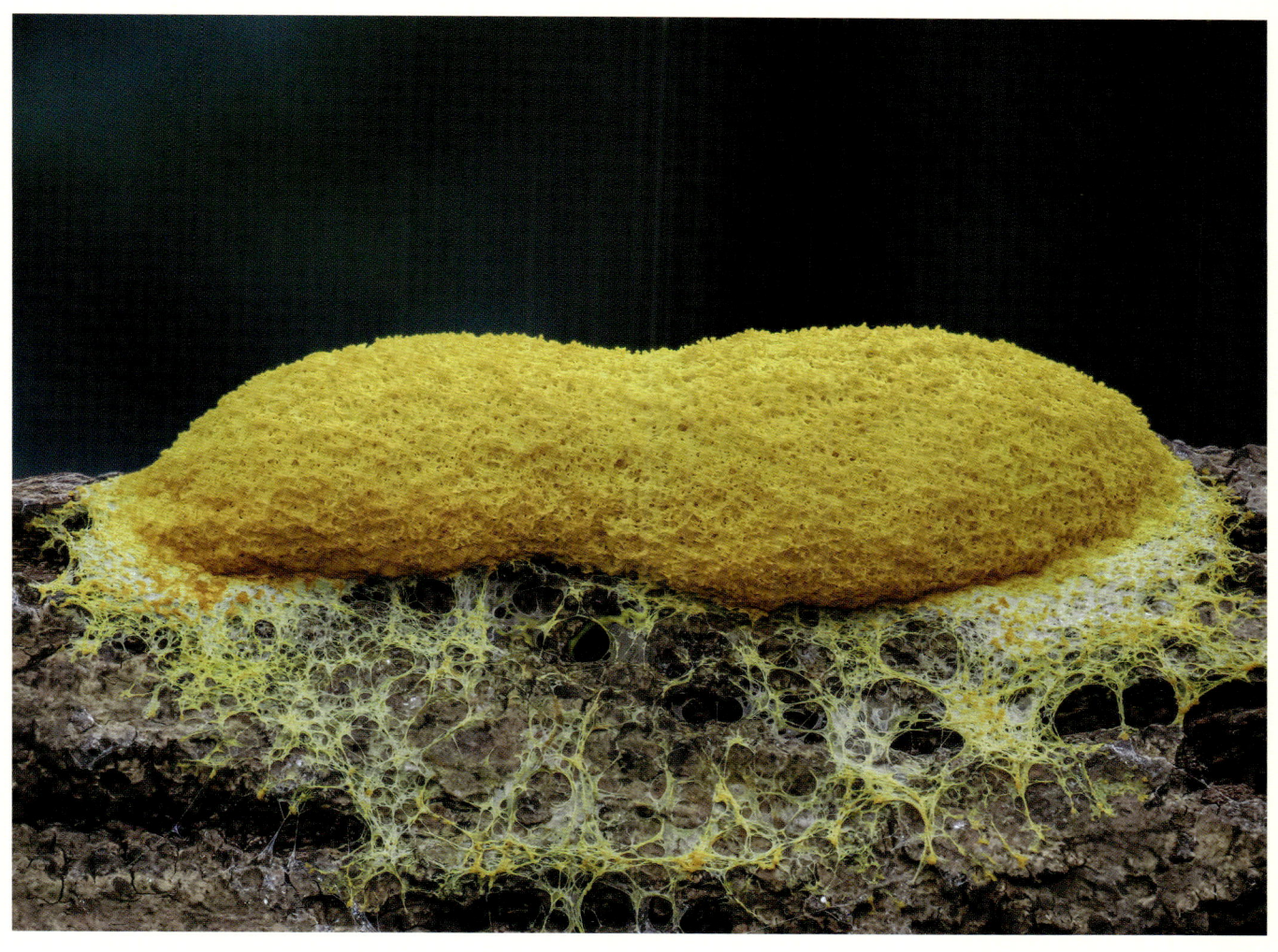

ABOVE: WOLF'S MILK (*LYCOGALA EPIDENDRUM*). OPPOSITE: DOG VOMIT SLIME MOLD (*FULIGO SEPTICA*).

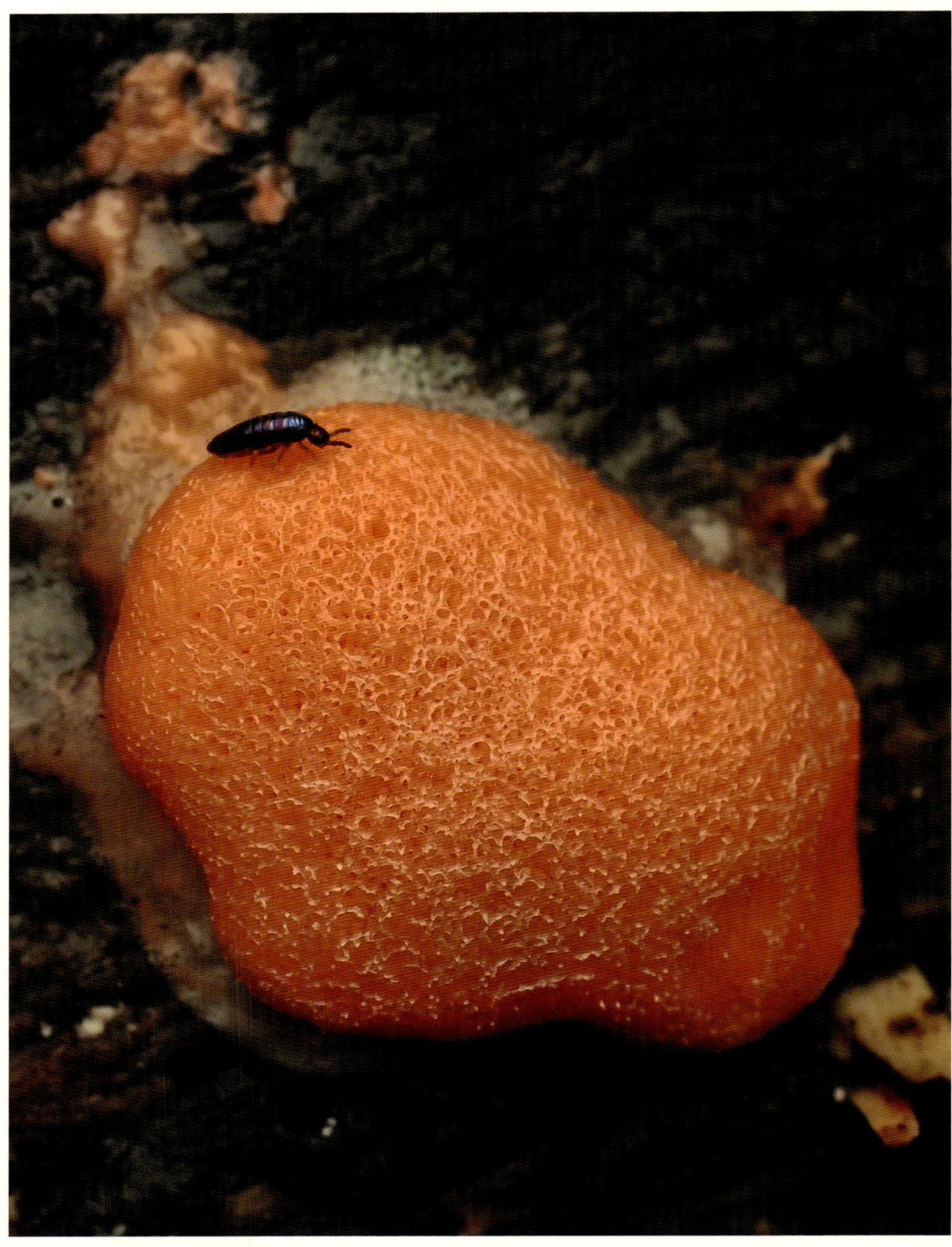

ABOVE AND OPPOSITE: *DICTYDIAETHALIUM PLUMBEUM.*

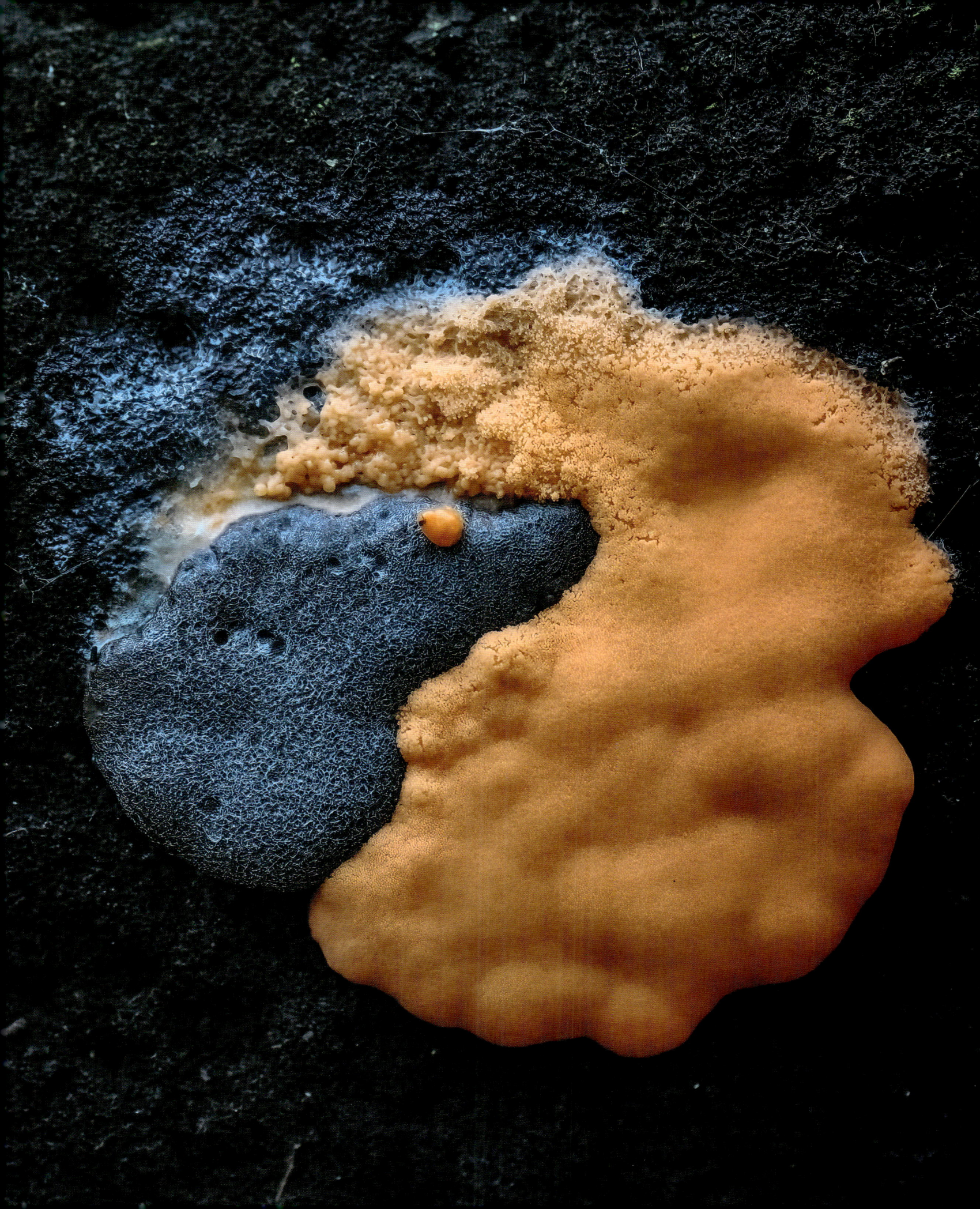

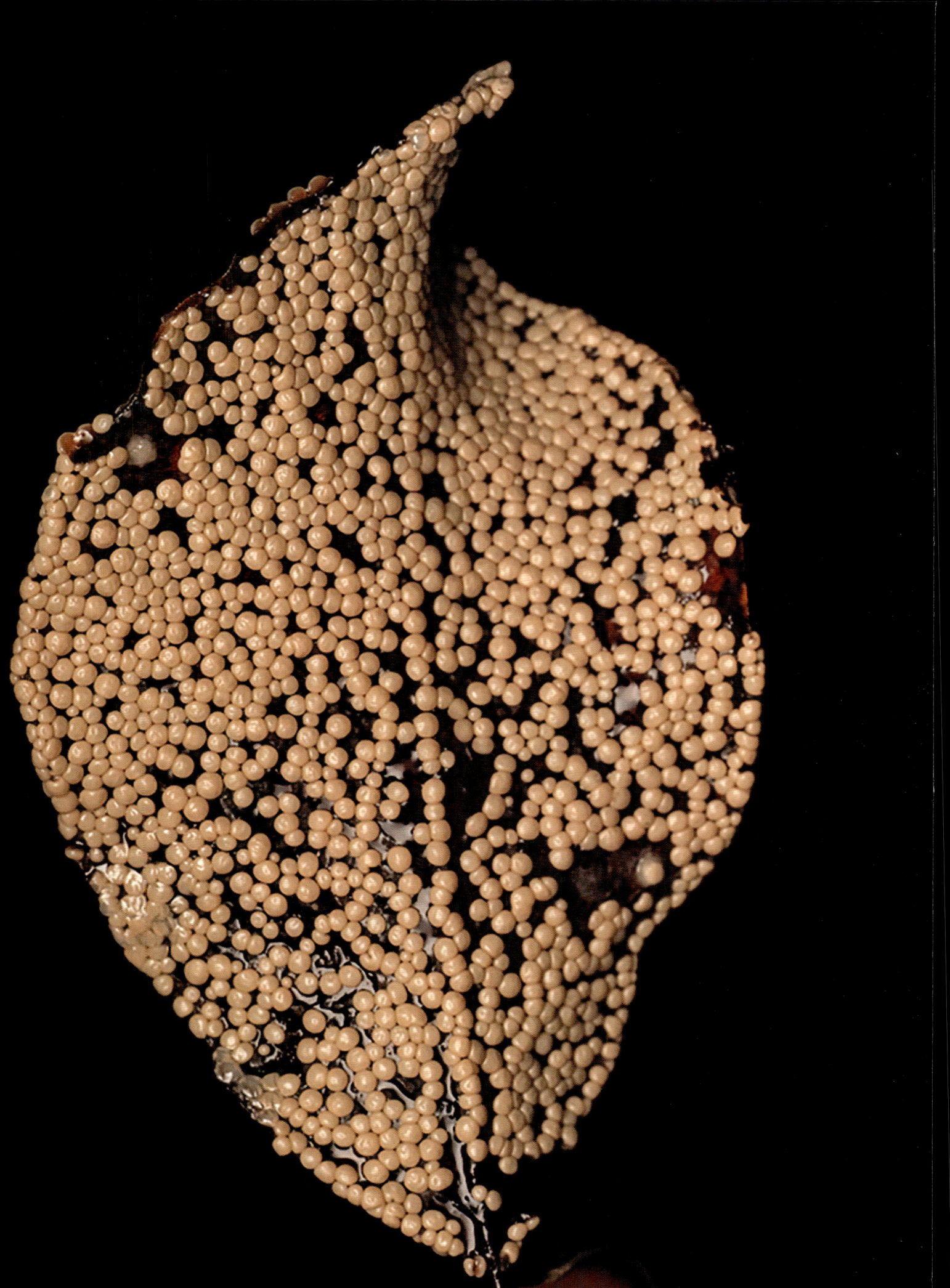

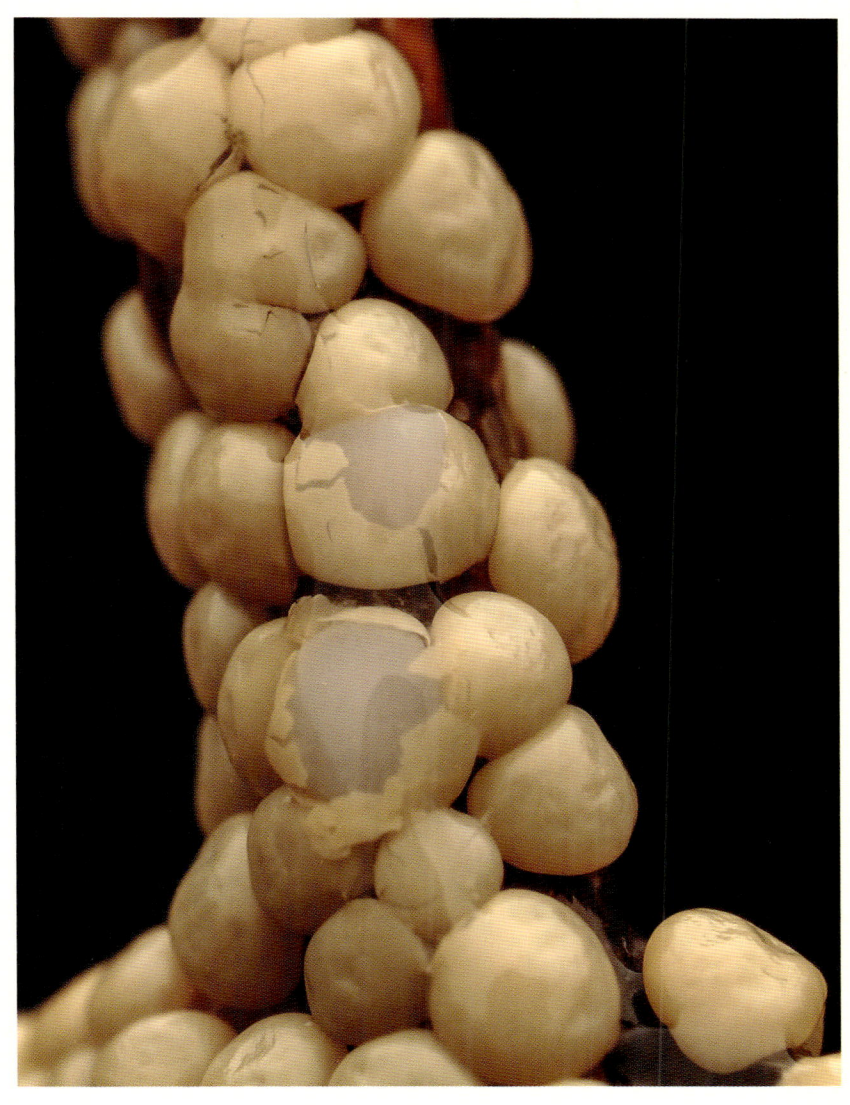

ABOVE AND OPPOSITE: *DIDERMA* SPP.

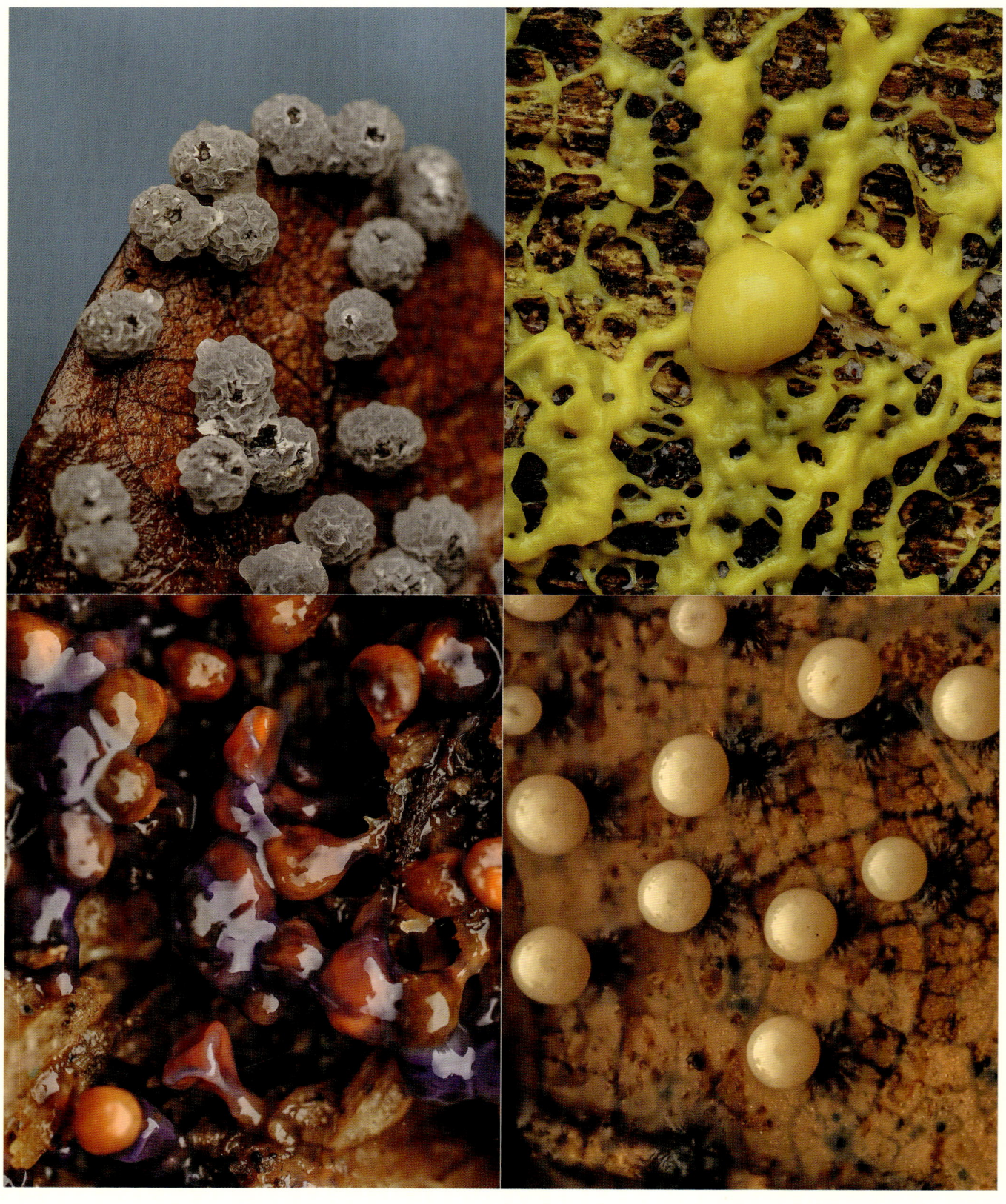

ABOVE LEFT: *DIDYMIUM* SPP. ABOVE RIGHT: *PLASMODIUM* SPP. BELOW LEFT AND BELOW RIGHT: UNKNOWN SLIME MOLD. OPPOSITE: UNKNOWN SLIME DEVELOPING.

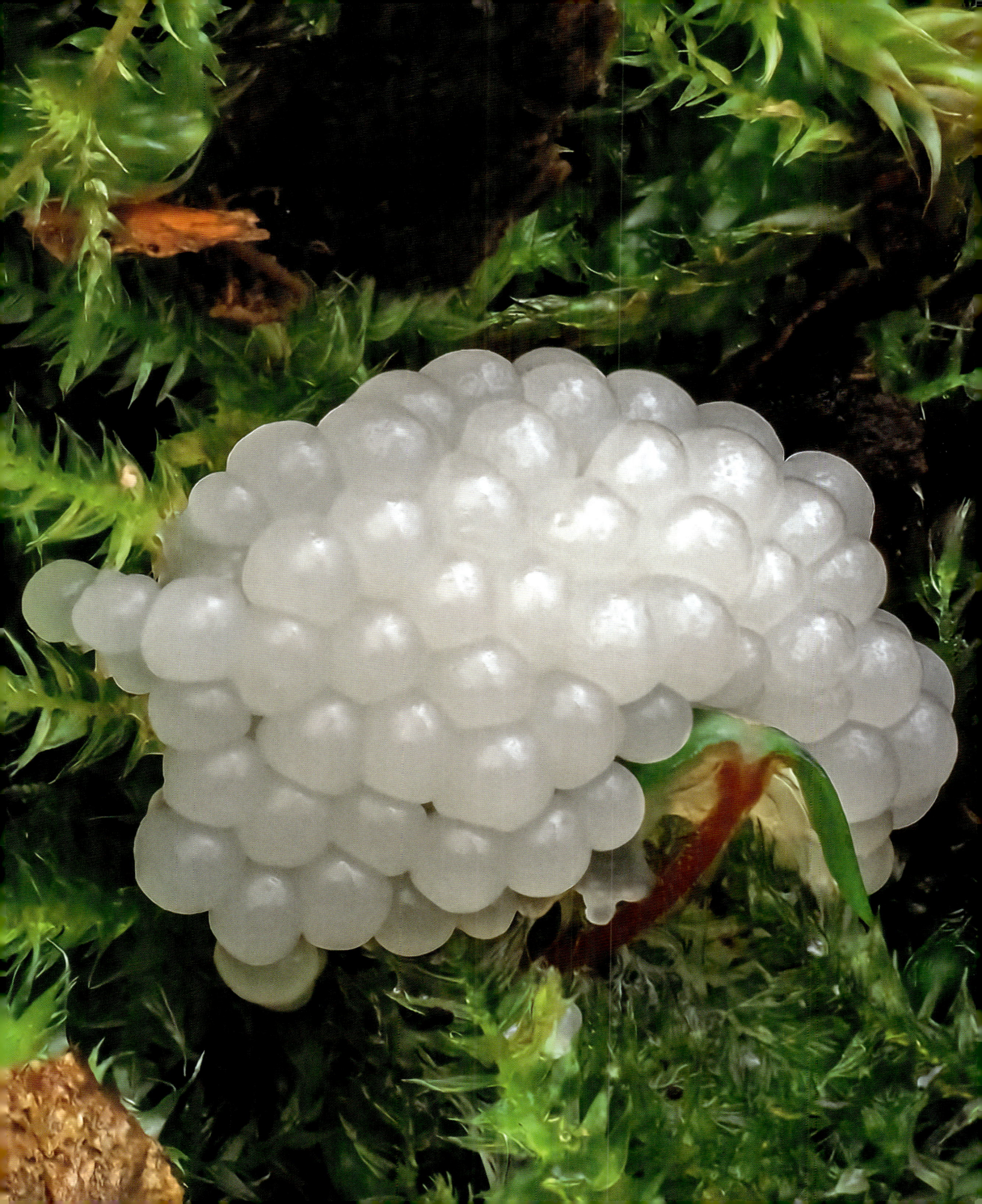

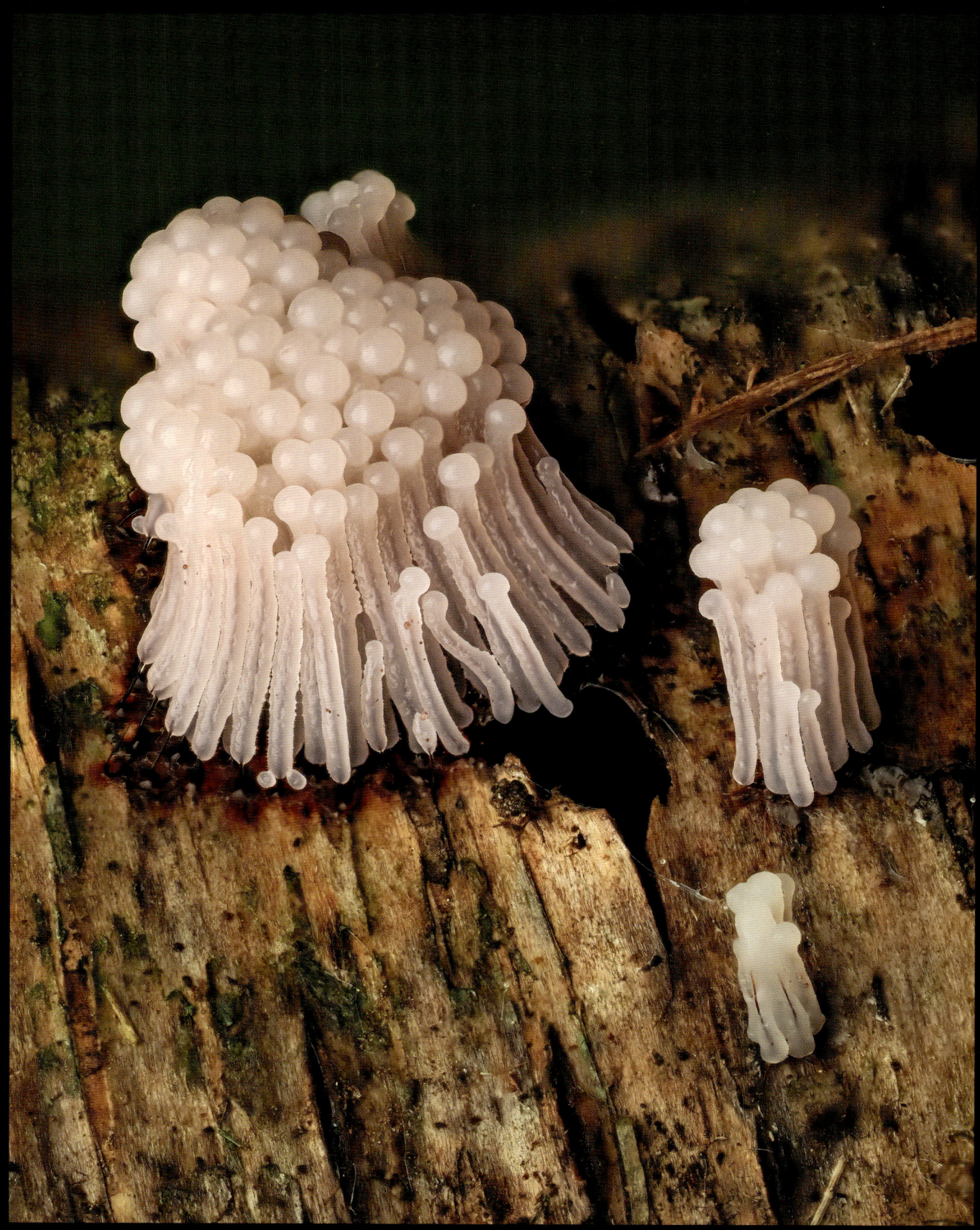

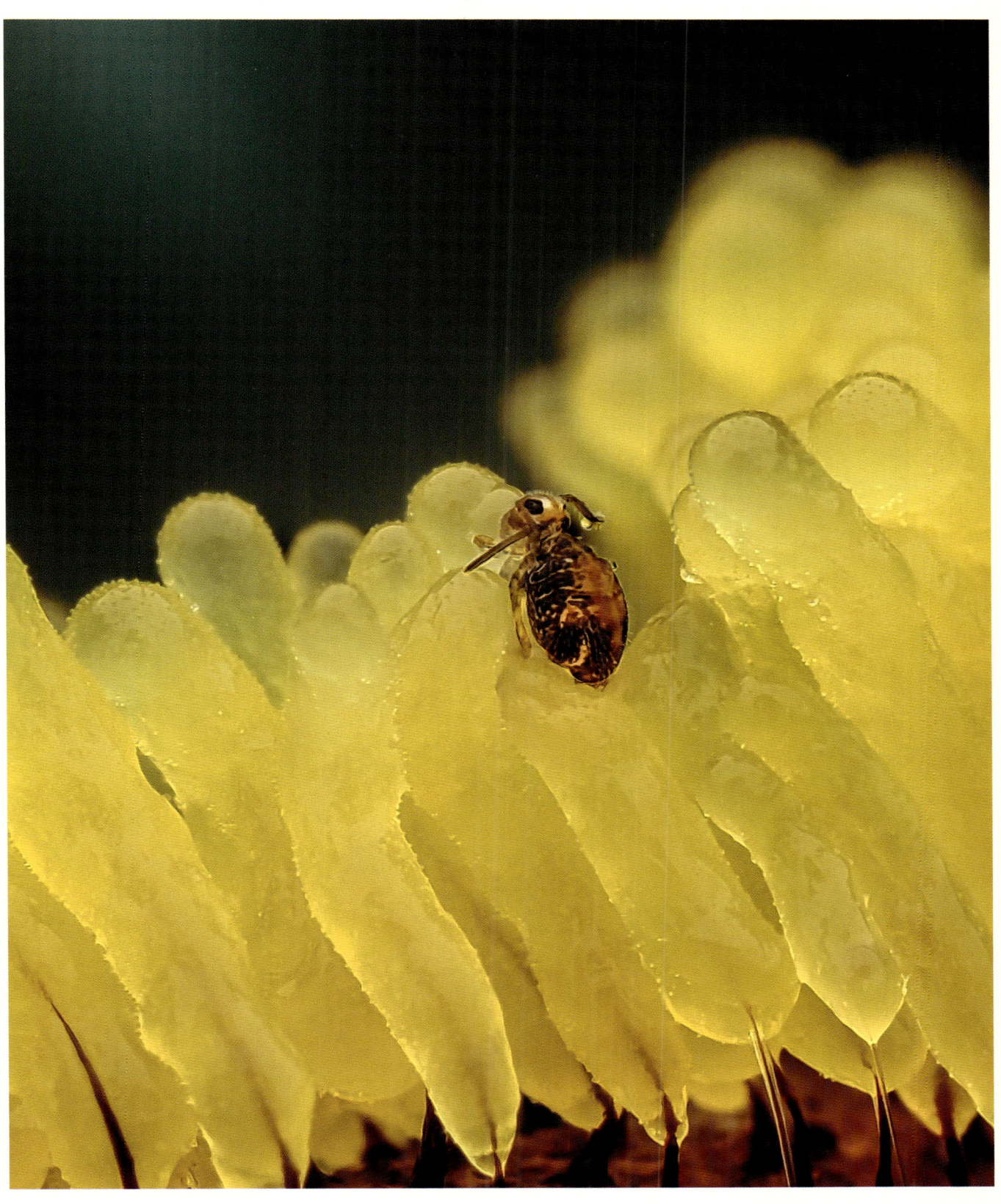

ABOVE: *STEMONITIS FLAVOGENITA* AND A GLOBULAR SPRINGTAIL. OPPOSITE: *STEMONITIS* SPP.

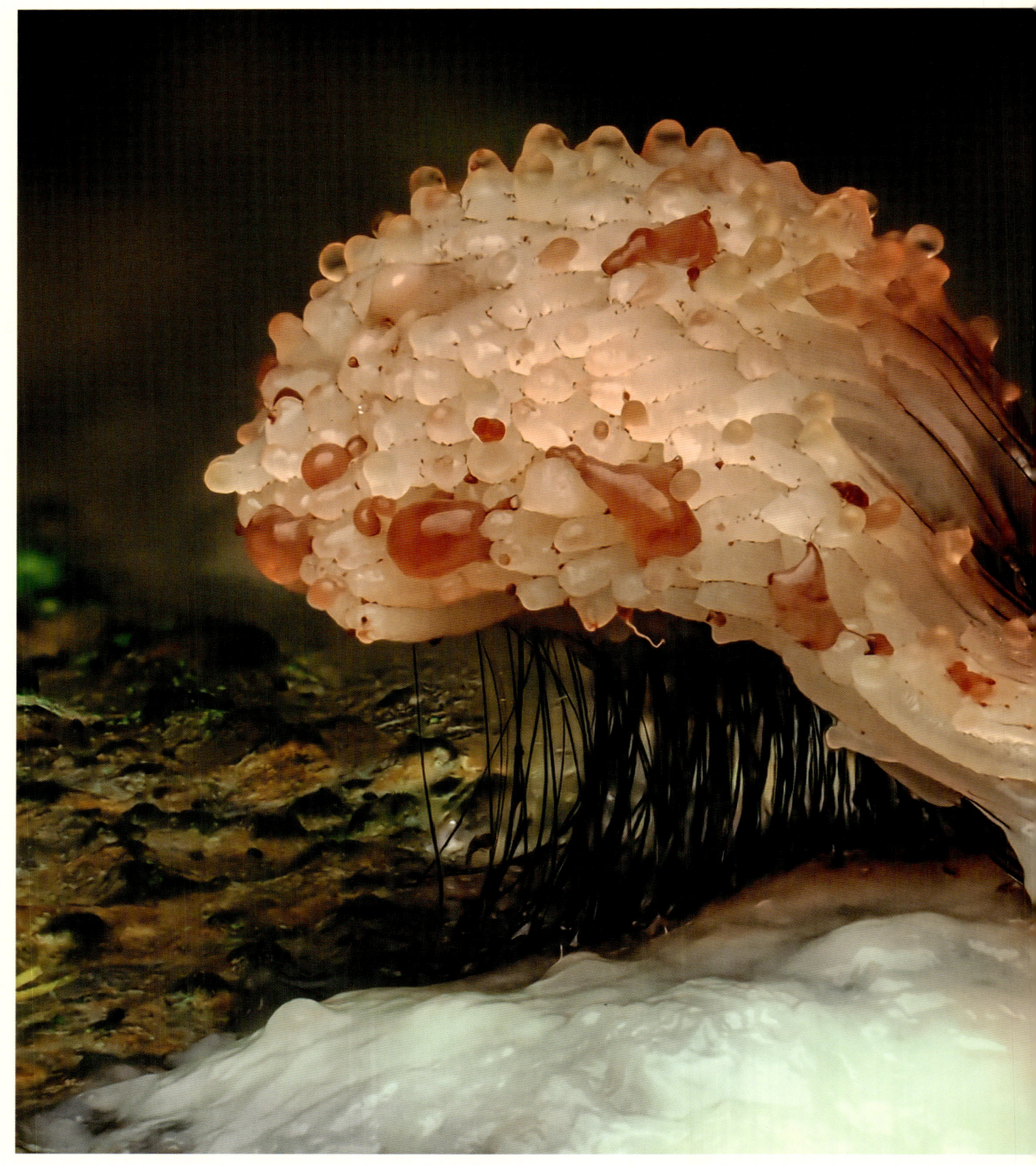

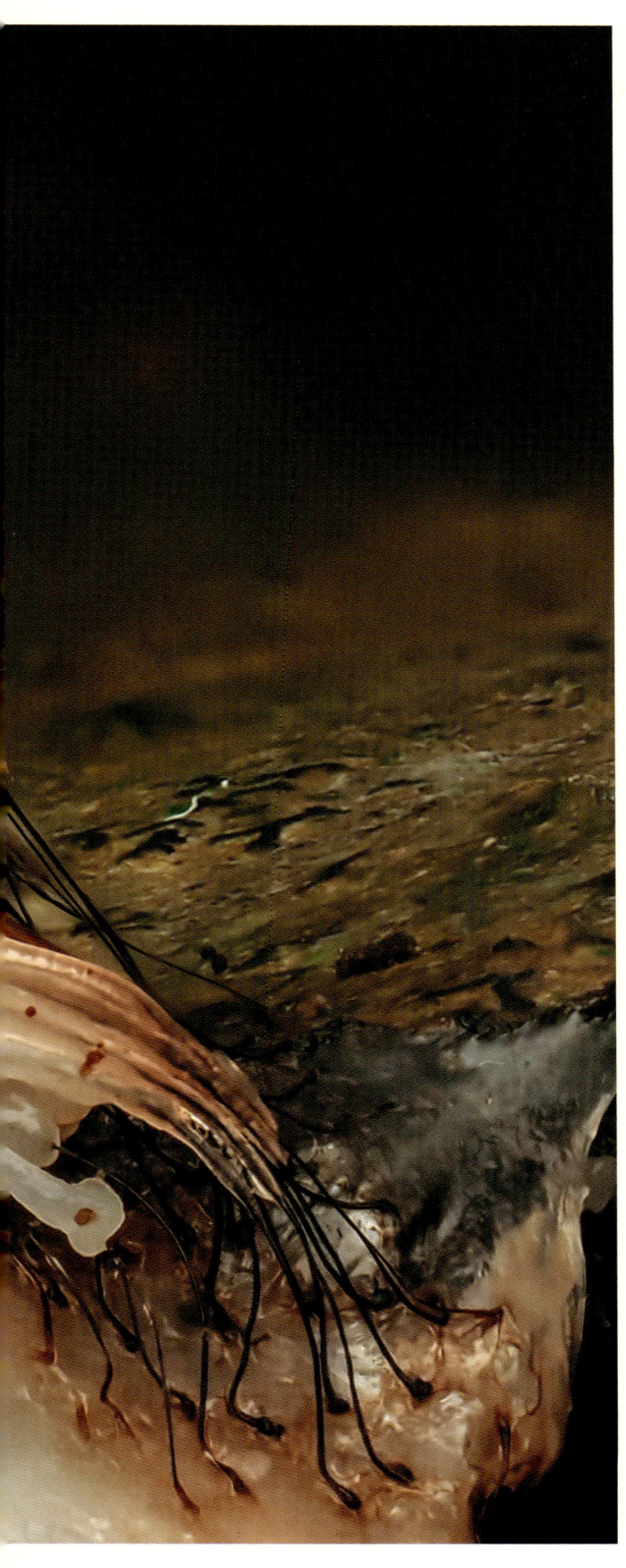

LEFT: *STEMONITIS* SPP.

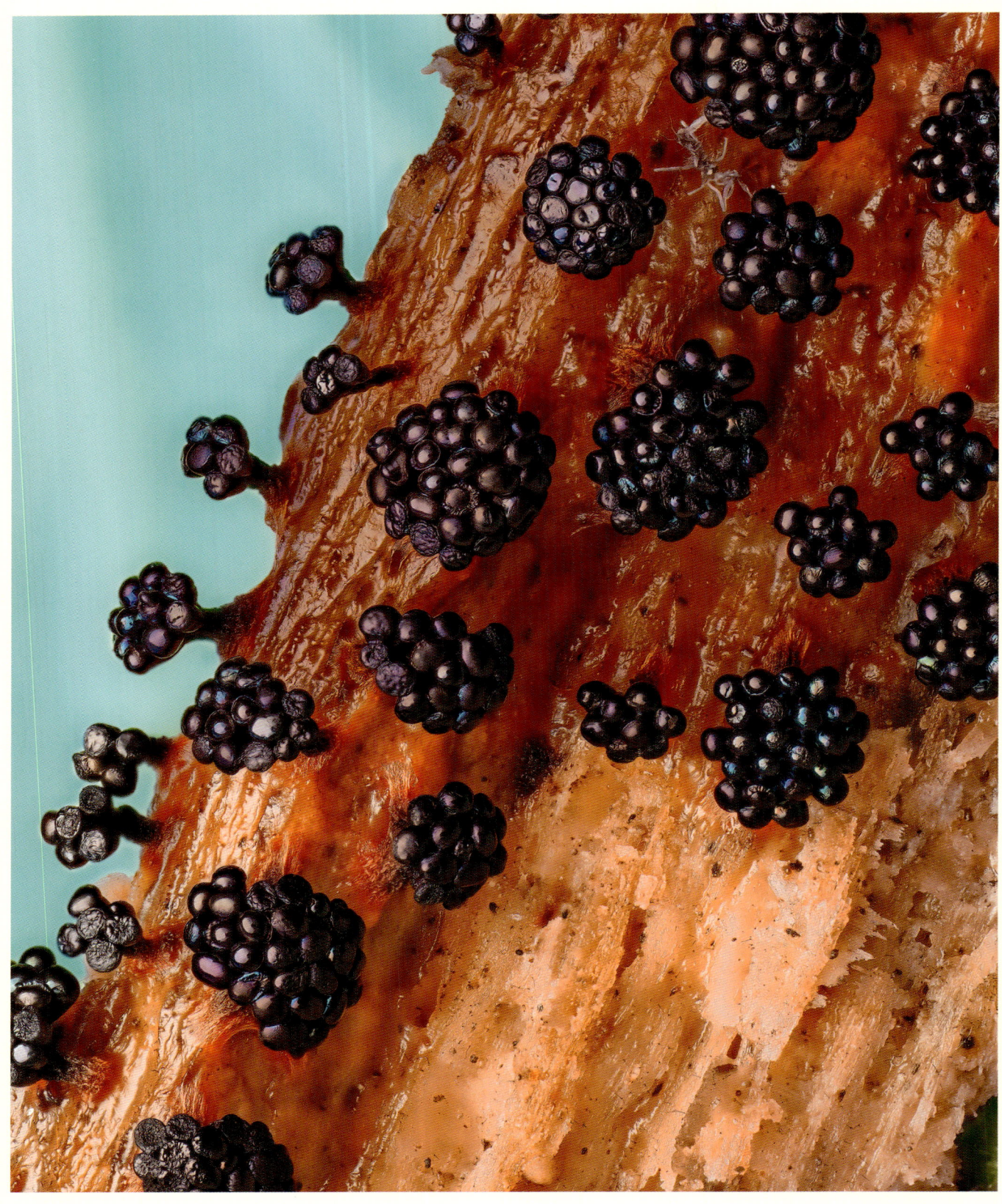

ABOVE: WASP'S NEST SLIME MOLD (*METATRICHIA VESPARIA*). OPPOSITE: *METATRICHIA FLORIFORMIS*.

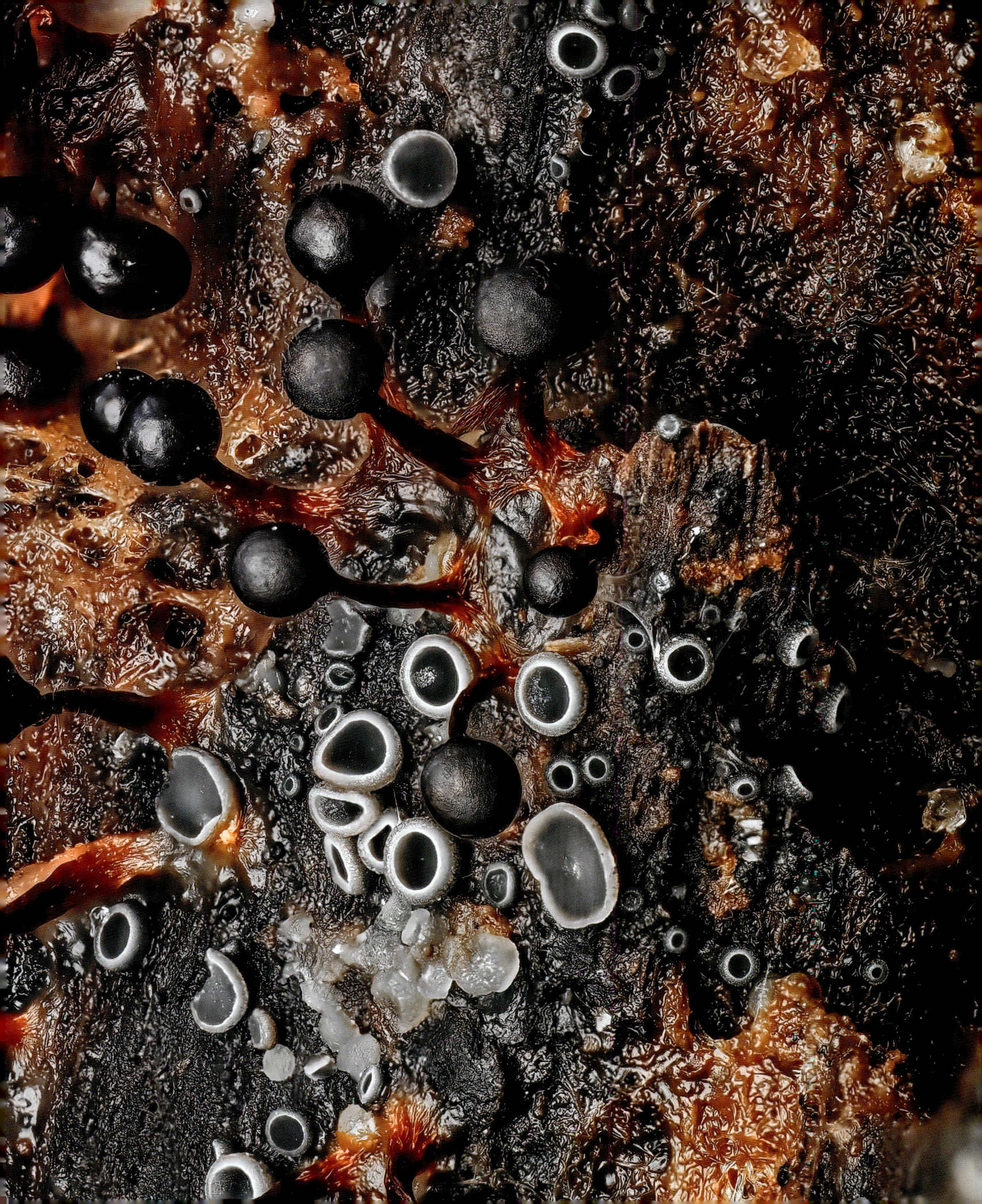

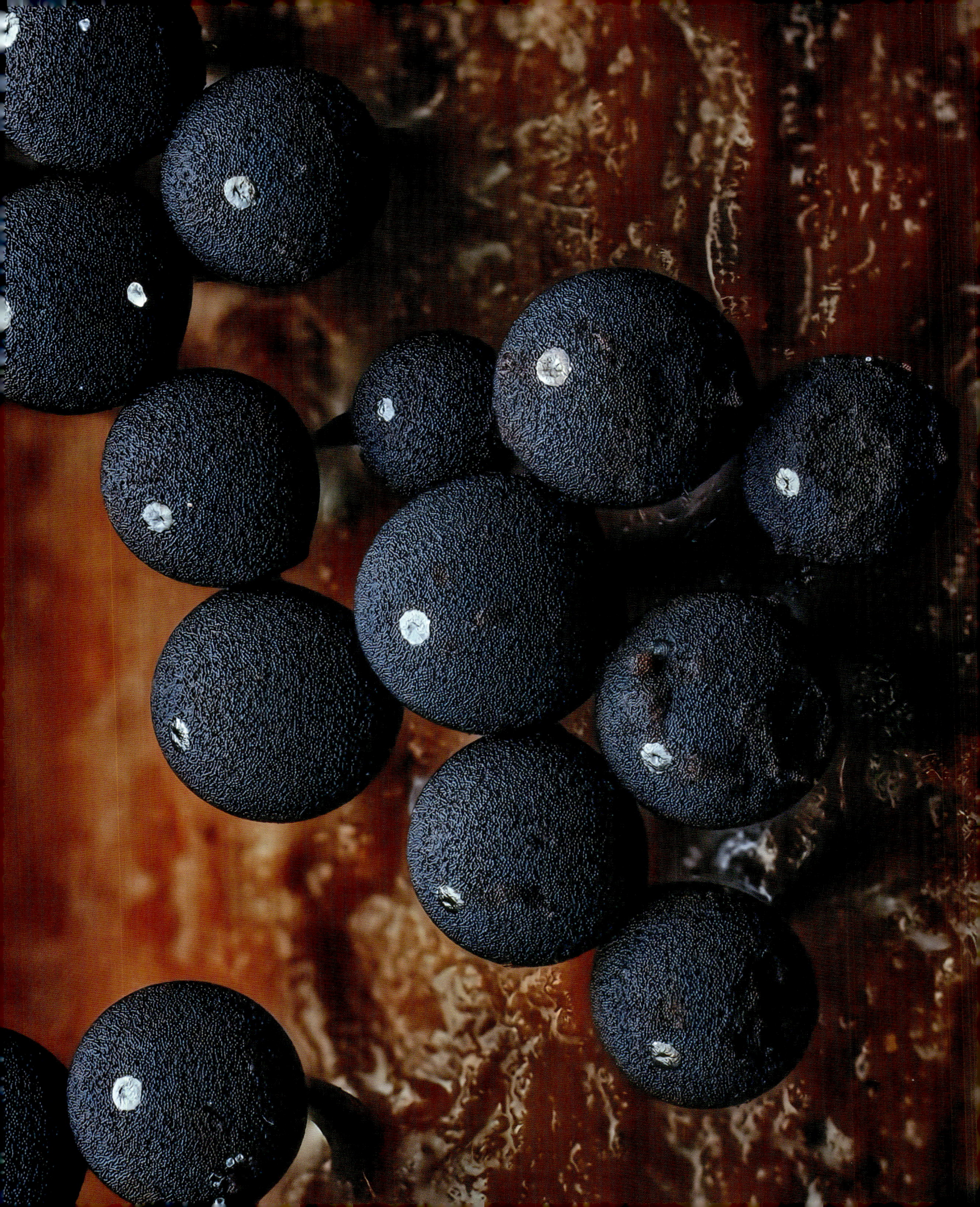

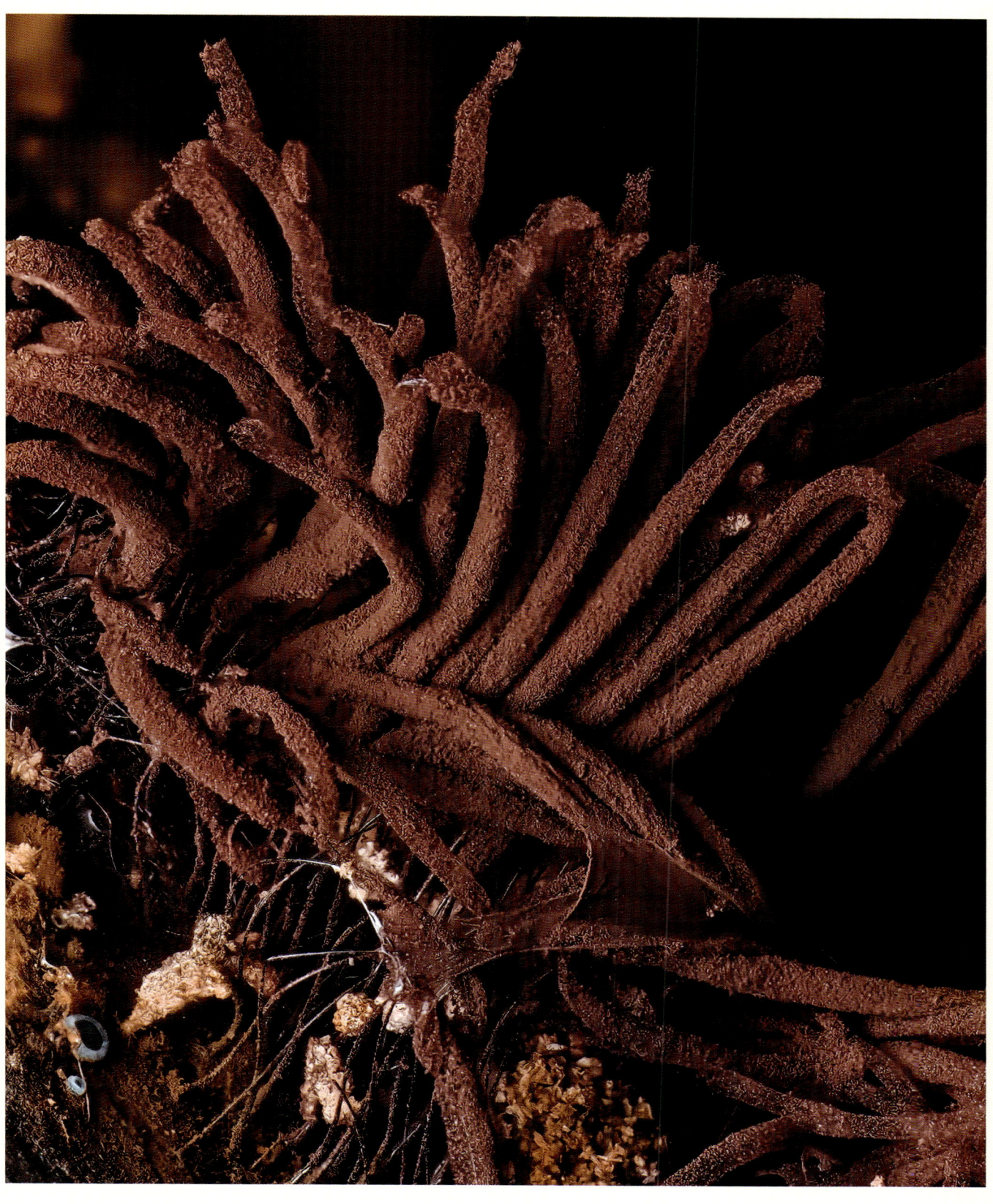

ABOVE: CHOCOLATE TUBE SLIME (*STEMONITIS SP.*). OPPOSITE: *ENERTHENEMA PAPILLATUM.*

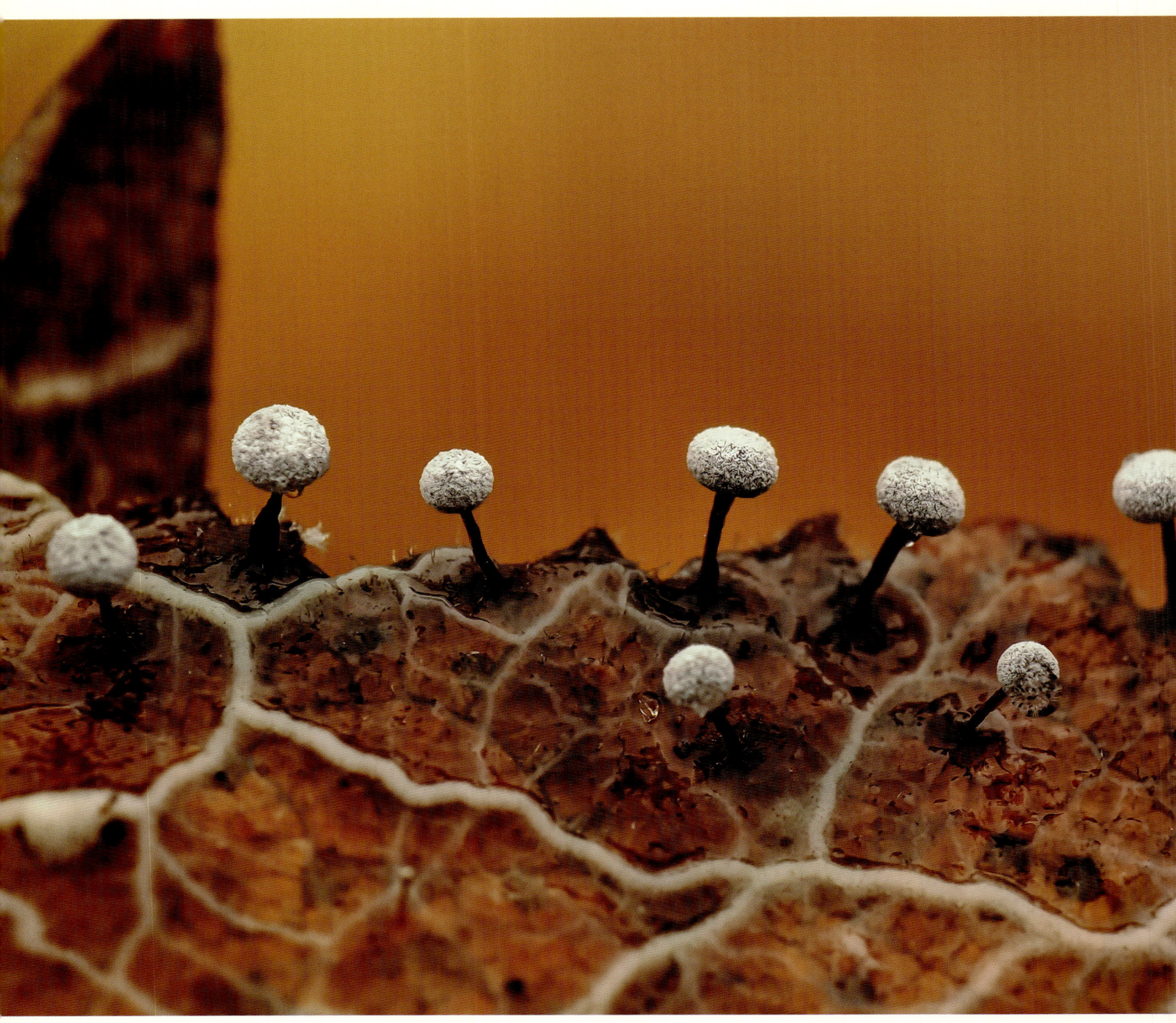

ABOVE AND OPPOSITE: *PHYSARUM* SPP.

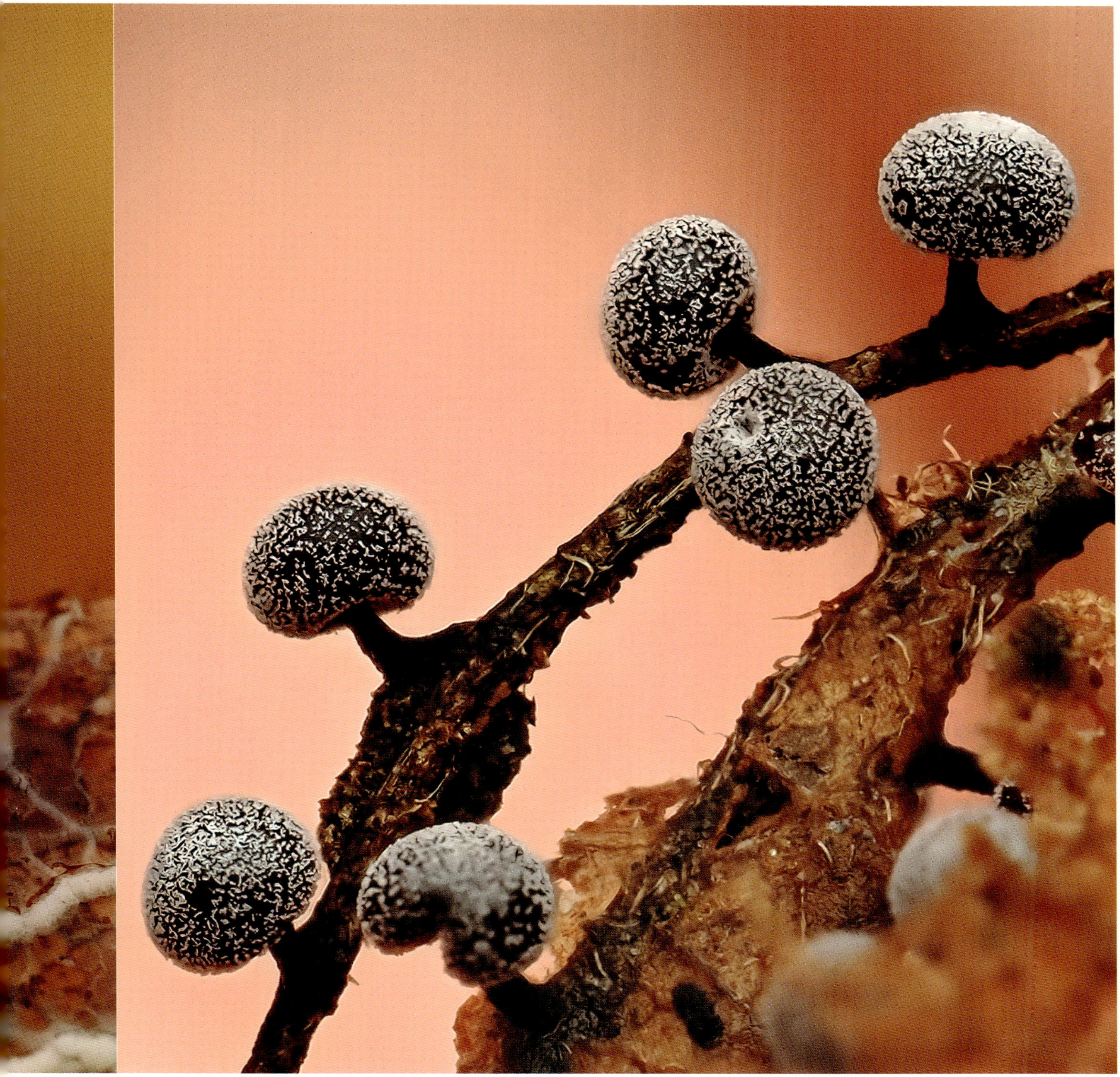

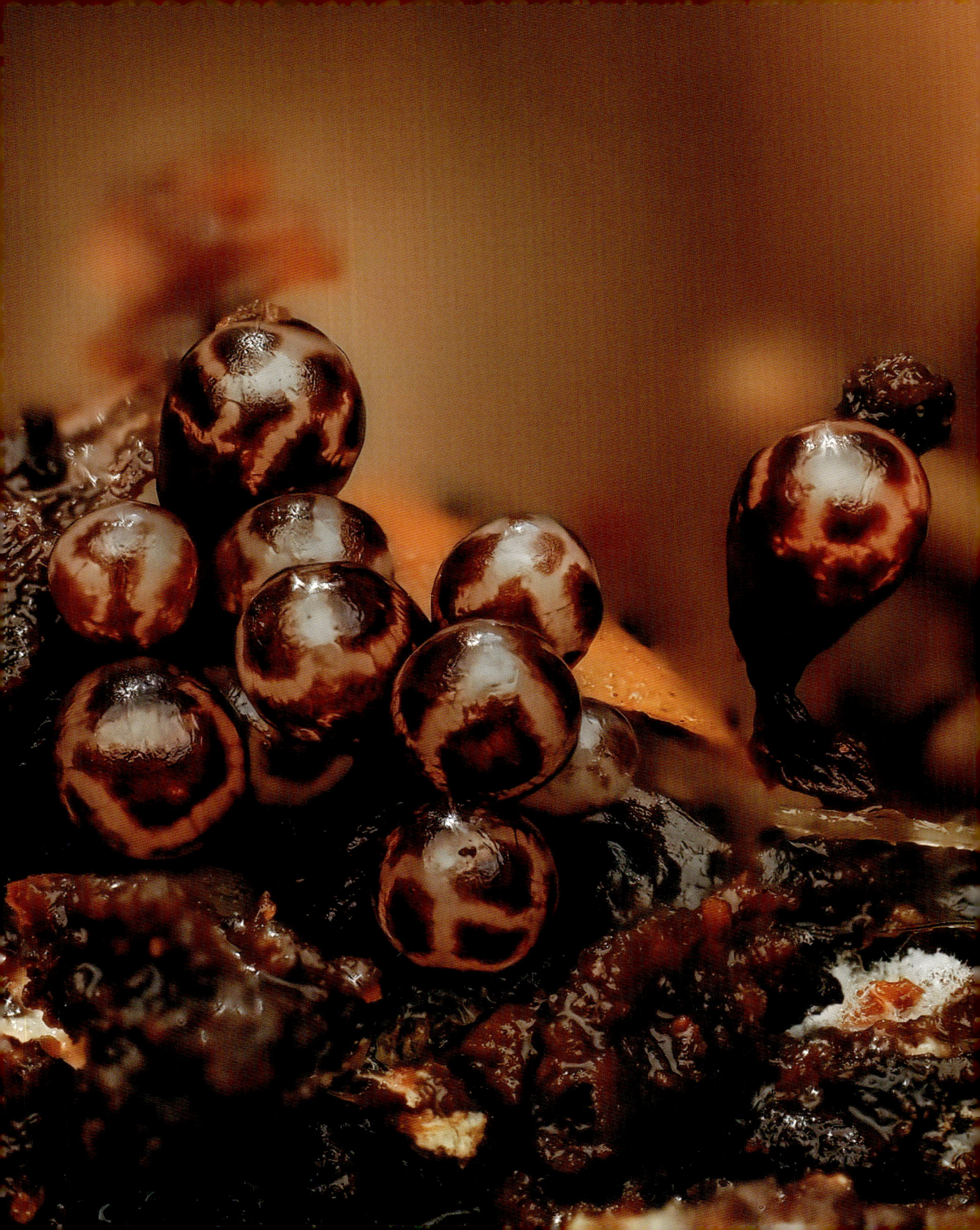

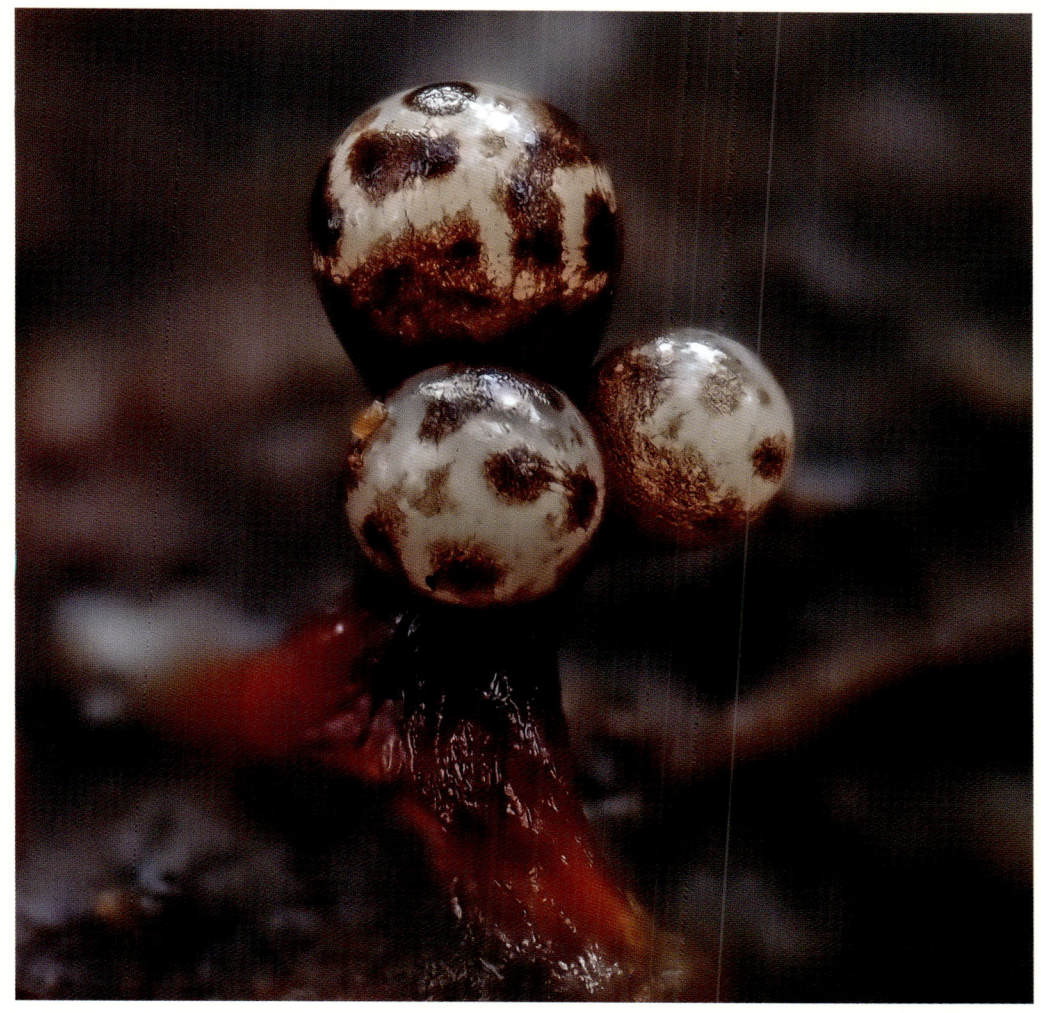

ABOVE AND OPPOSITE: TRICHIA BOTRYTIS COMPLEX.

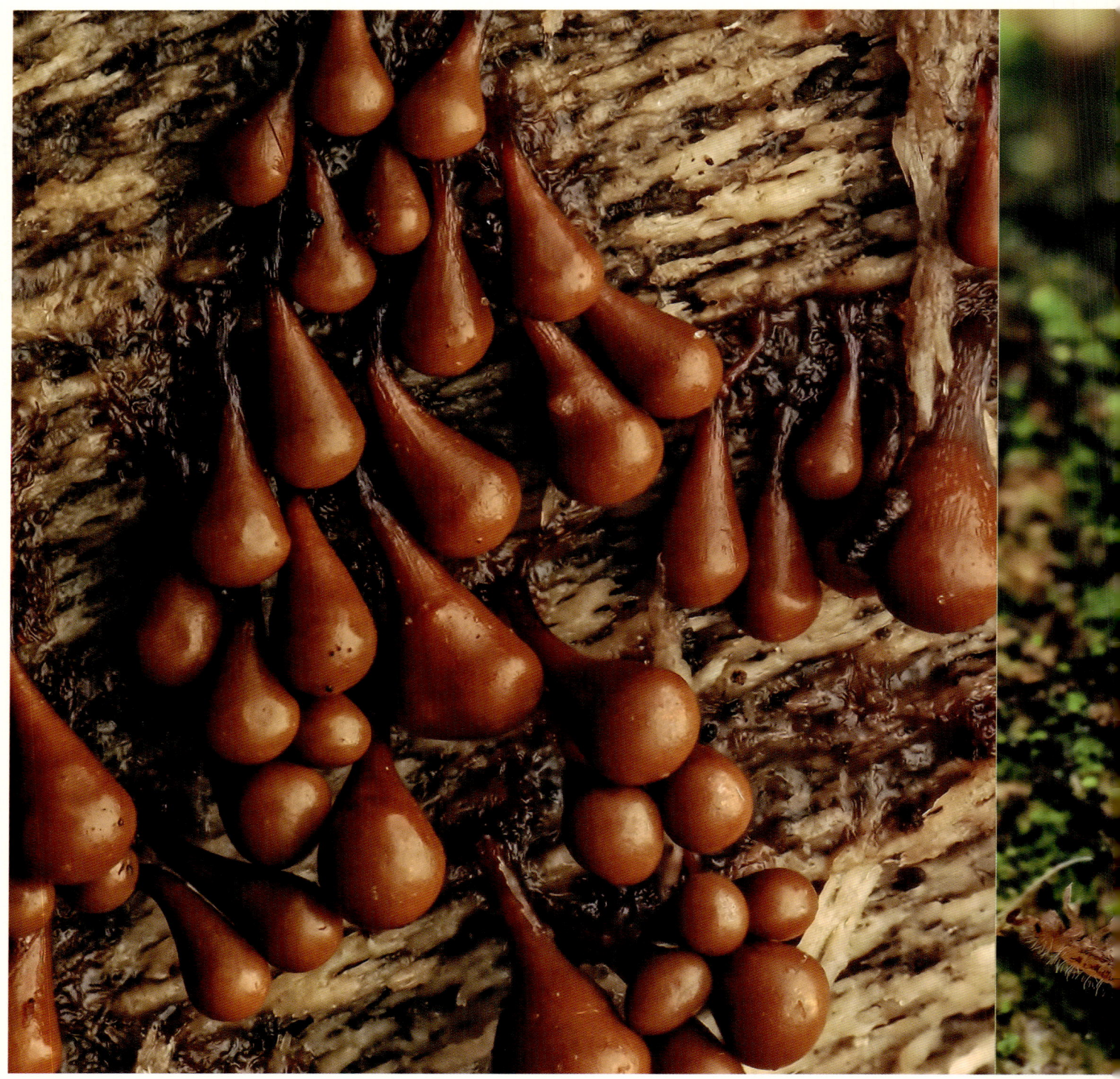

ABOVE: INSECT-EGG SLIME (*LEOCARPUS FRAGILIS*). OPPOSITE: *WILLKOMMLANGEA RETICULATA.*

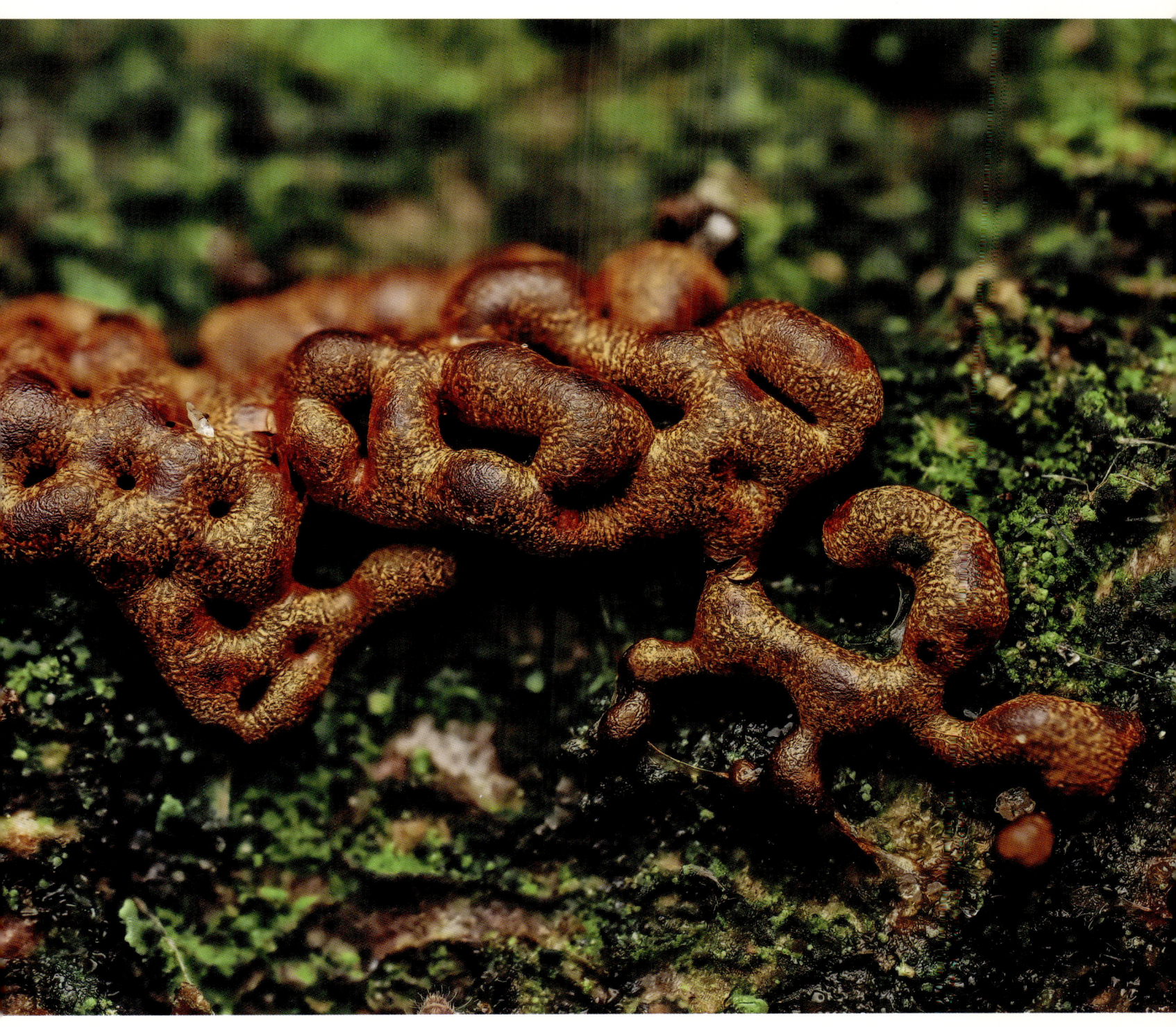

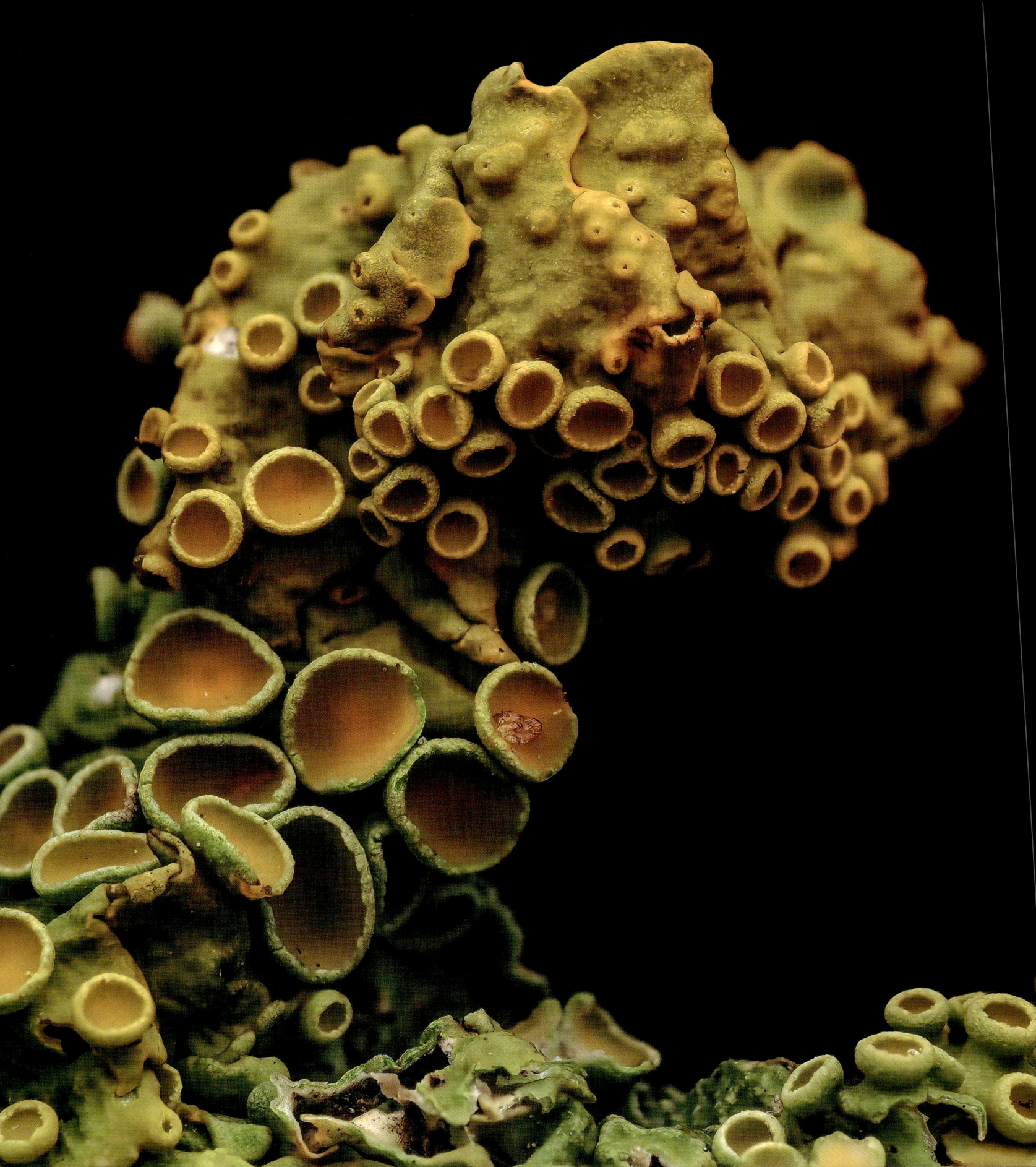

Lichens

There are over two thousand species of lichen where I live. With as much time as I spend searching the forest floor for gorgeous treasures to photograph, I find and photograph *a lot* of lichen.

The word "lichen" comes from the Greek word *leichens*, which means "licker." Most lichens adhere closely to the surfaces they grow on, almost as if they are licking the rock or tree bark! Lichens have existed for 400 million years, predating the dinosaurs by at least 150 million years. Fossil evidence suggests that early lichens played a role in soil formation and the colonization of land by plants.

Despite its incredible history and ecological significance, lichen is often an overlooked organism that is just as fantastical as slime. Lichens are often thought of as plants, but they are not; they are several things combined! Lichens are actually a complex symbiosis between fungi and algae, or sometimes cyanobacteria. Together, they create an entity that can endure extreme conditions where neither would survive alone.

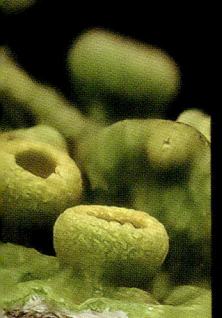

OPPOSITE AND FOLLOWING SPREAD: GOLDEN SHIELD LICHEN (*XANTHORIA PAR ETINA*).

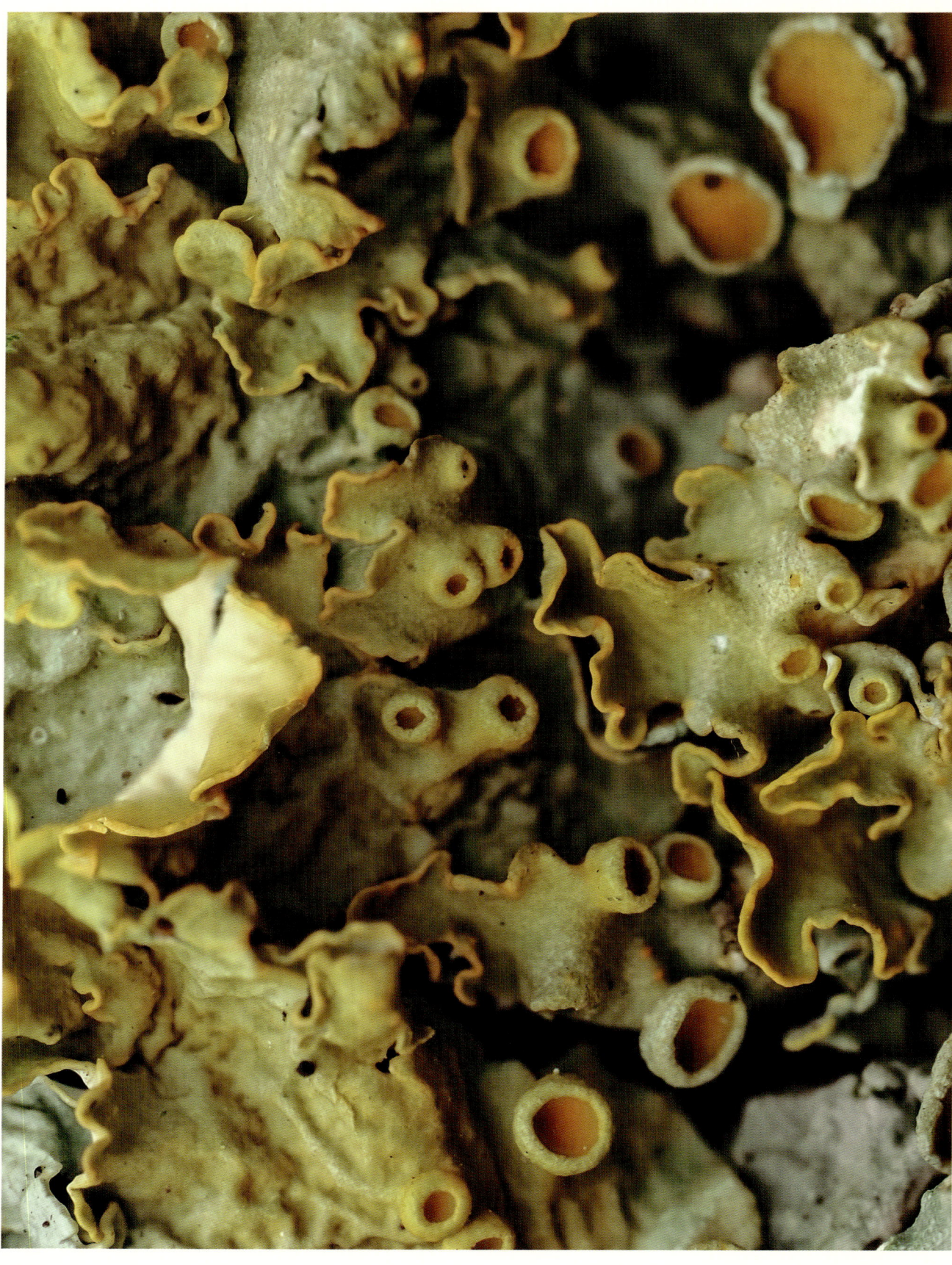

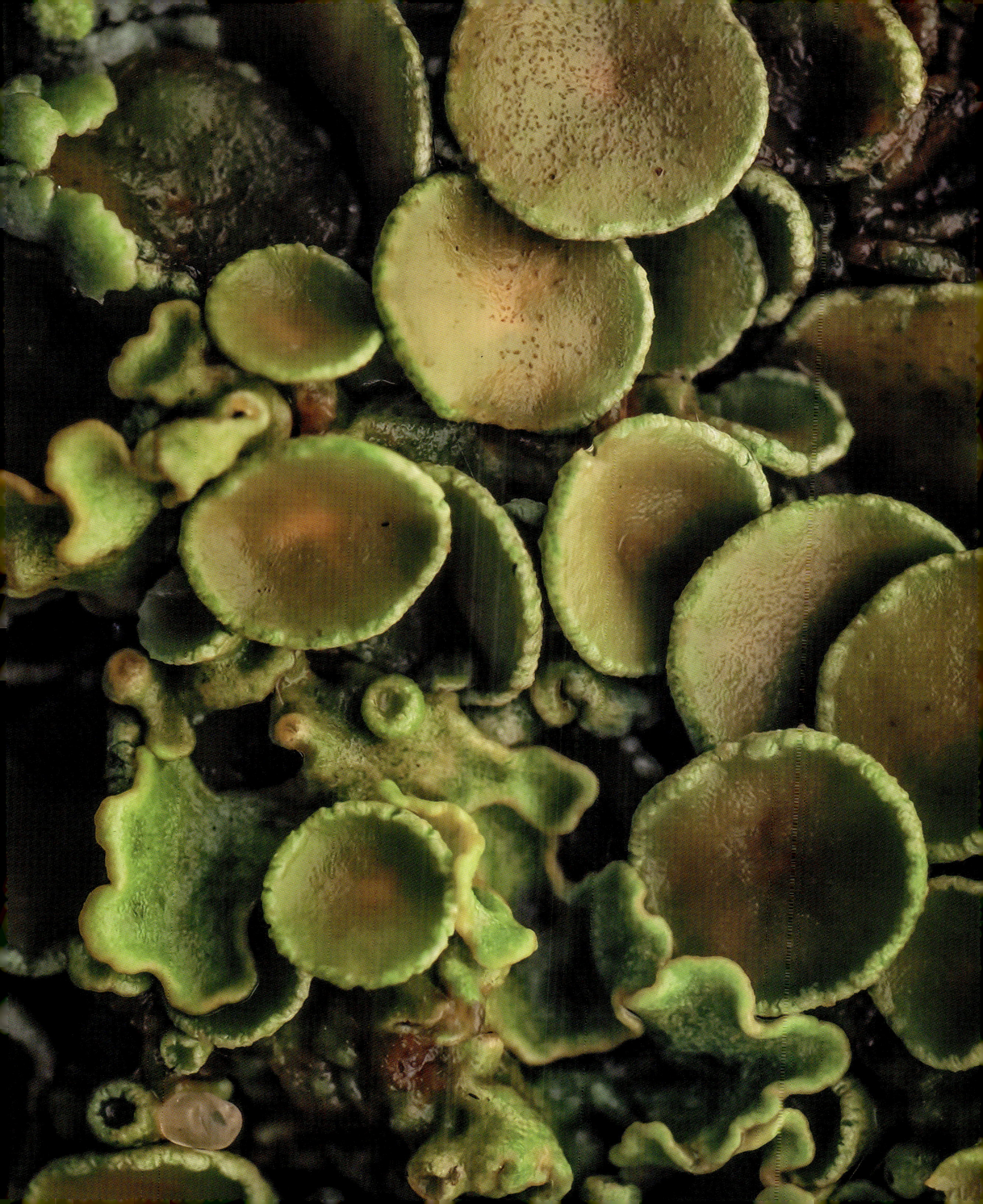

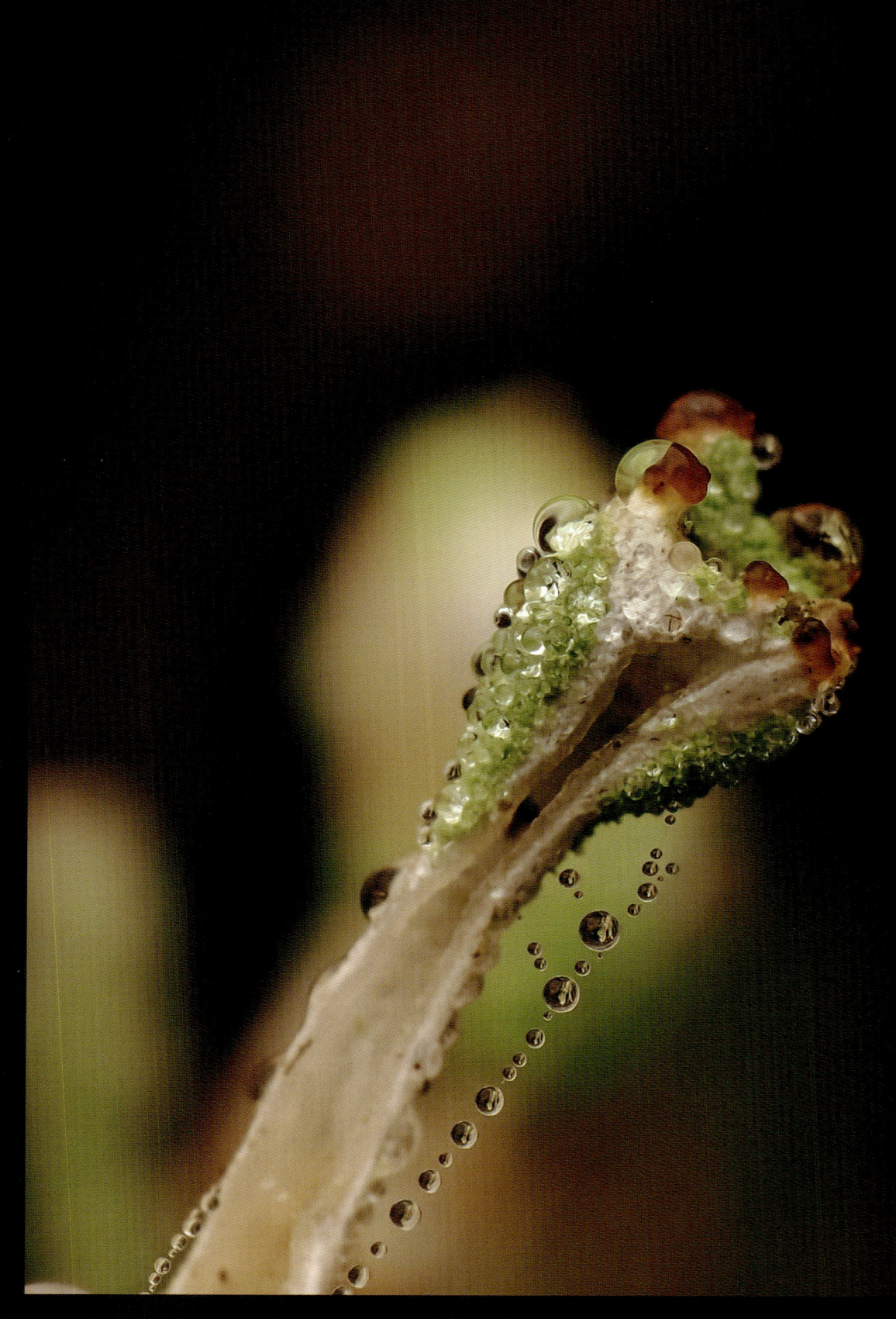

Lichens are often pioneer species, the first organisms to colonize bare and harsh environments such as rocks, deserts, and tundra. By breaking down these surfaces, they contribute to soil formation, creating a path for other plants to grow. In this way, lichens are a strong indicator of environmental health, and air quality as well. They take in water and nutrients from the air and are very sensitive to pollution. In polluted areas, some types of lichens might be missing. Most of the lichen I photograph are found on tiny twigs, fallen branches, old fence posts, and ancient rocks. They prefer environments where they can receive plenty of sunlight and air.

Many animals, from tiny insects to big mammals, use lichens for food and shelter. Like fungi, the layers and textures of lichens also provide small critters great hiding spots from predators. However, they also serve as places for predatory insects to hunt. I rarely pick up a lichen-covered stick without seeing a myriad of soil animals having already made it their home. I've been known to find a stick encrusted with lichens and then sit in the forest photographing it for an hour because there is just so much to capture!

Lichens can lose almost all their water content and enter a state of dormancy, which allows them to survive in extremely dry conditions. Some lichen species can survive for several months without water, while others have been known to endure dry conditions for years. Their ability to withstand drought depends on the species and the environmental conditions. Nearly every summer in London, we have a heatwave. Some recent years have been so brutal that all the grass dies and trees start dropping leaves two months early. Despite the harsh conditions, lichens are still in abundance and often the only thing aside from bees that I can reliably find to photograph in the later summer months.

Aesthetically speaking, the physical appearance of lichen is mesmerizing and strange. Lichen comes in a variety of bright colors, from deep greens to radiant reds, neon yellows, minty blues, pinks, and even black. They speckle their habitats with these splashes of colors, adding a fairytale beauty to both natural and urban landscapes. After a fresh rain, they plump up a bit and their already vibrant colors brighten further.

Unlike plants, lichens do not have roots, stems, or leaves. Their structure is formed by the fungal component, which can take on various forms such as crusty, leafy, or shrubby. Among its many forms are intricate squid-like cups—some lichens lie flat but are a bit crinkly like papier mâché, resembling a craggy landscape. Lichens can also be fluffy and sprout from tree branches with long papery tendrils or form thin tubular structures similar to fingers A few of the most magical lichens are *Cladonia*, some of which have a goblet structure that catches raindrops and transforms dew into a perfect sphere, making for sensational photography.

127

OPPOSITE AND FOLLOWING PAGES: *CLADONIA* SPP.

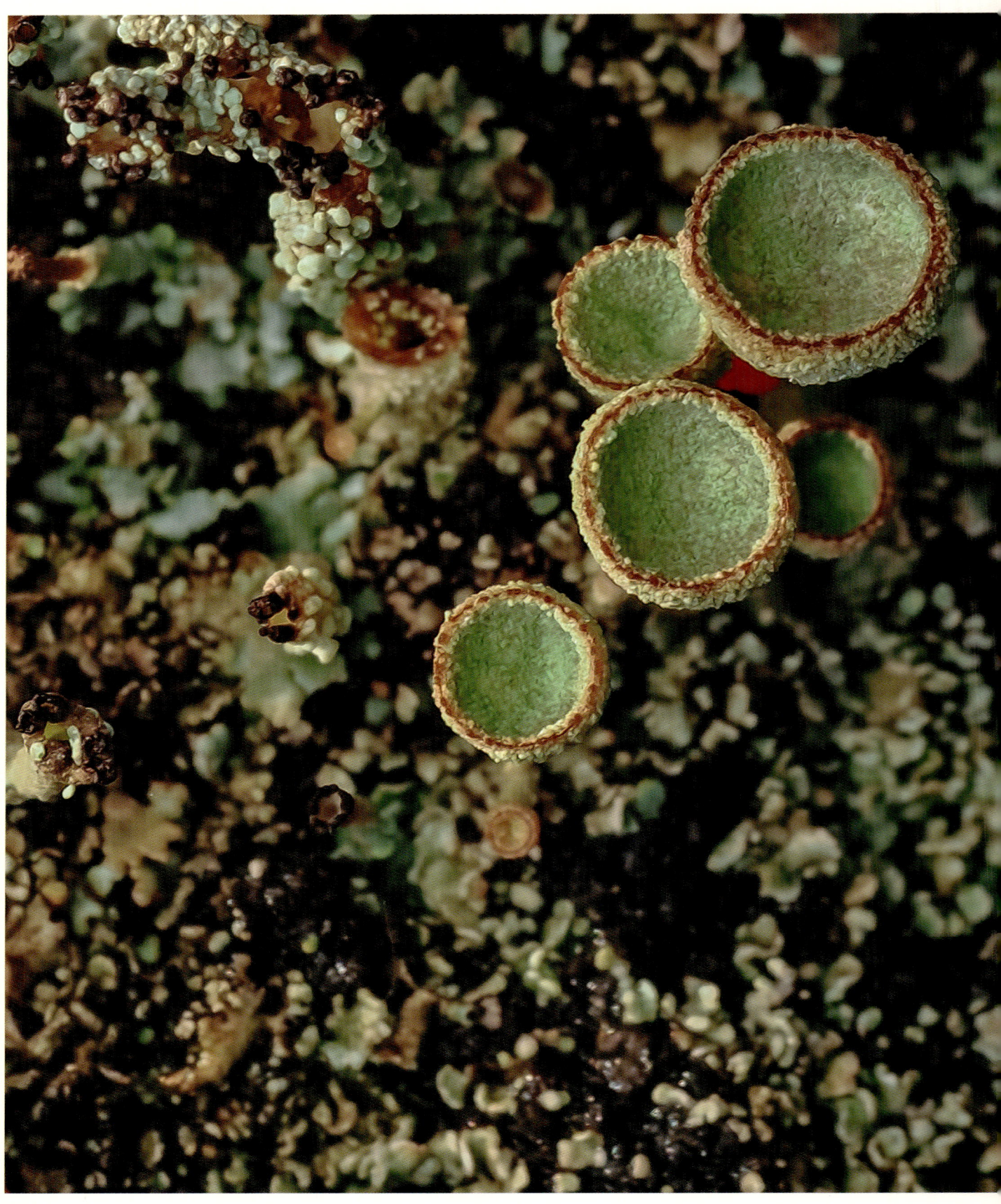

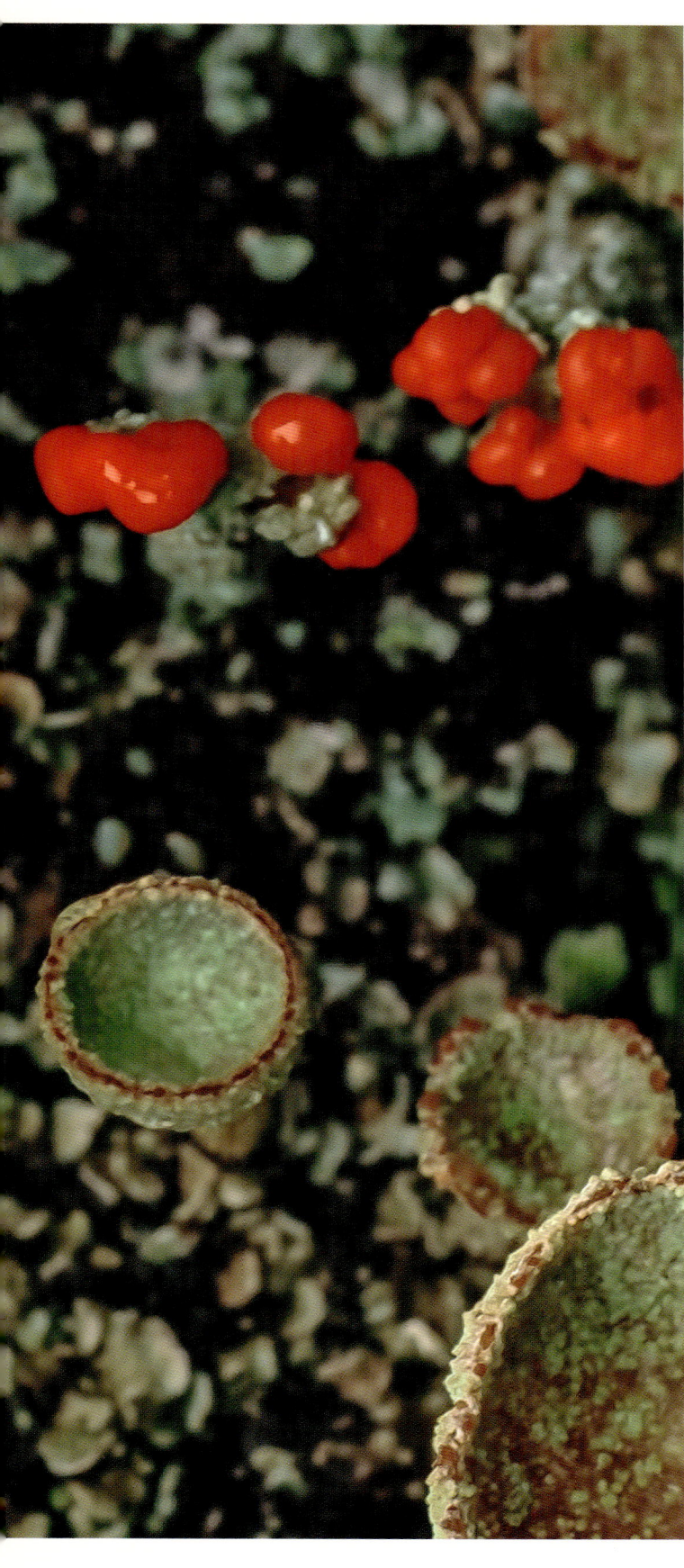

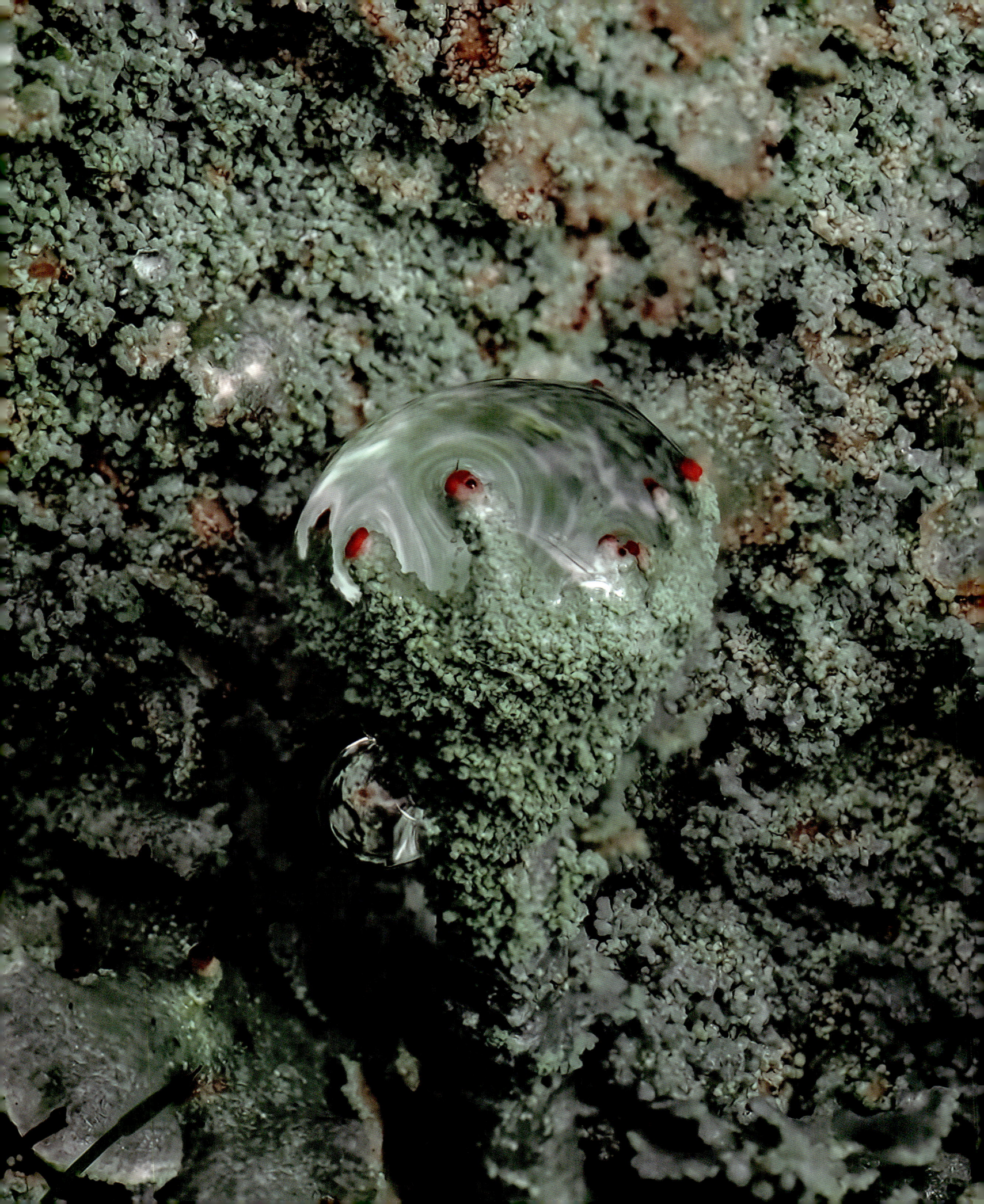

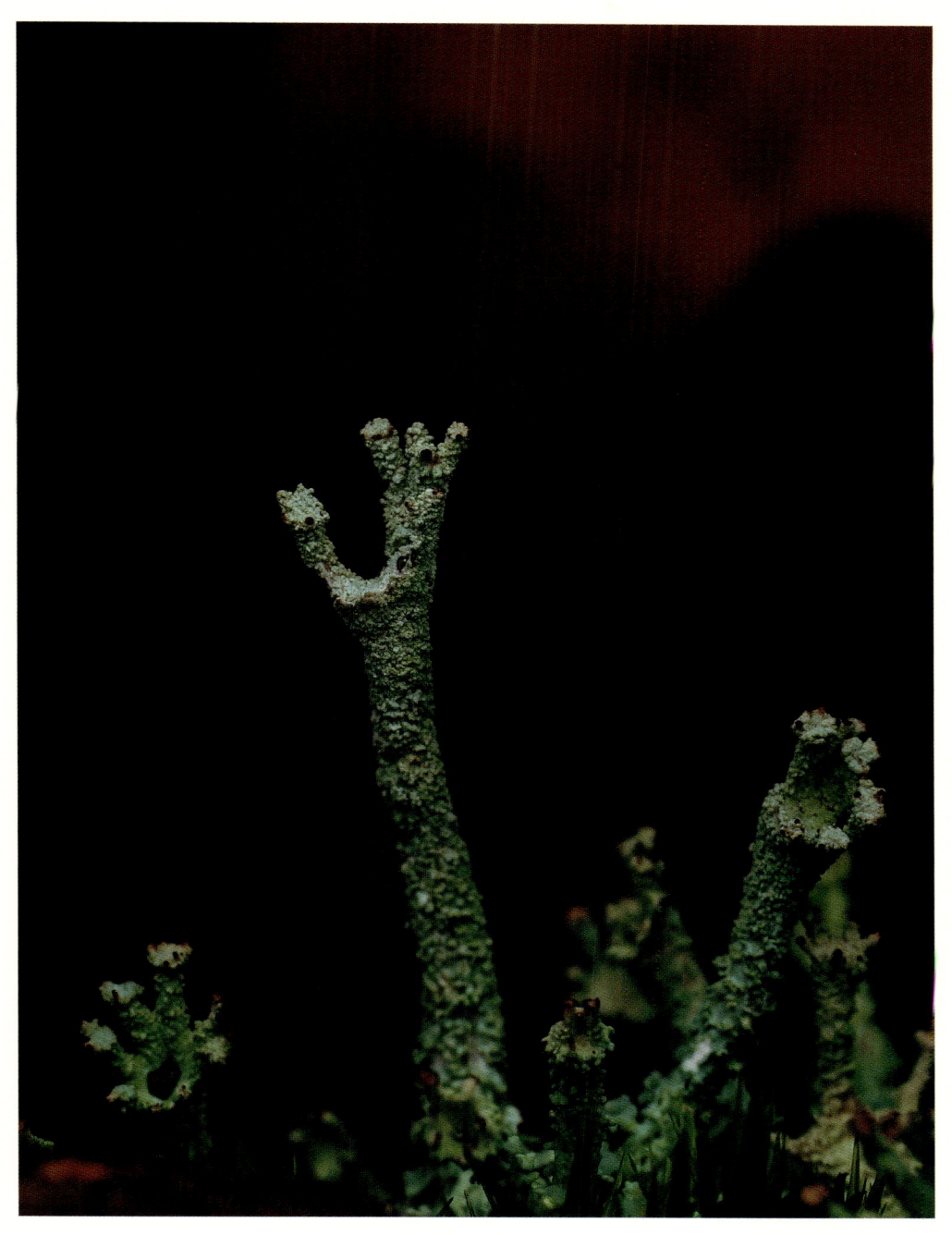

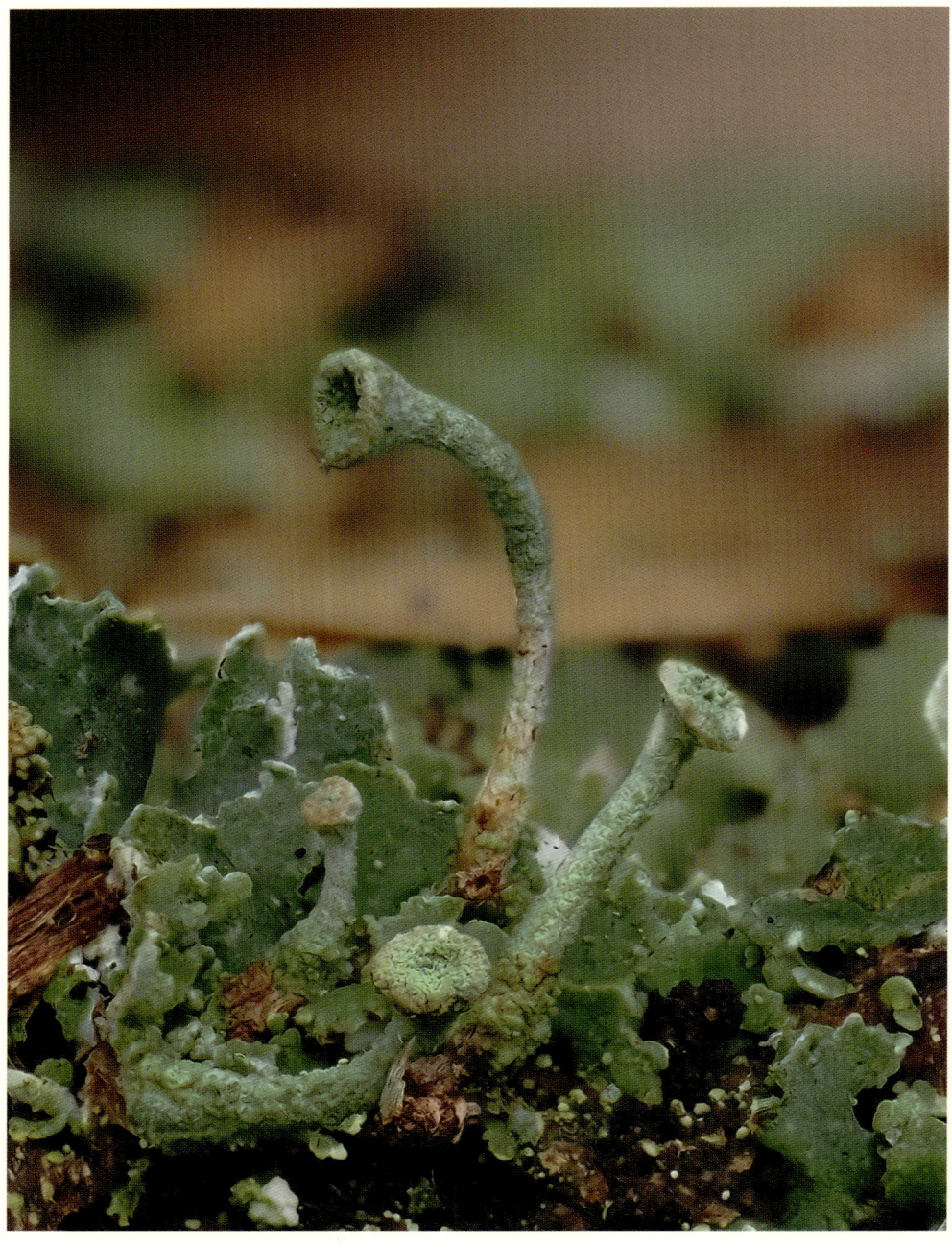

ABOVE: *CLADONIA* SPP. OPPOSITE: LIPSTICK POWDERHORN (*CLADONIA MACILENTA*).

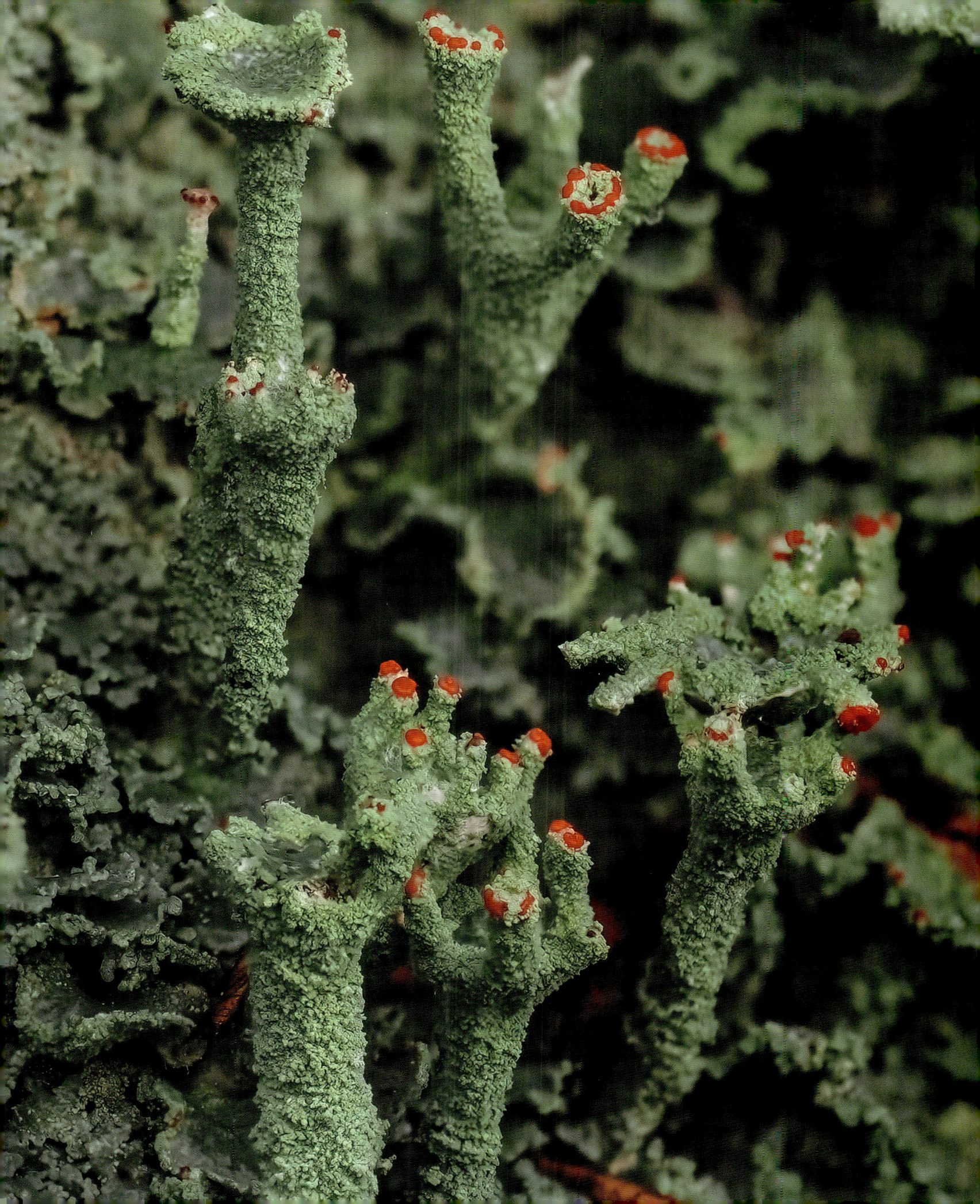

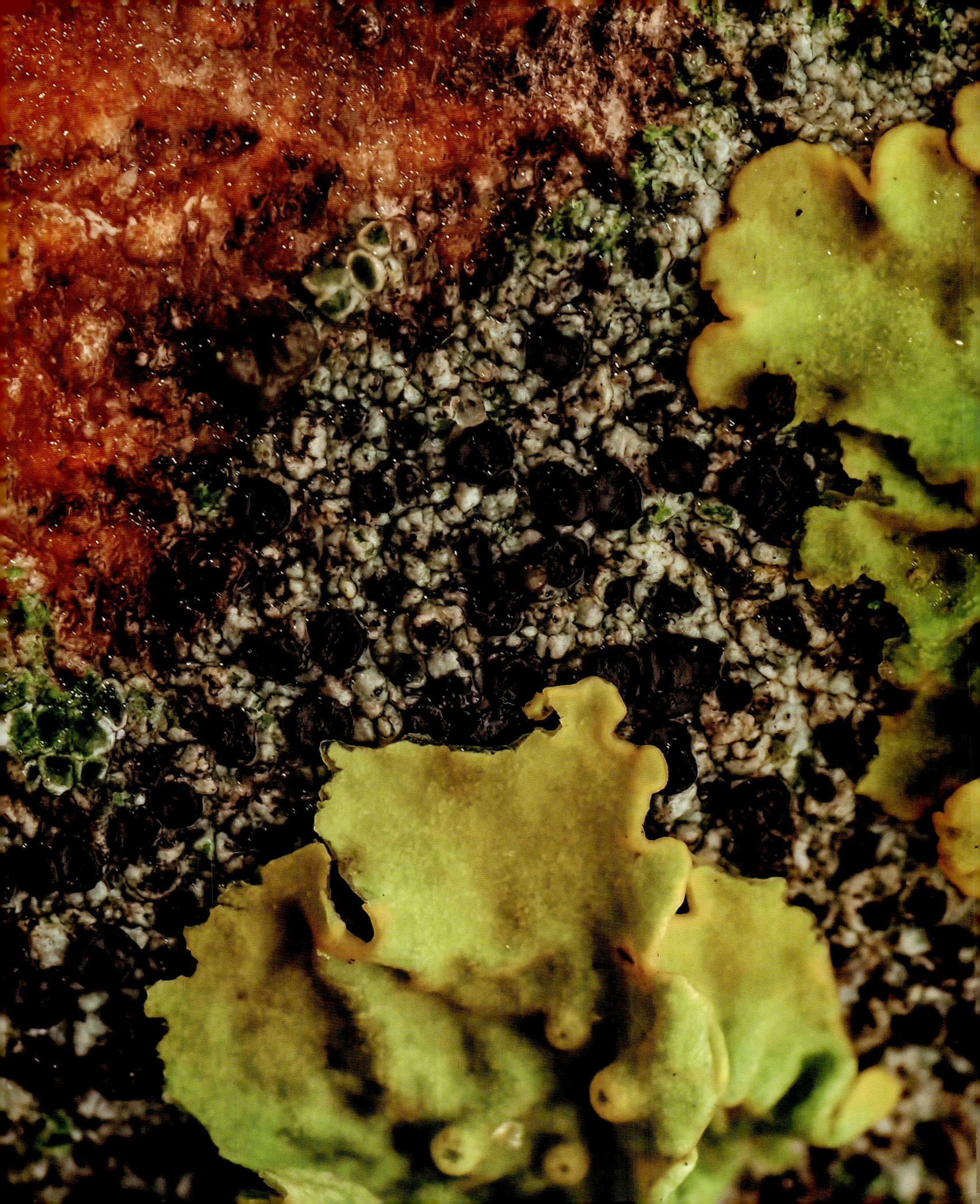

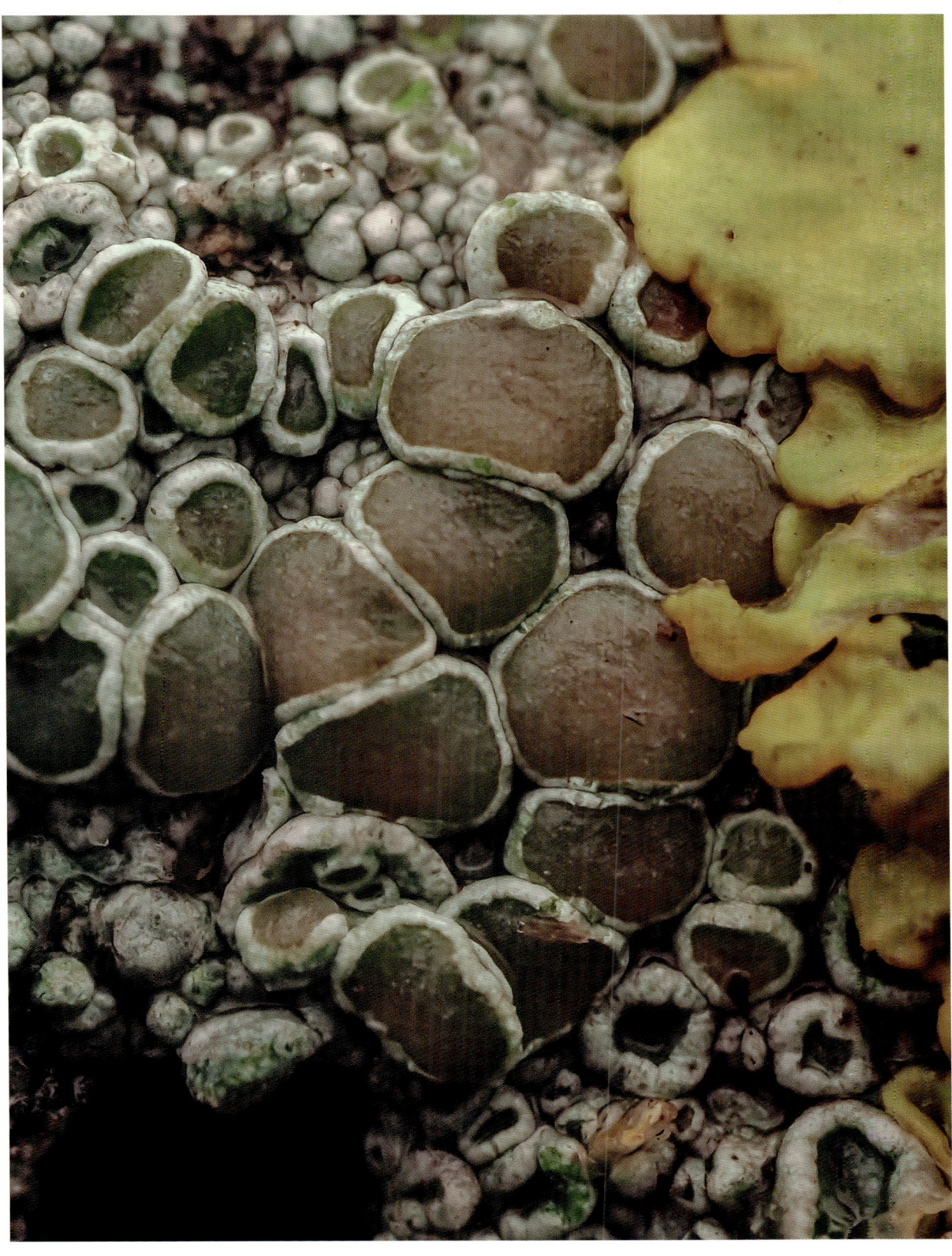

ABOVE AND OPPOSITE: UNKNOWN LICHENS.

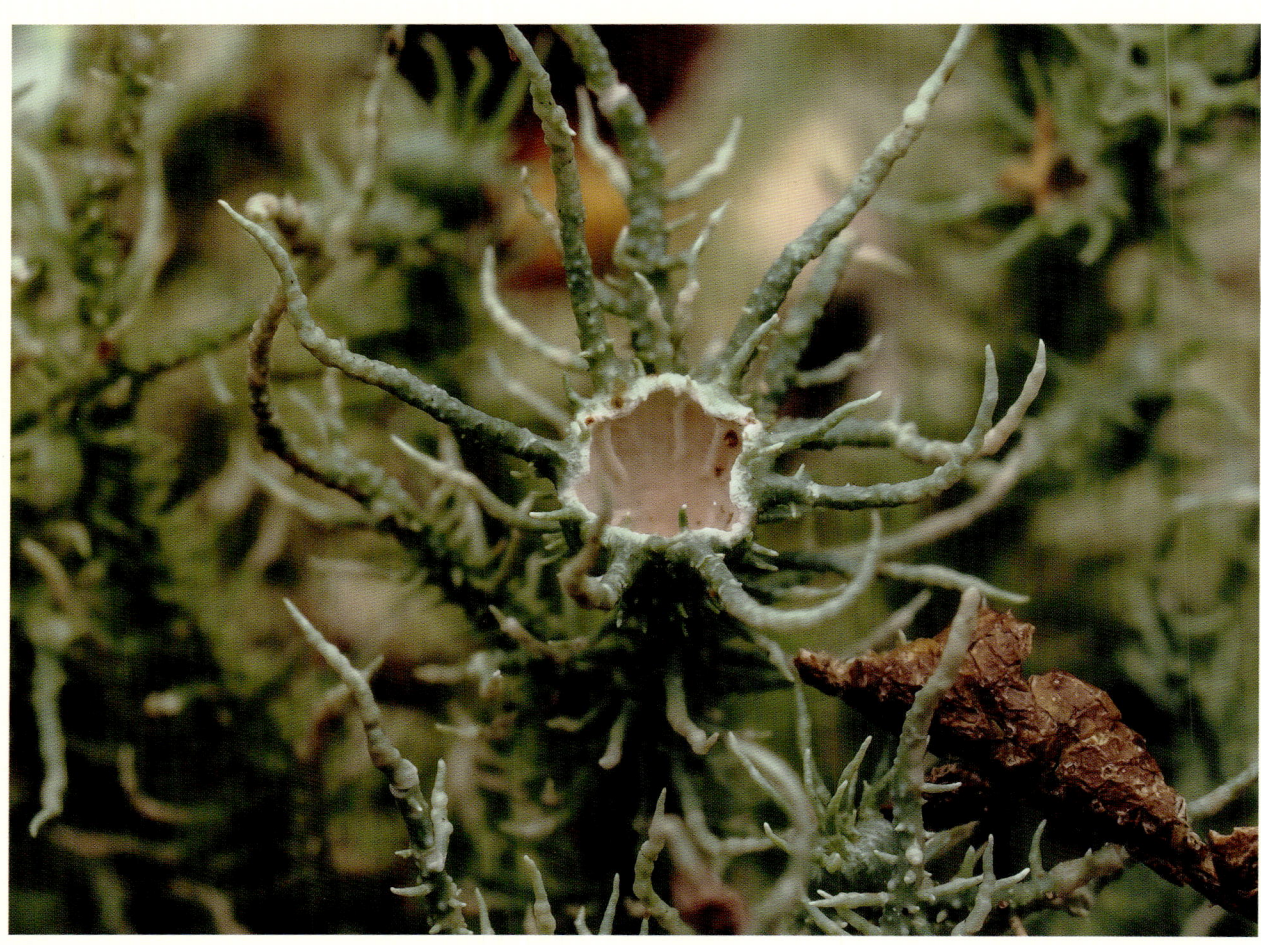

ABOVE AND OPPOSITE: BEARD LICHENS (*USNEA* SPP.).

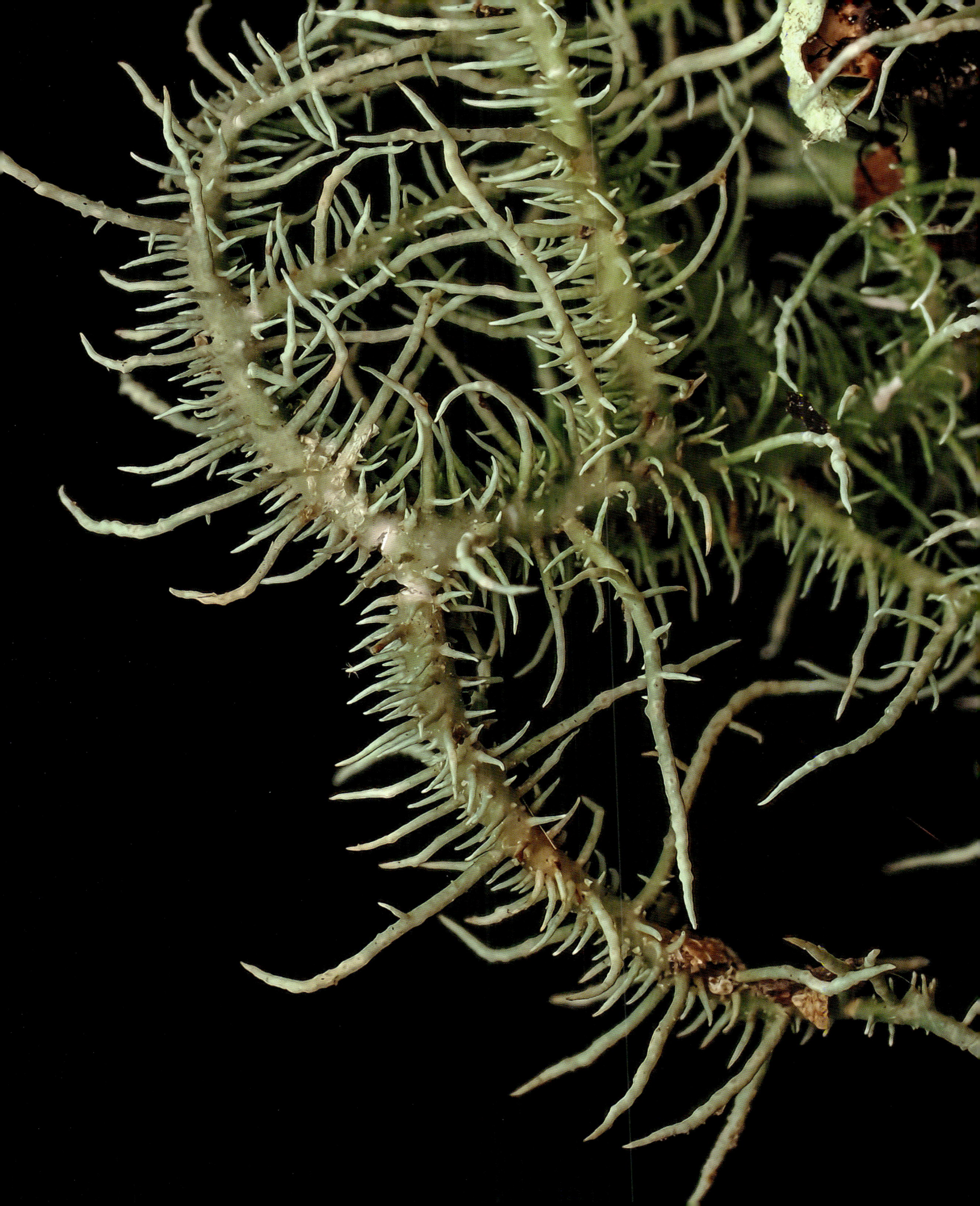

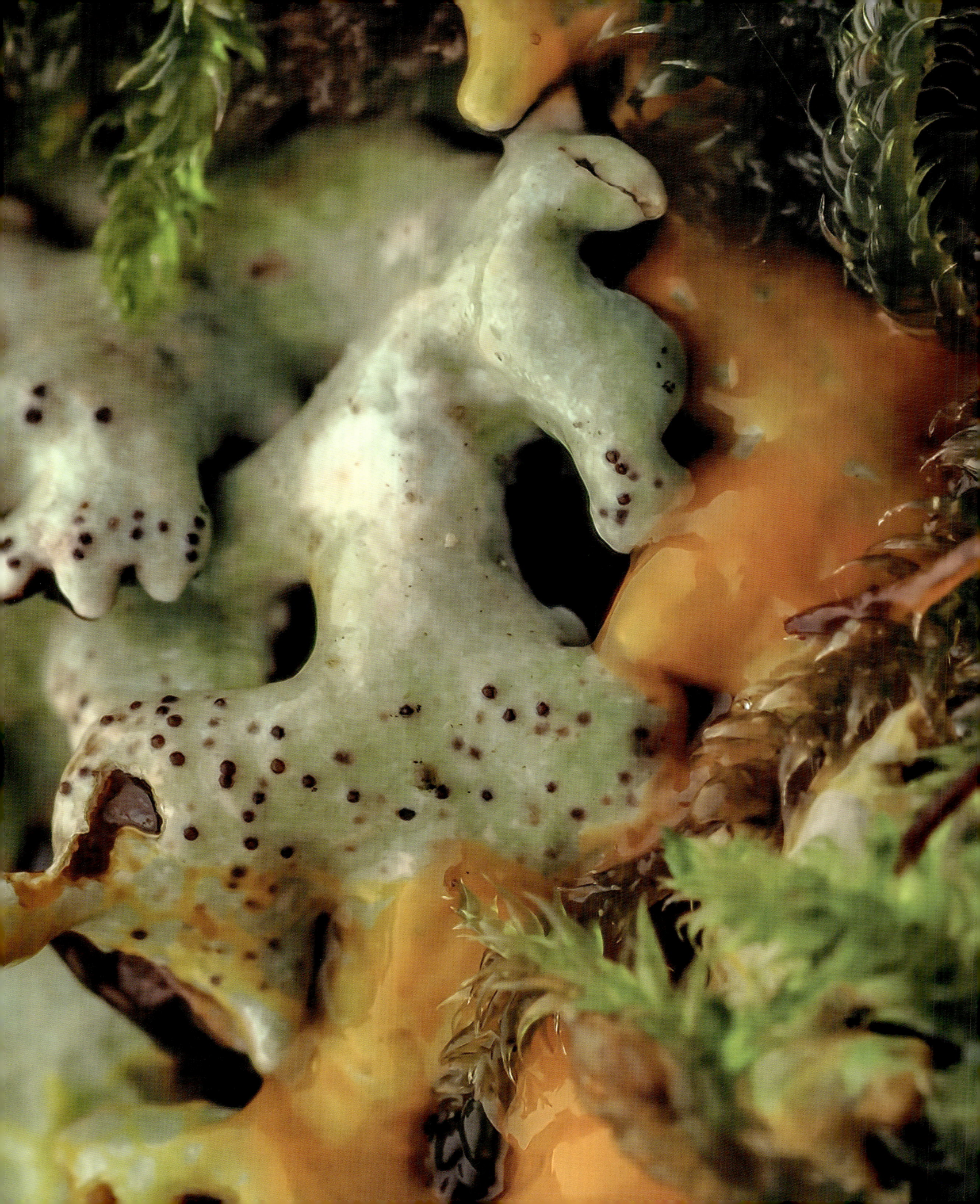

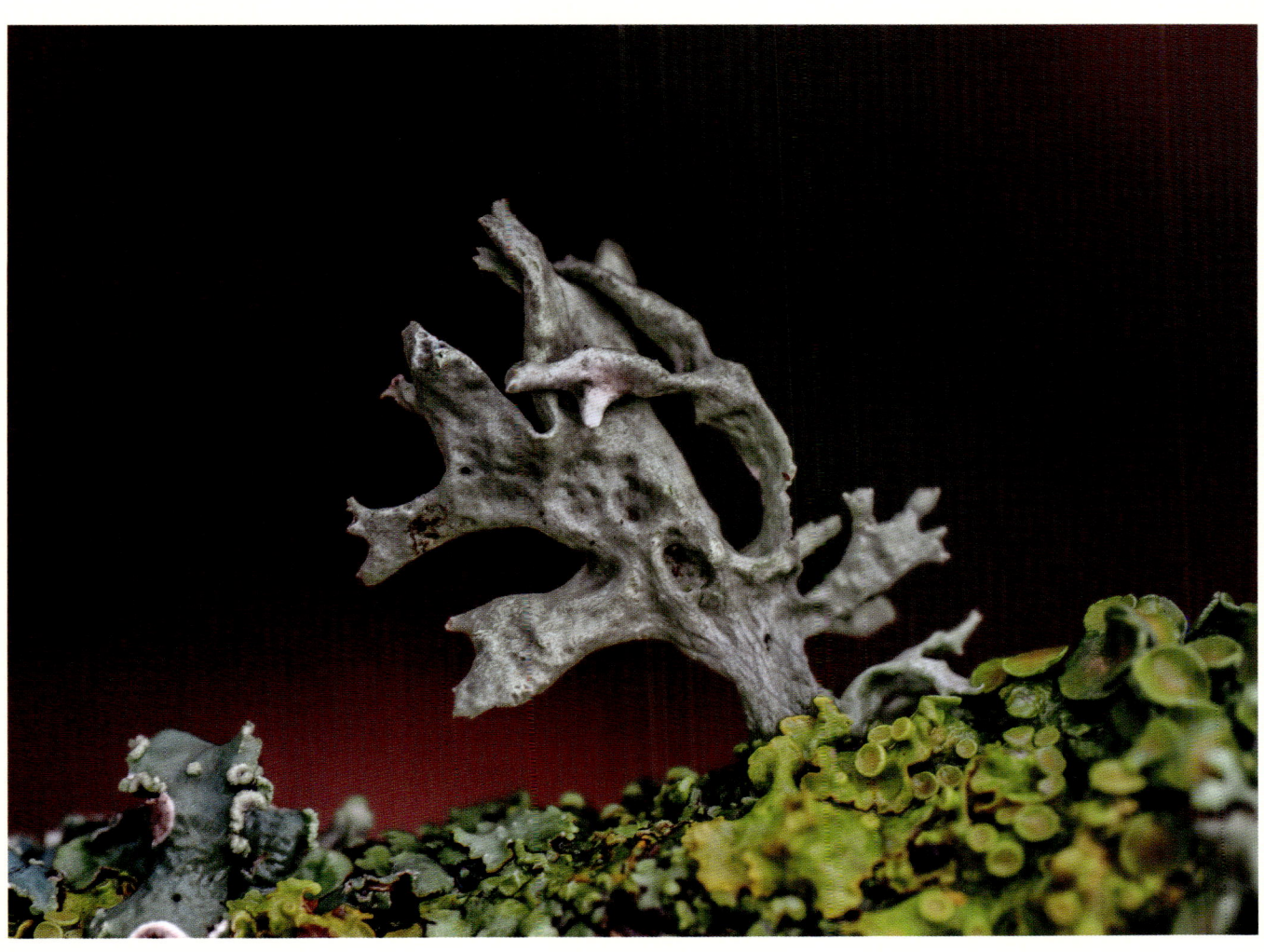

ABOVE AND OPPOSITE: UNKNOWN LICHENS.

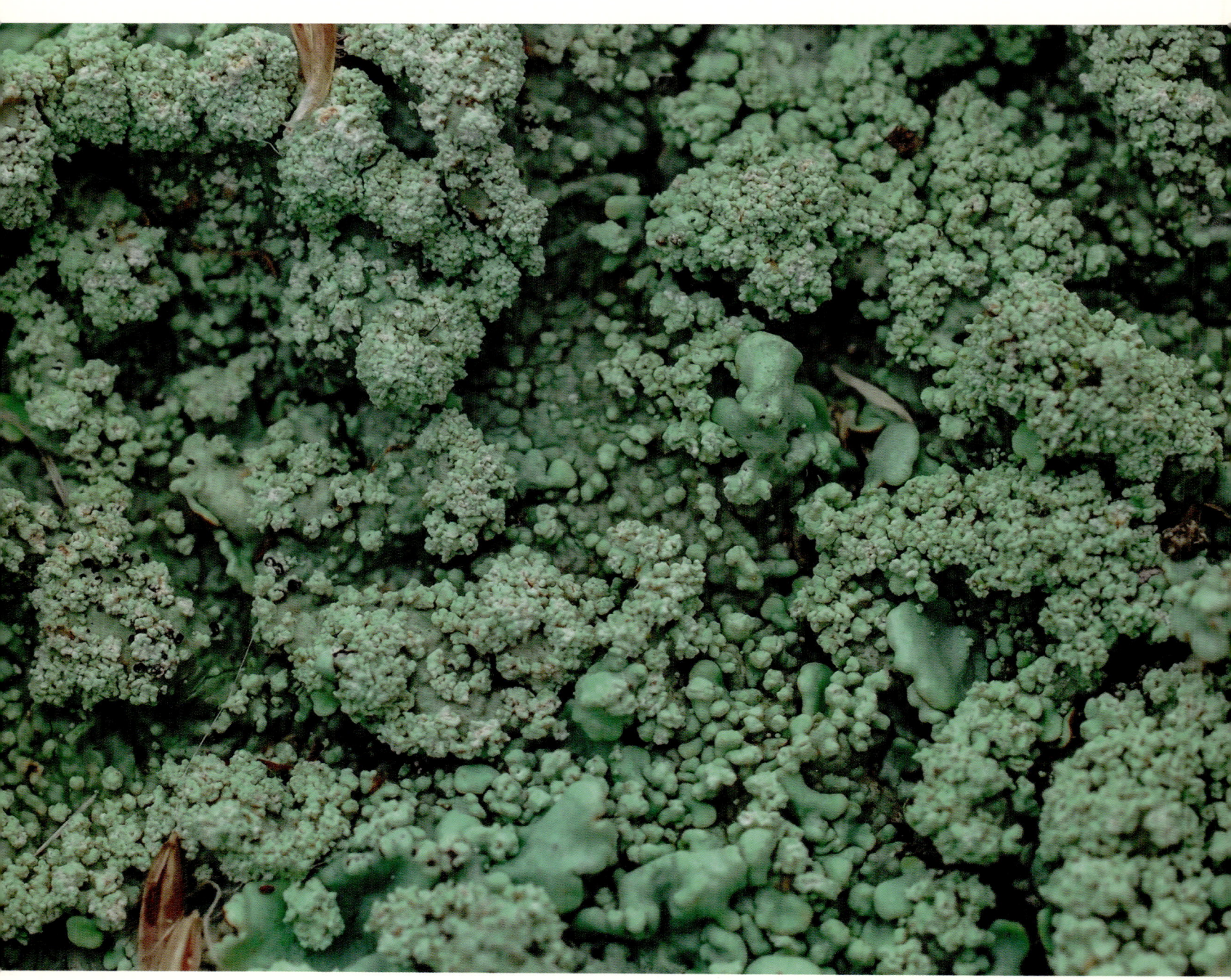

ABOVE AND OPPOSITE: UNKNOWN LICHENS.

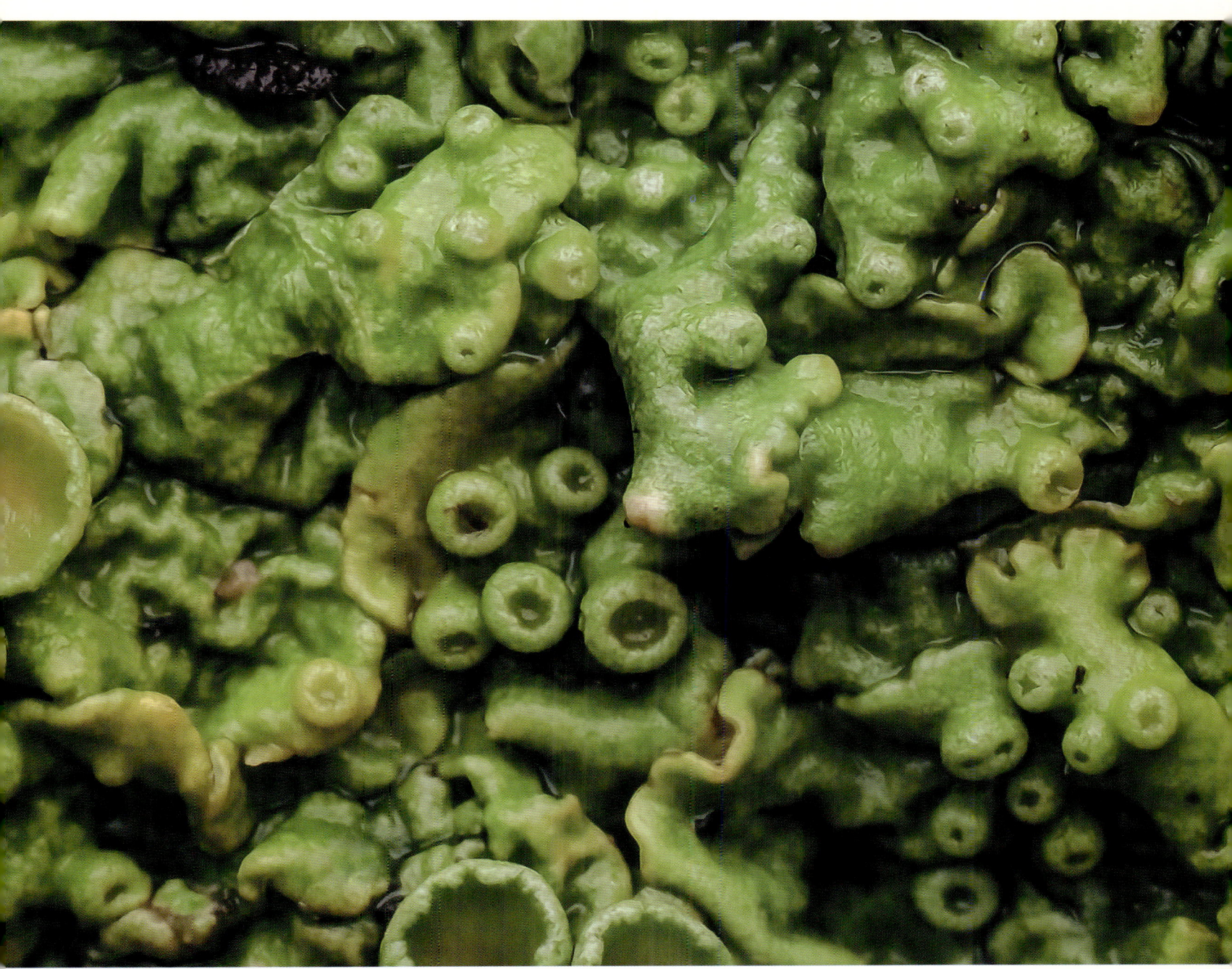

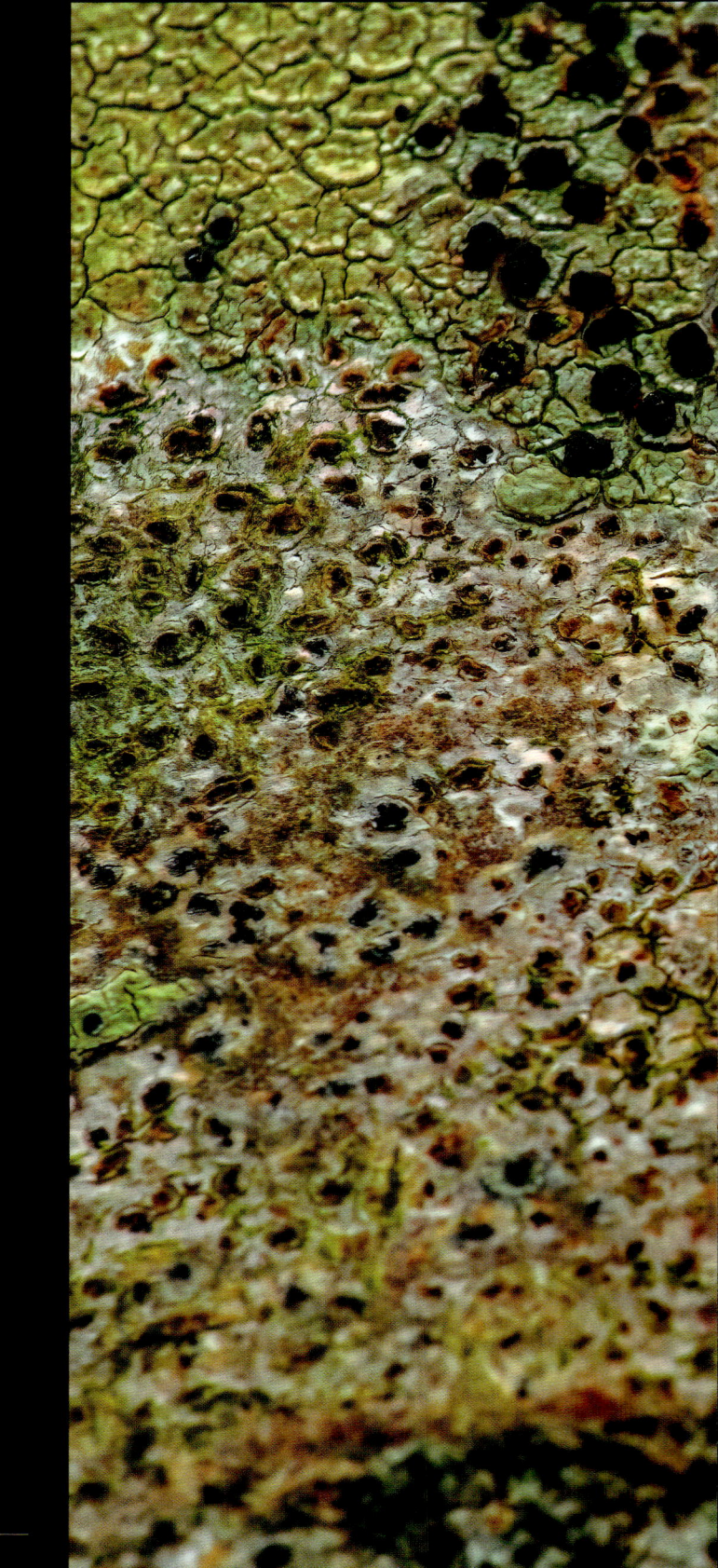

RIGHT: DISC LICHENS (*LECIDELLA* SPP.)

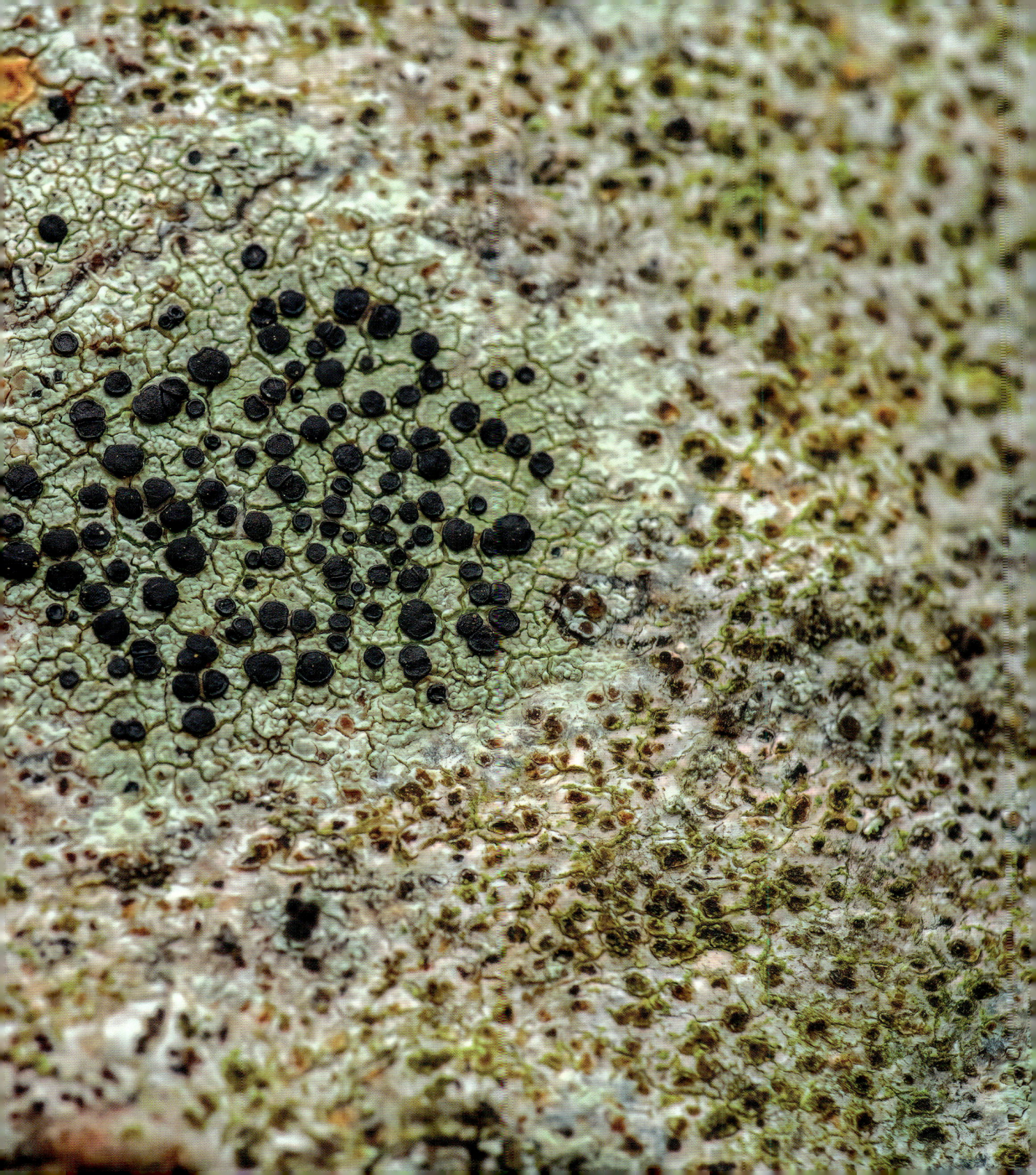

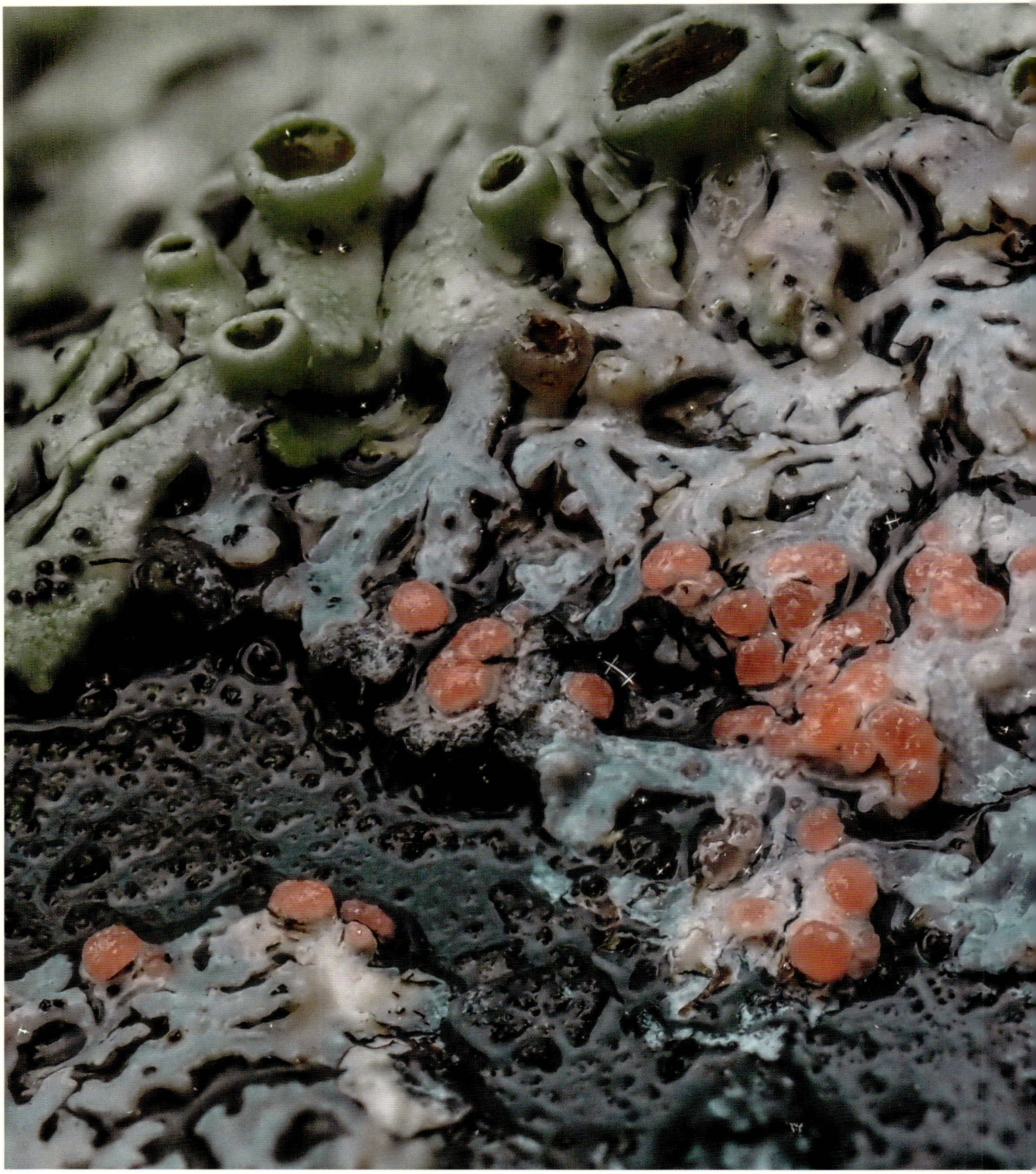

144

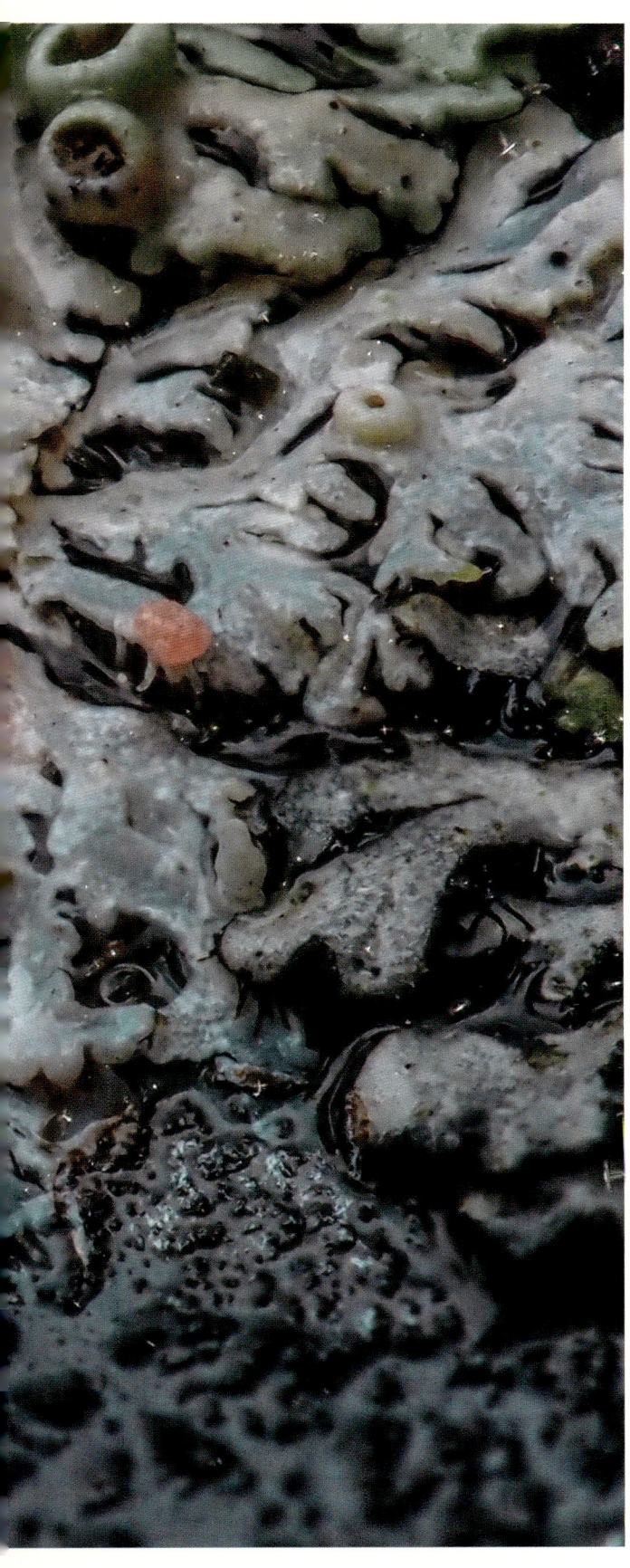

LEFT: *ERYTHRICIUM AURANTIACUM.*

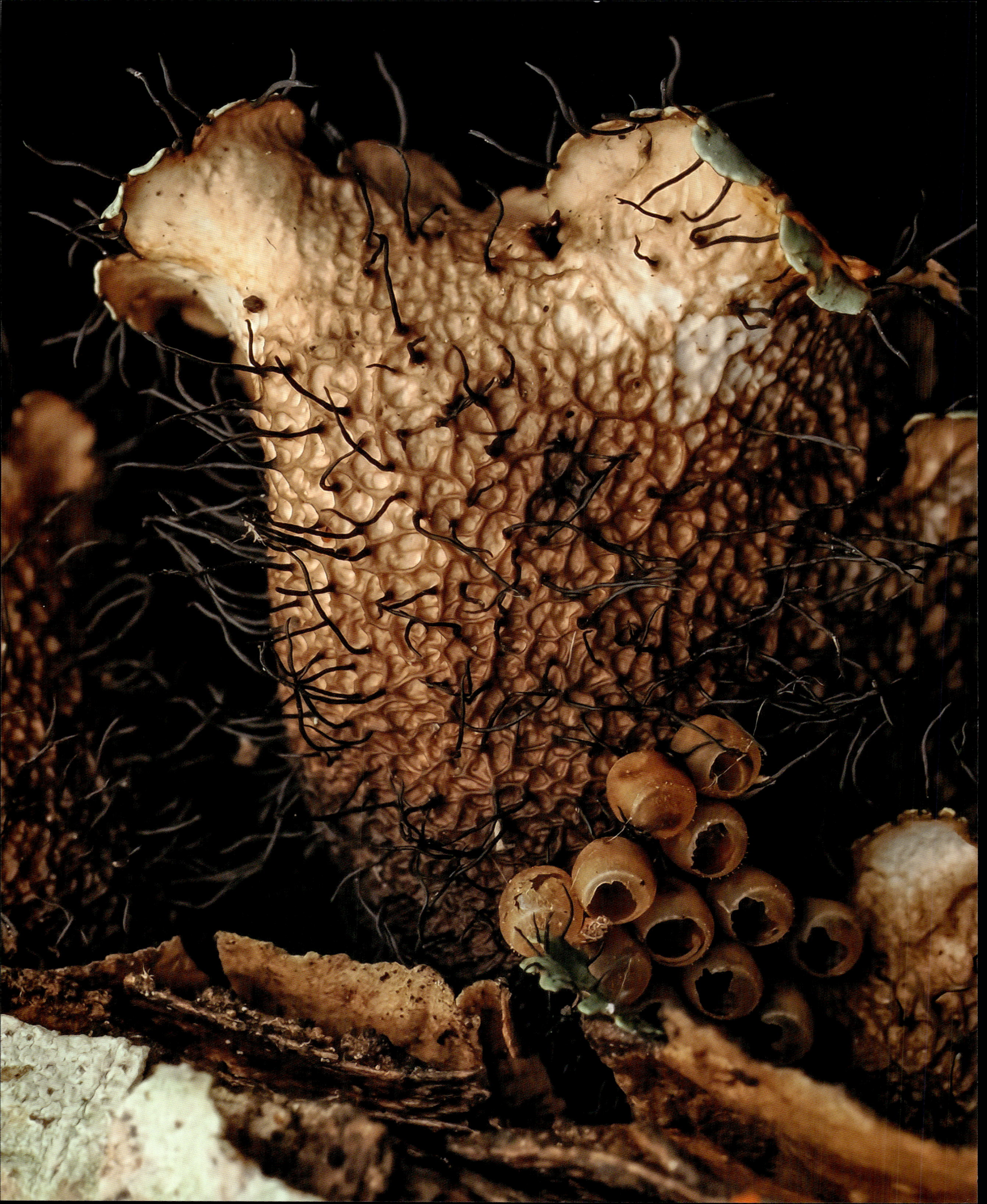

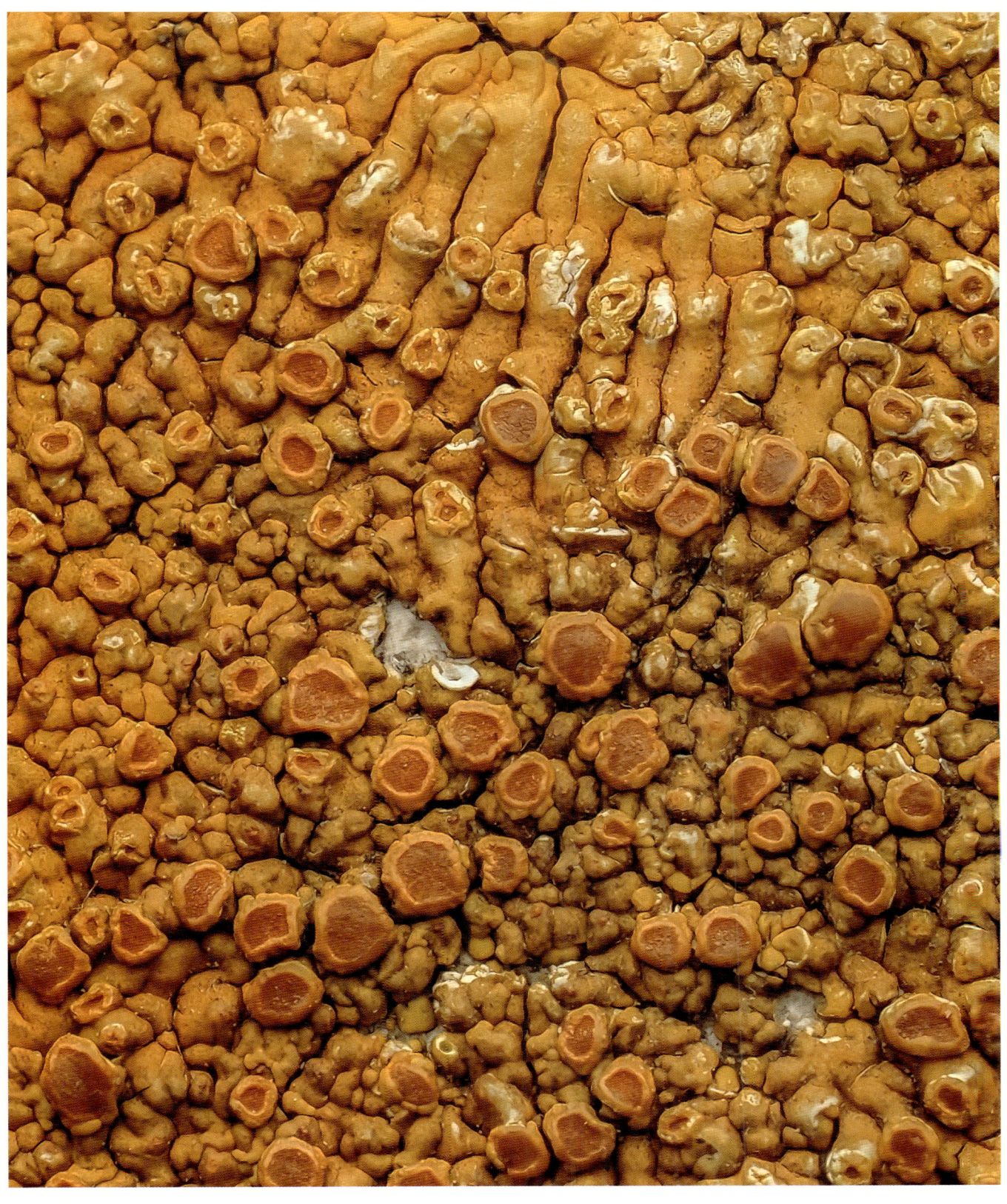

ABOVE: *VARIOSPORA* SPP. OPPOSITE: FOLIOSE LICHENS WITH CILIA AND INSECT EGG CASINGS.

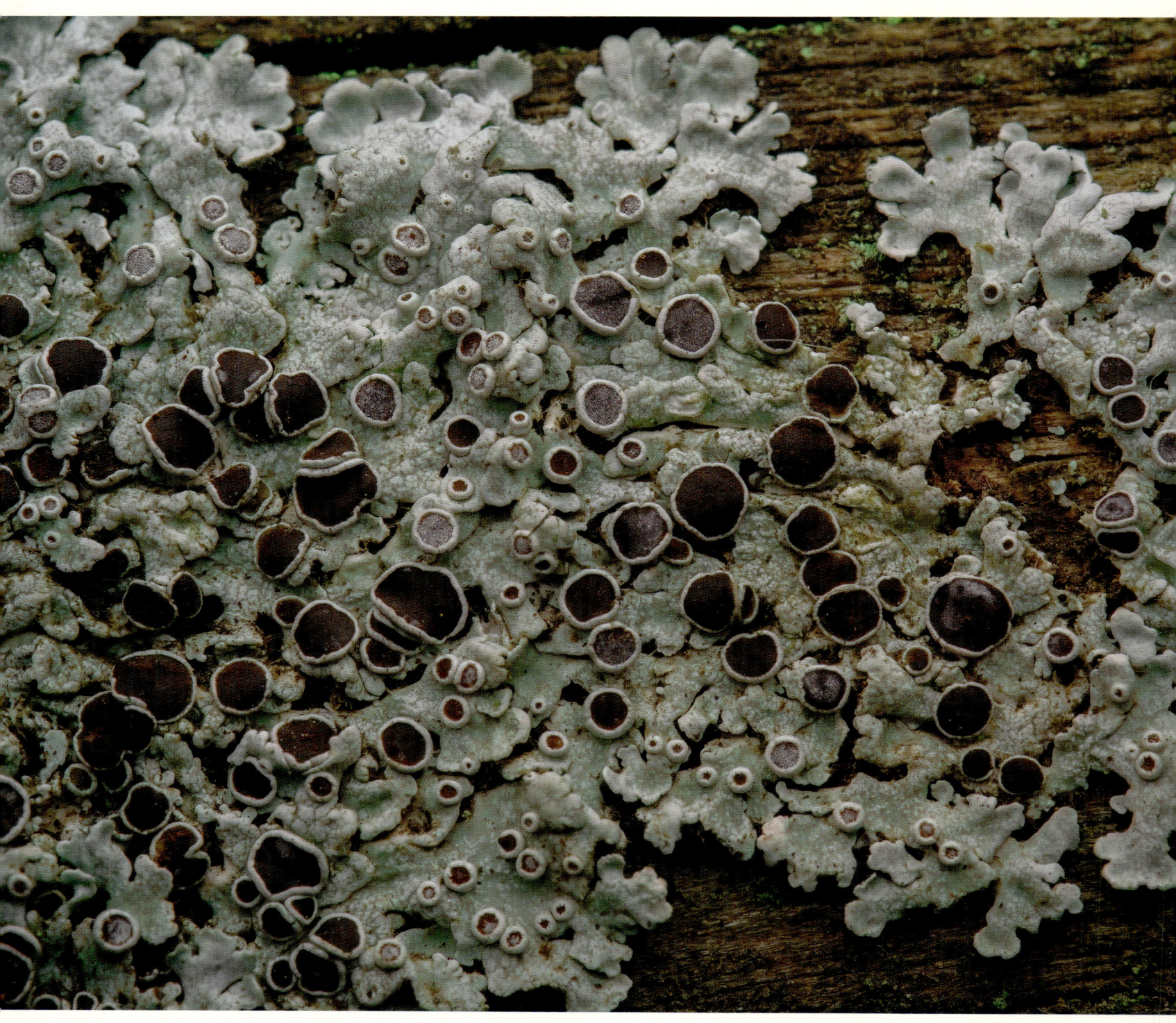

ABOVE AND OPPOSITE: ROSETTE LICHENS (*PHYSCIA* SPP.).

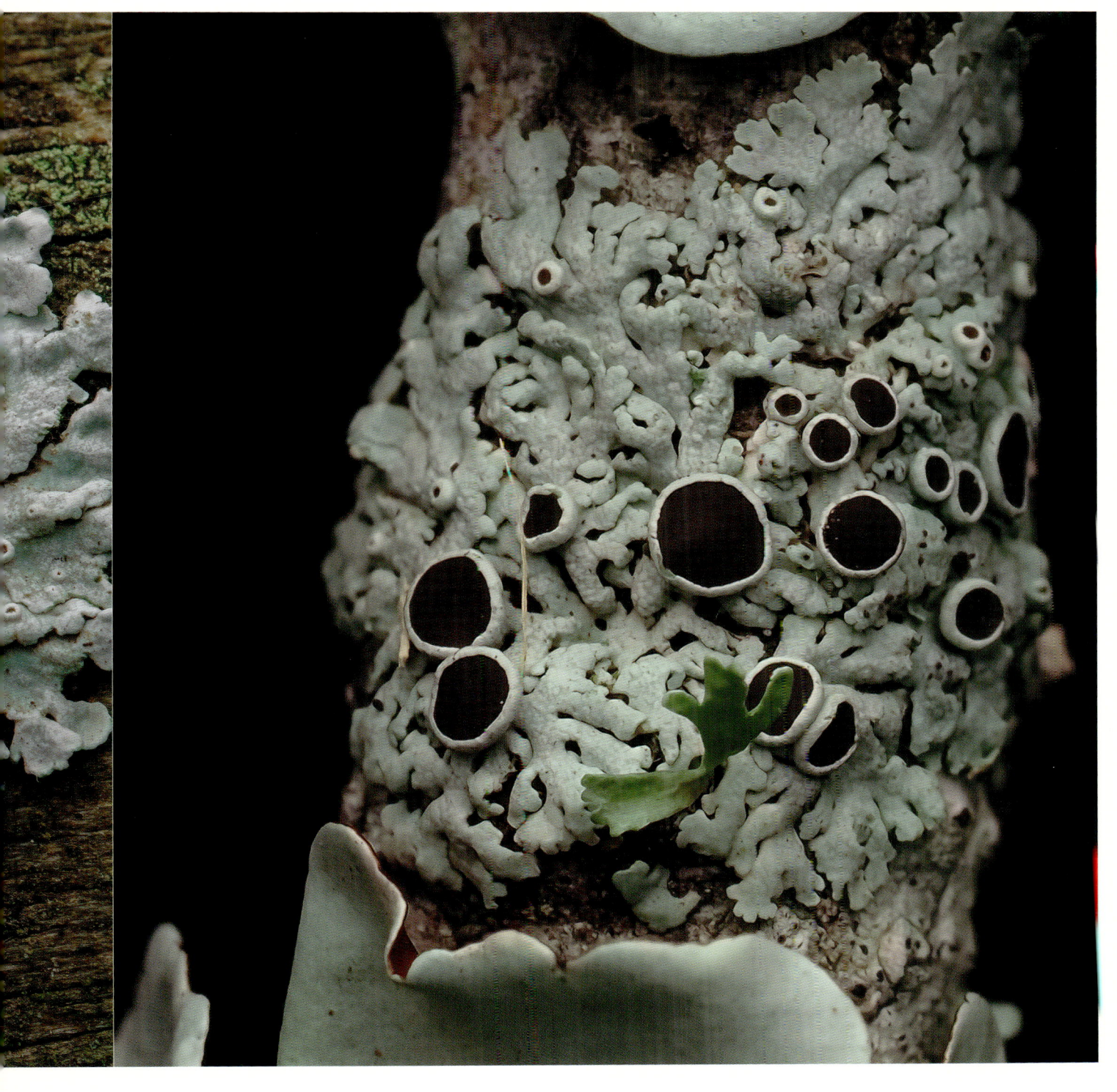

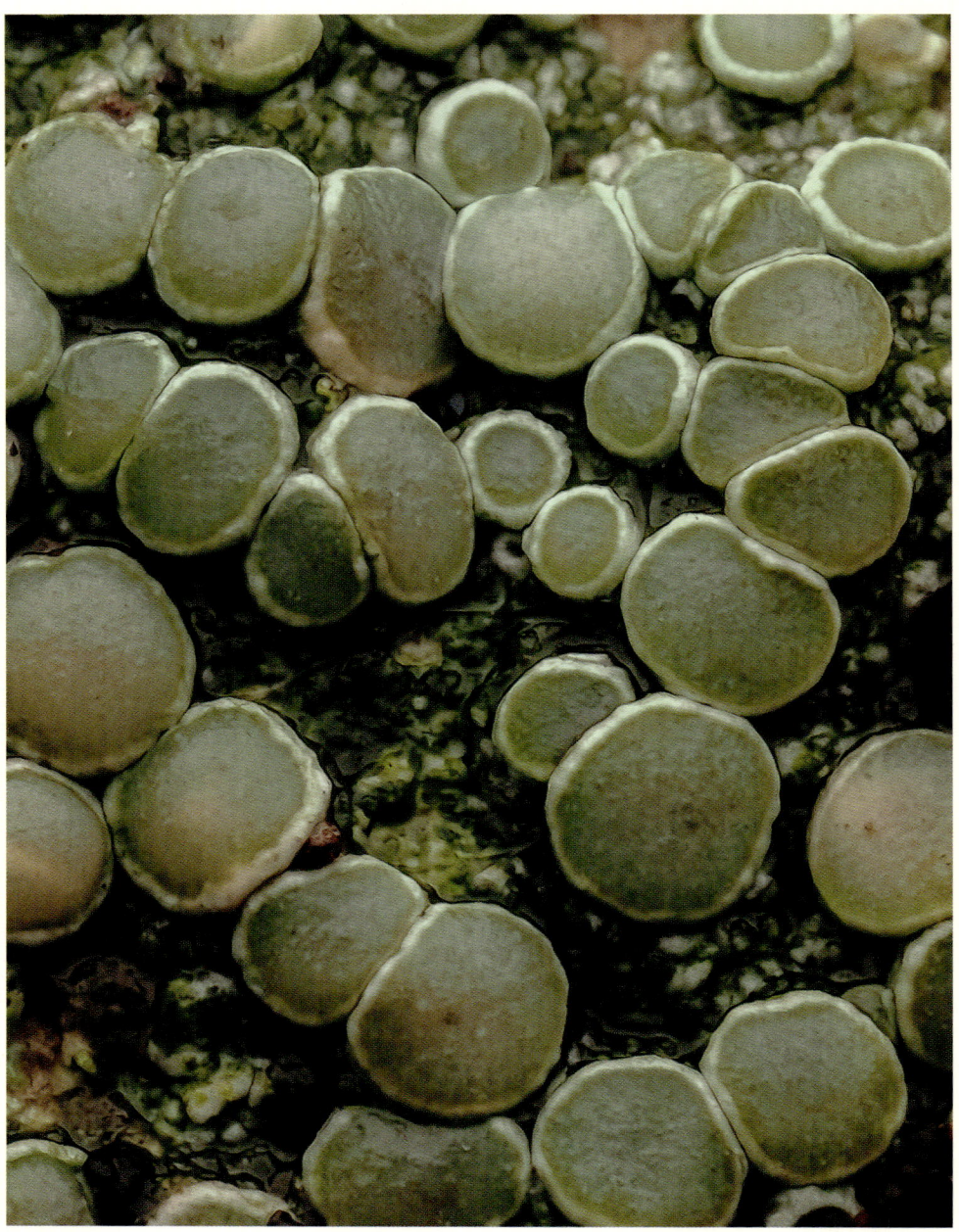

ABOVE AND OPPOSITE: RIM LICHENS (*LECANORA* SPP.).

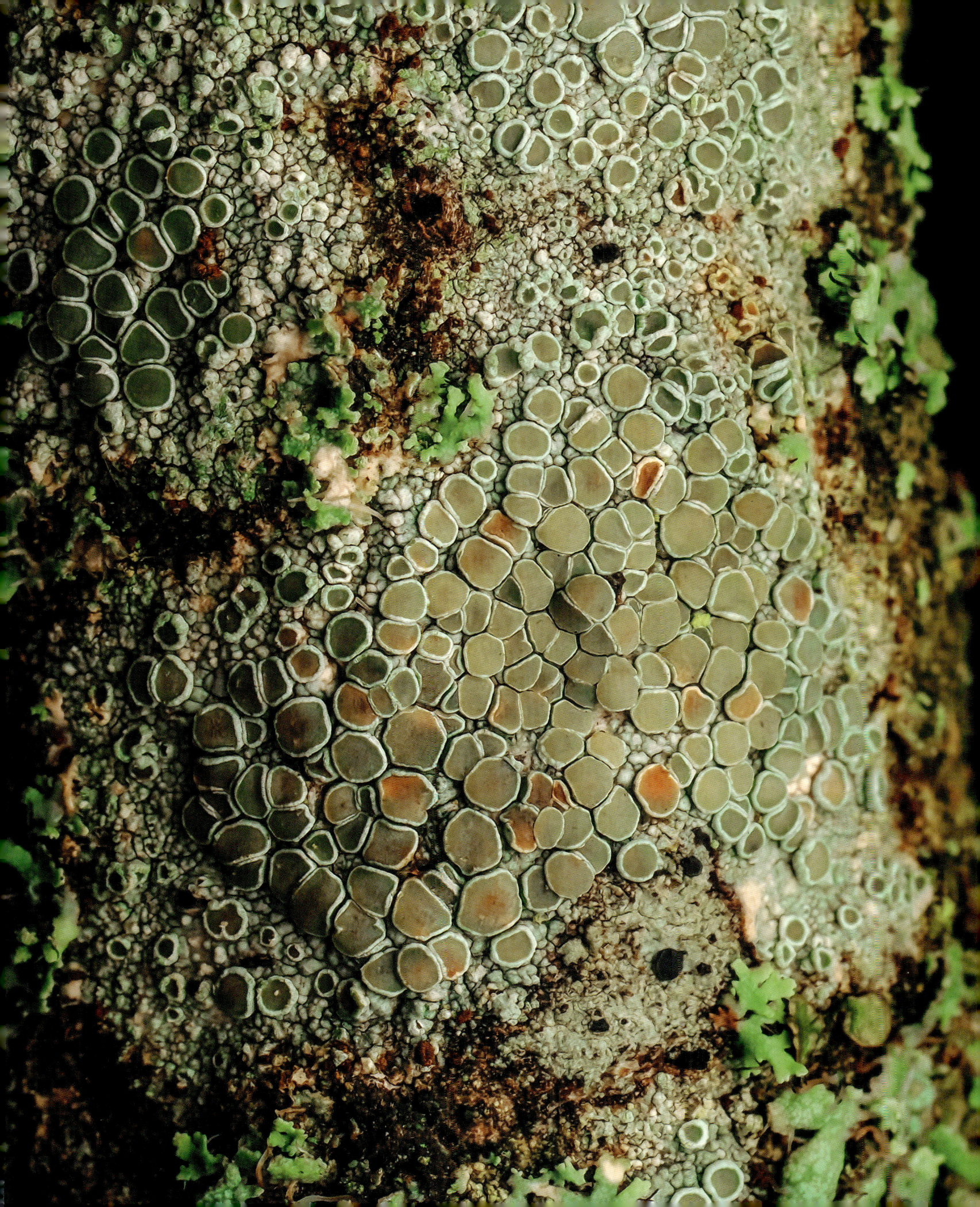

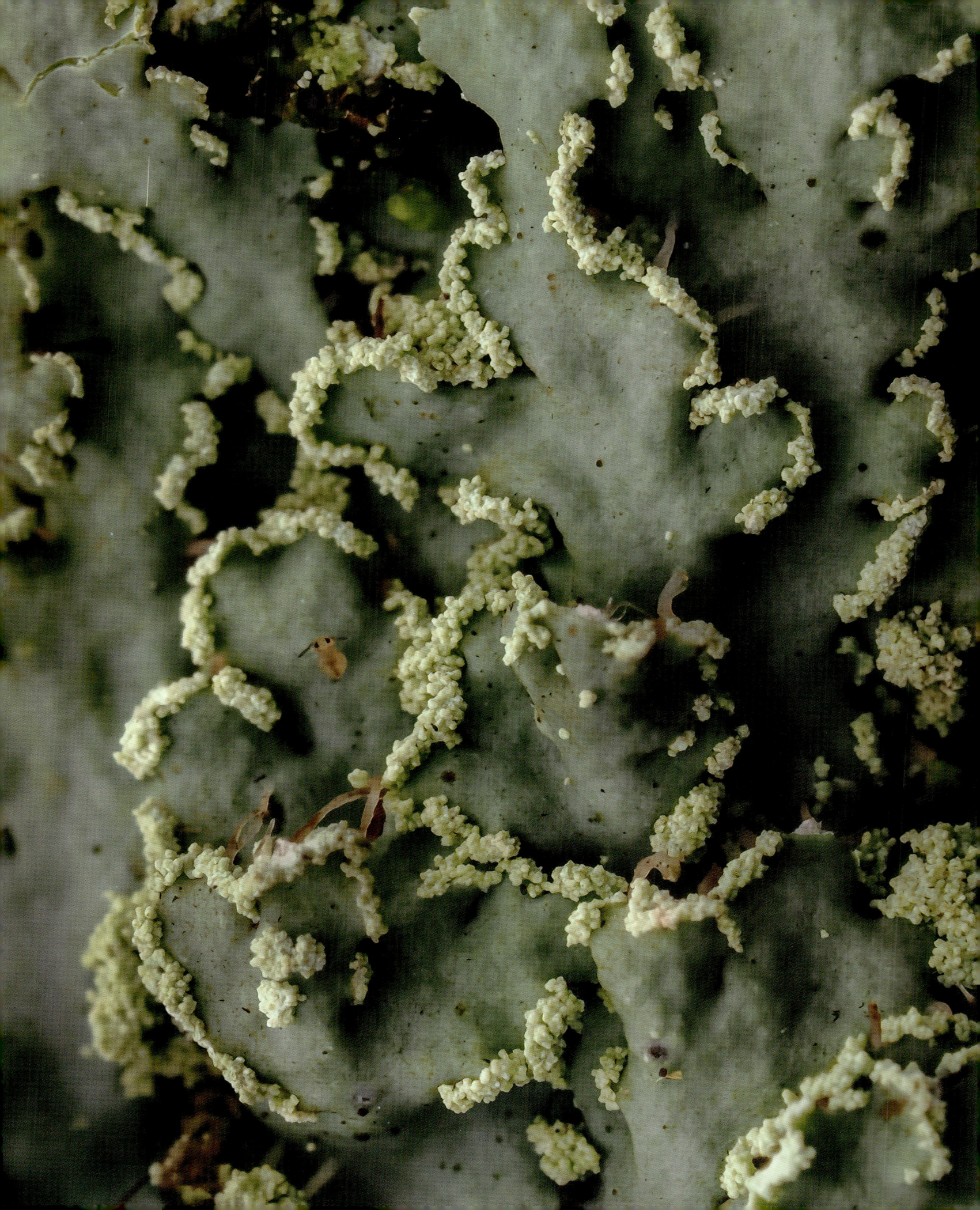

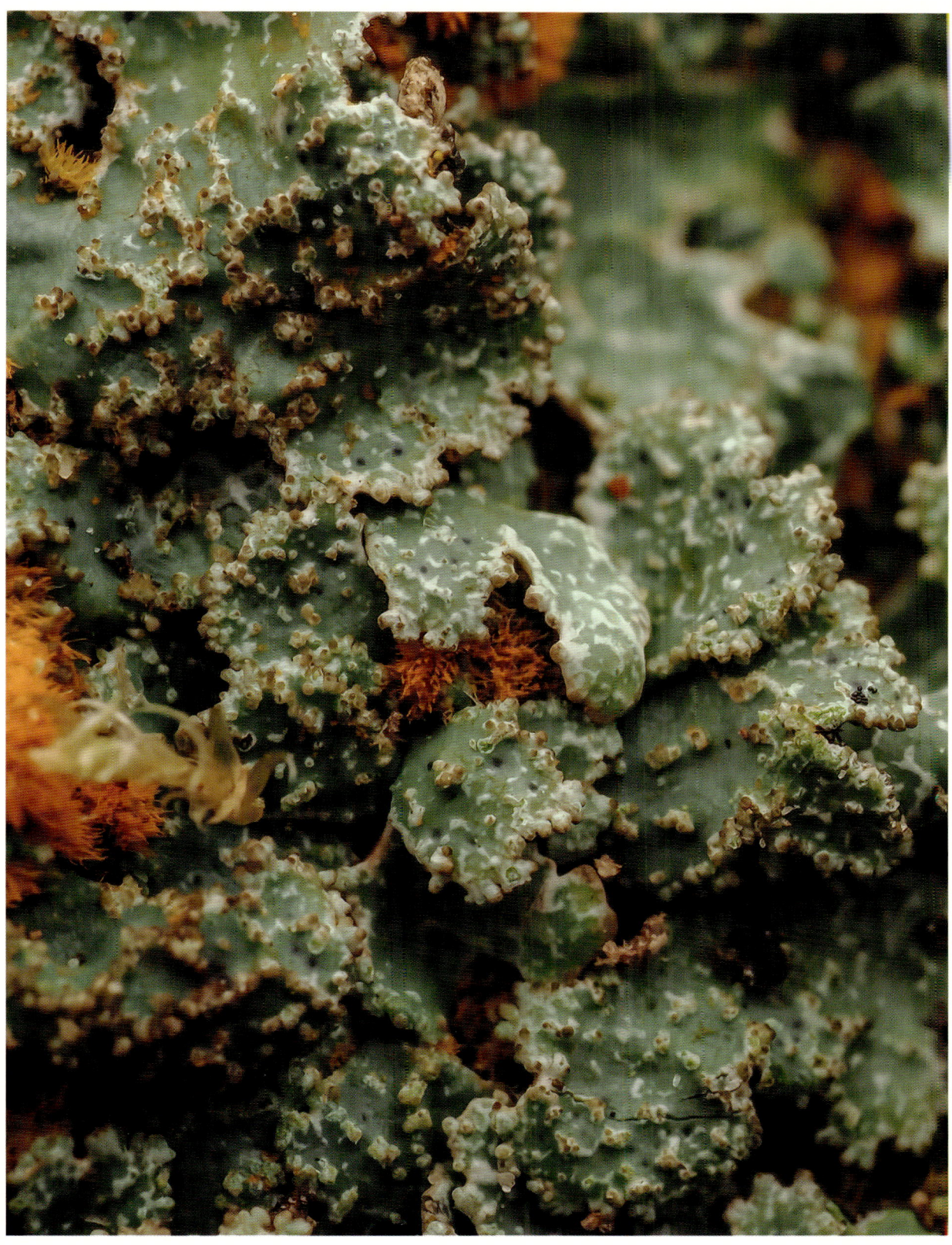

ABOVE: TYPICAL SHIELD LICHEN (SUBFAMILY *PARMELIOIDEAE*). OPPOSITE: POWDERED SPECKLED SHIELD LICHEN (*PUNCTELIA* SPP.).

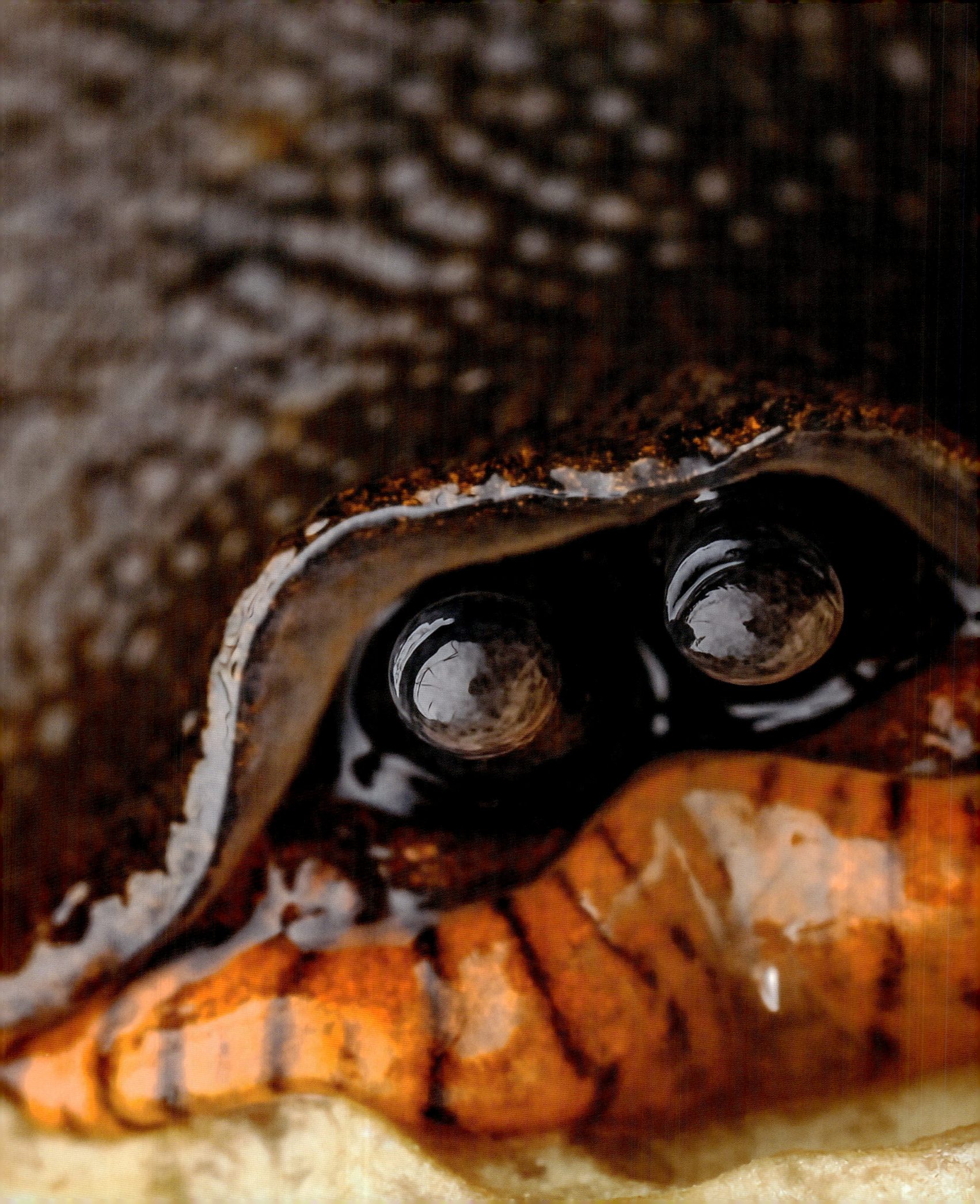

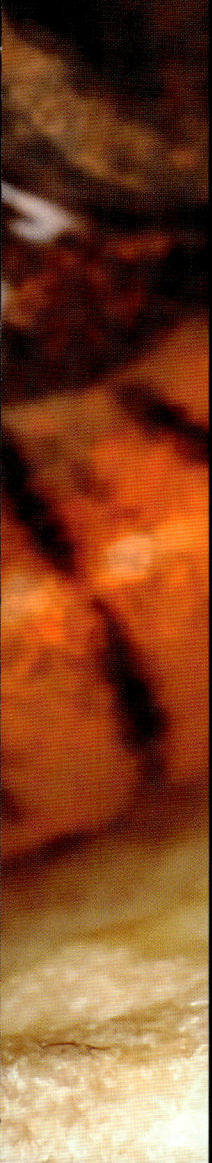

underside of their b

trail of goo in their w

where they've been.

Maybe if I had a

vegetables that I've t

and care, I would ha

slugs and snails . . . b

around and eating a

paths, being impossi

silly faces, taking vid

down leaves; using t

their very own minia

thousands of teeny t

They grind their foo

can see pass down t

OPPOSITE AND FOLLOWING PAGES:

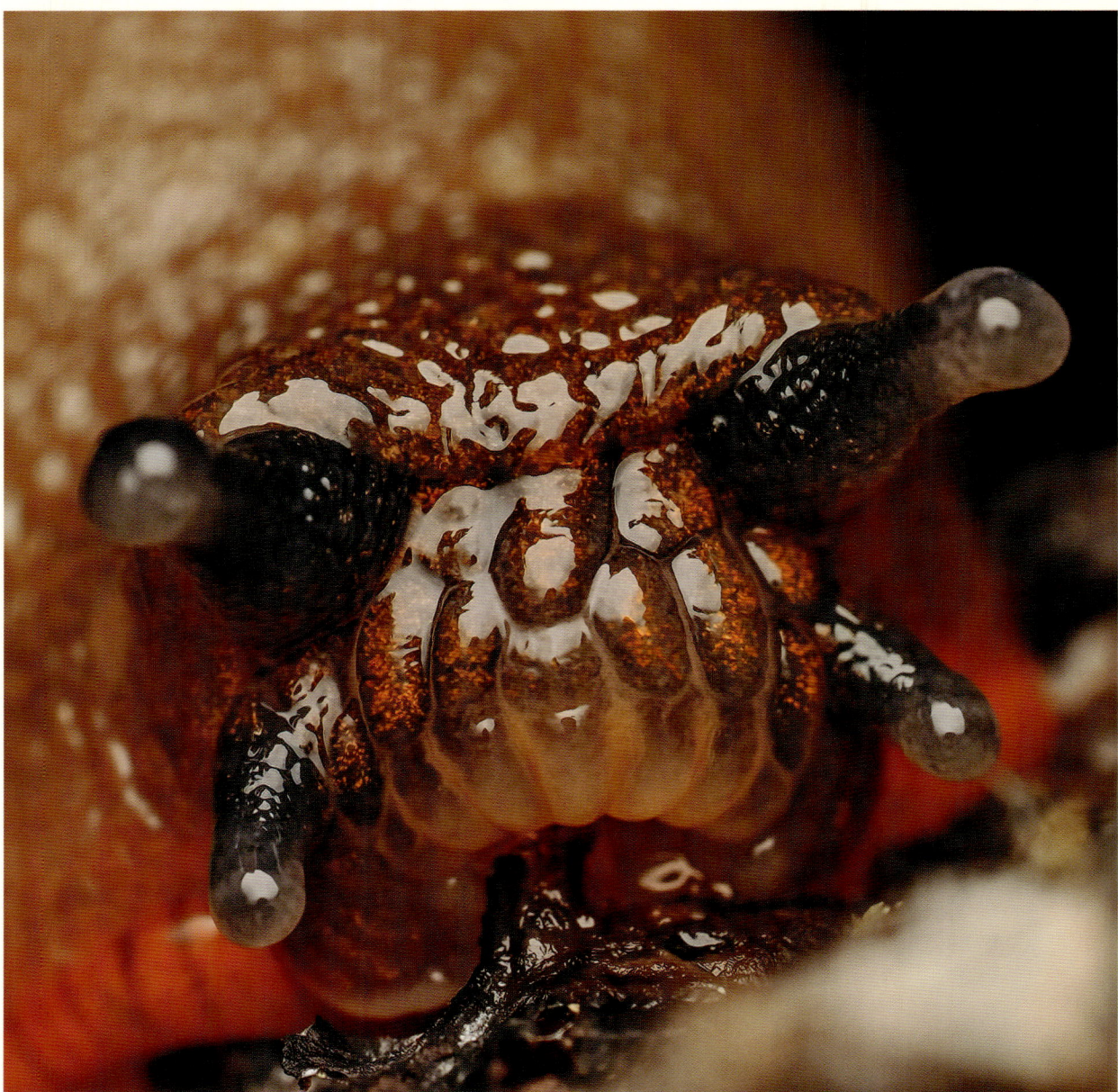

You can't deny the joy of seeing a snail release its optical tentacles, or "eye stalks." They are like the periscopes of the gastropod world. Gastropods certainly can't see in HD (high definition), but they can tell light from dark, spot basic shapes, and detect movement enough to consistently turn away from me right when I'm getting the best close-up photographs.

What's also cool about these tentacles is that they can move around independently.

They can swivel their eyes in all directions, so if danger approaches—say, an enthusiastic macrophotographer—they can pull those eye stalks back into their head, like retracting into a little safe house. Crazily enough, should they lose a stalk in a forest mishap, they can actually grow them back.

Although there are over 60,000 species of gastropods spread out across the globe, in the United Kingdom (where I do most of my

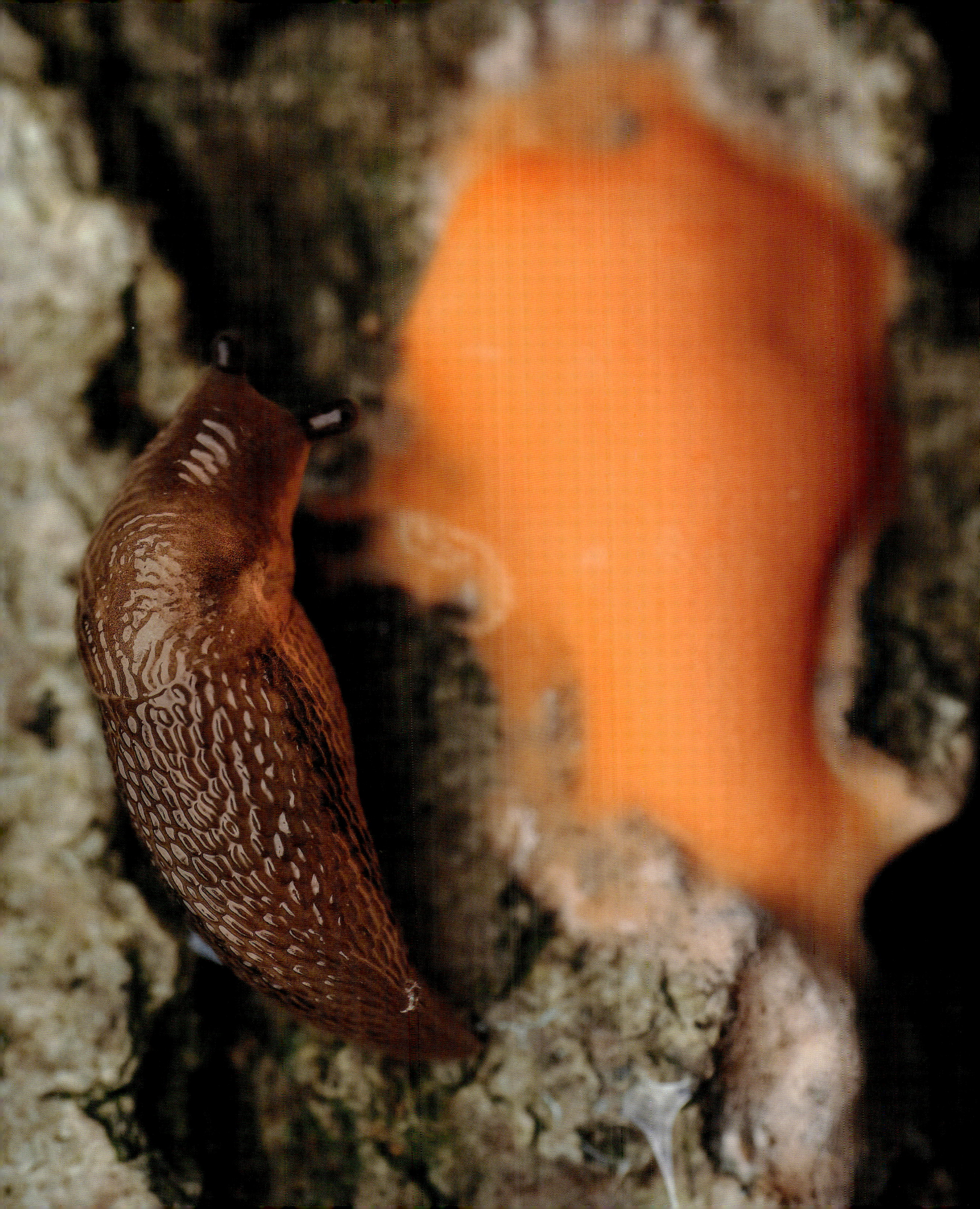

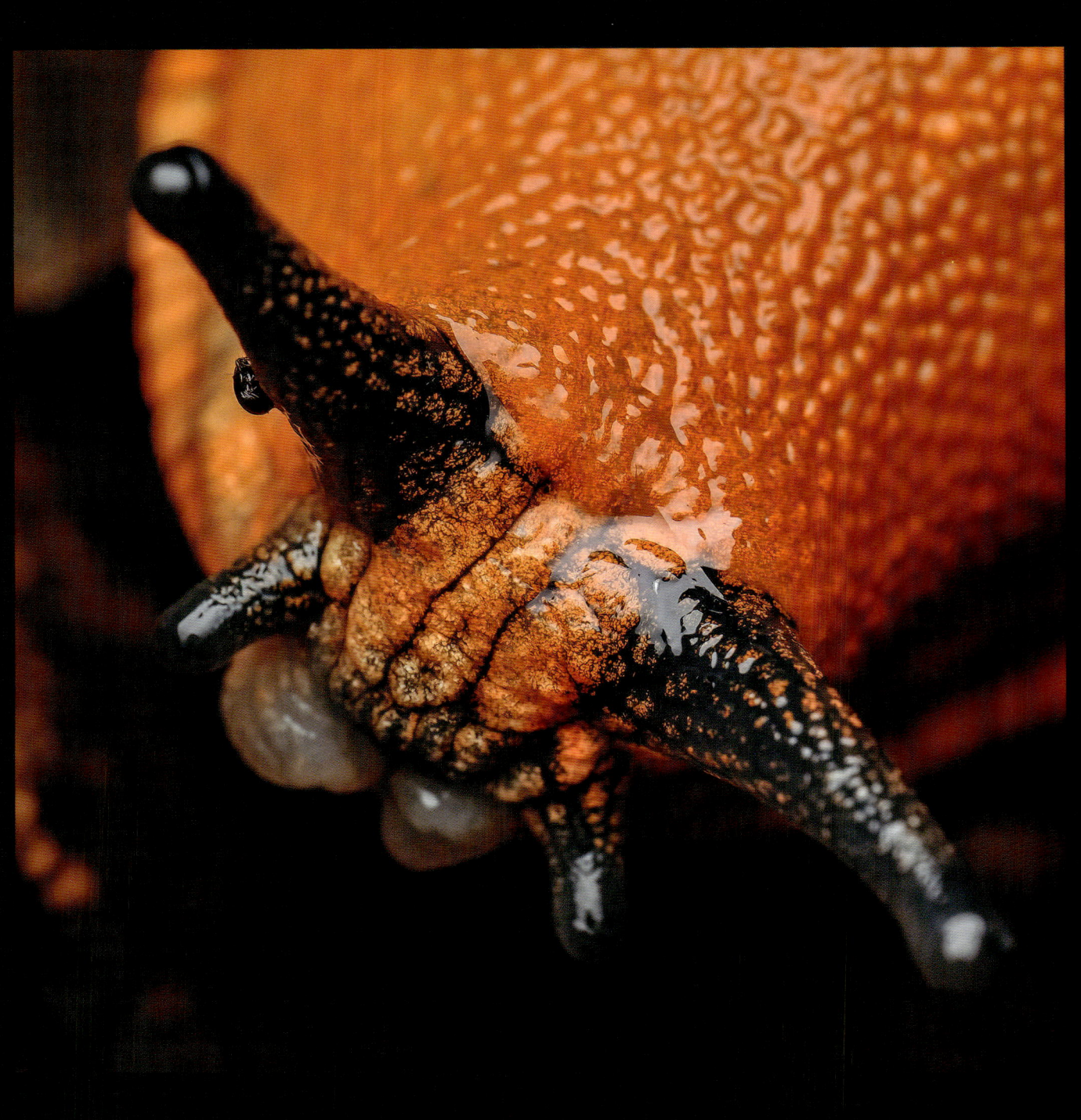

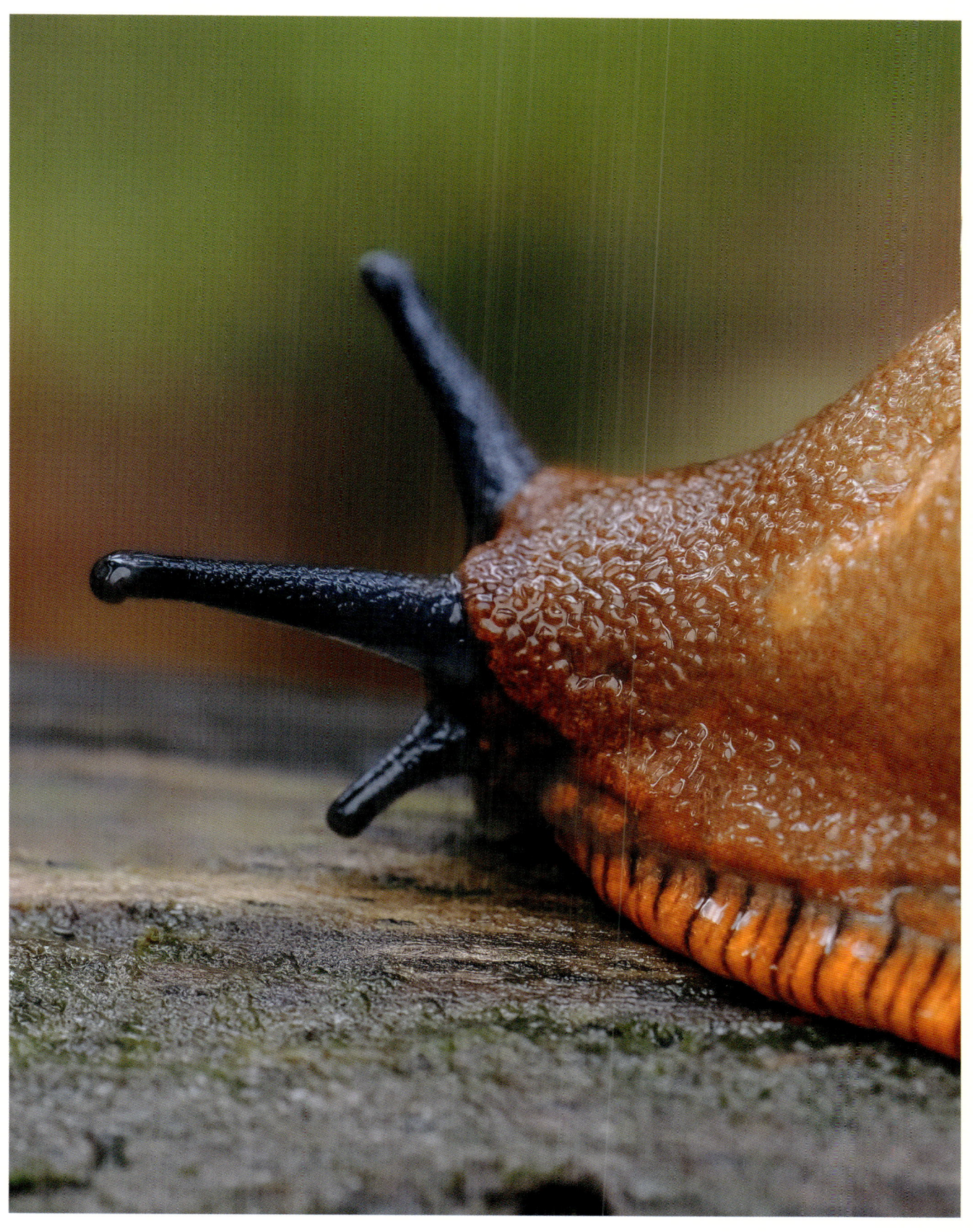

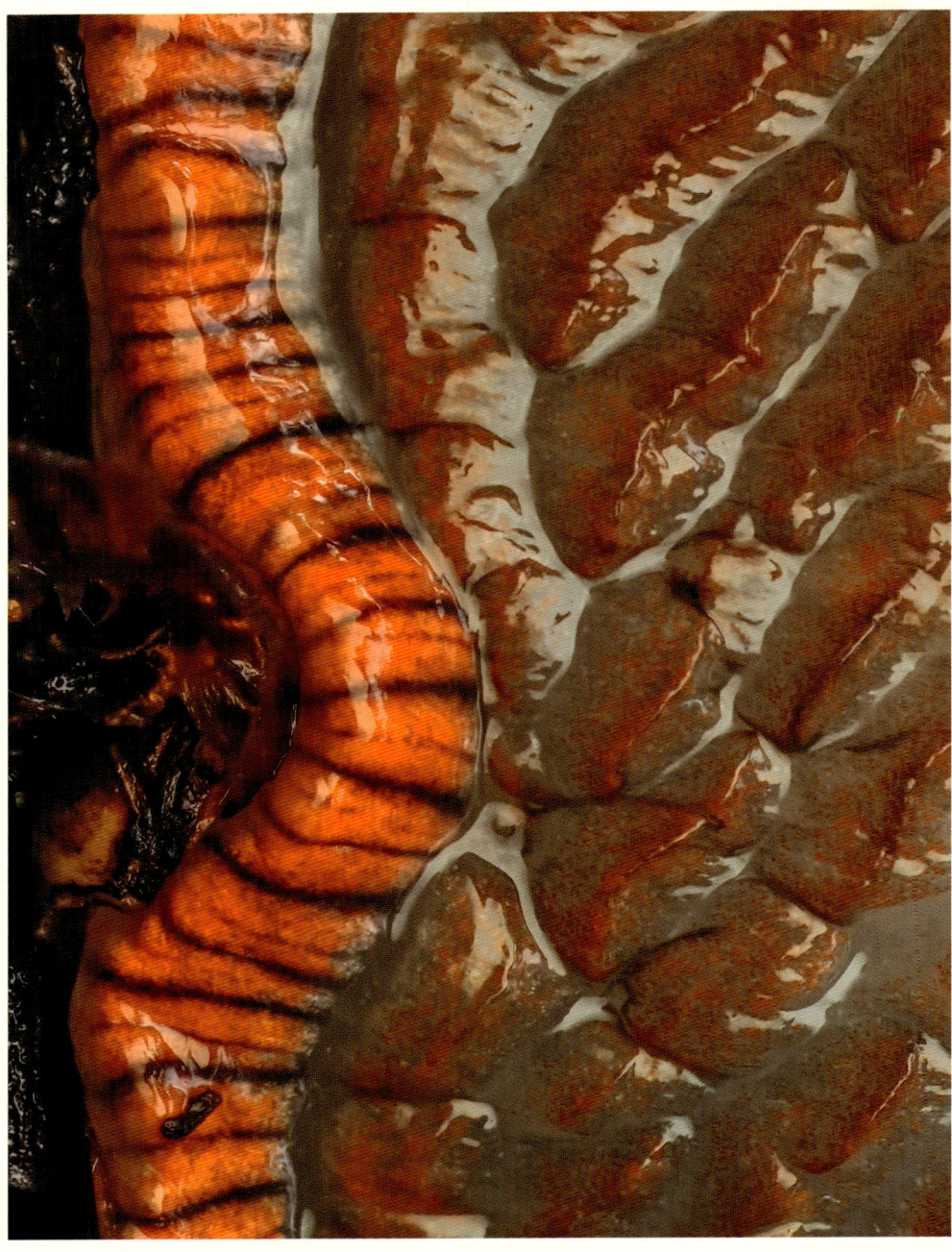

photography) there are just over 100 species of land snails and around forty species of slugs. And I love them all! Every squishy, delightful, blobby one of them.

Slugs and snails have a well-earned negative reputation among most gardeners. They are often seen as pests due to their voracious appetites and tendency to absolutely demolish gardens. They do have natural predators like hedgehogs and birds, but the gastropod population, in London at least, is most certainly not at threat. In the summer, after a heavy rain, they can be seen climbing walls, trees, and putting their lives at risk by clogging up footpaths. Despite their bad reputation, they actually do some good things too!

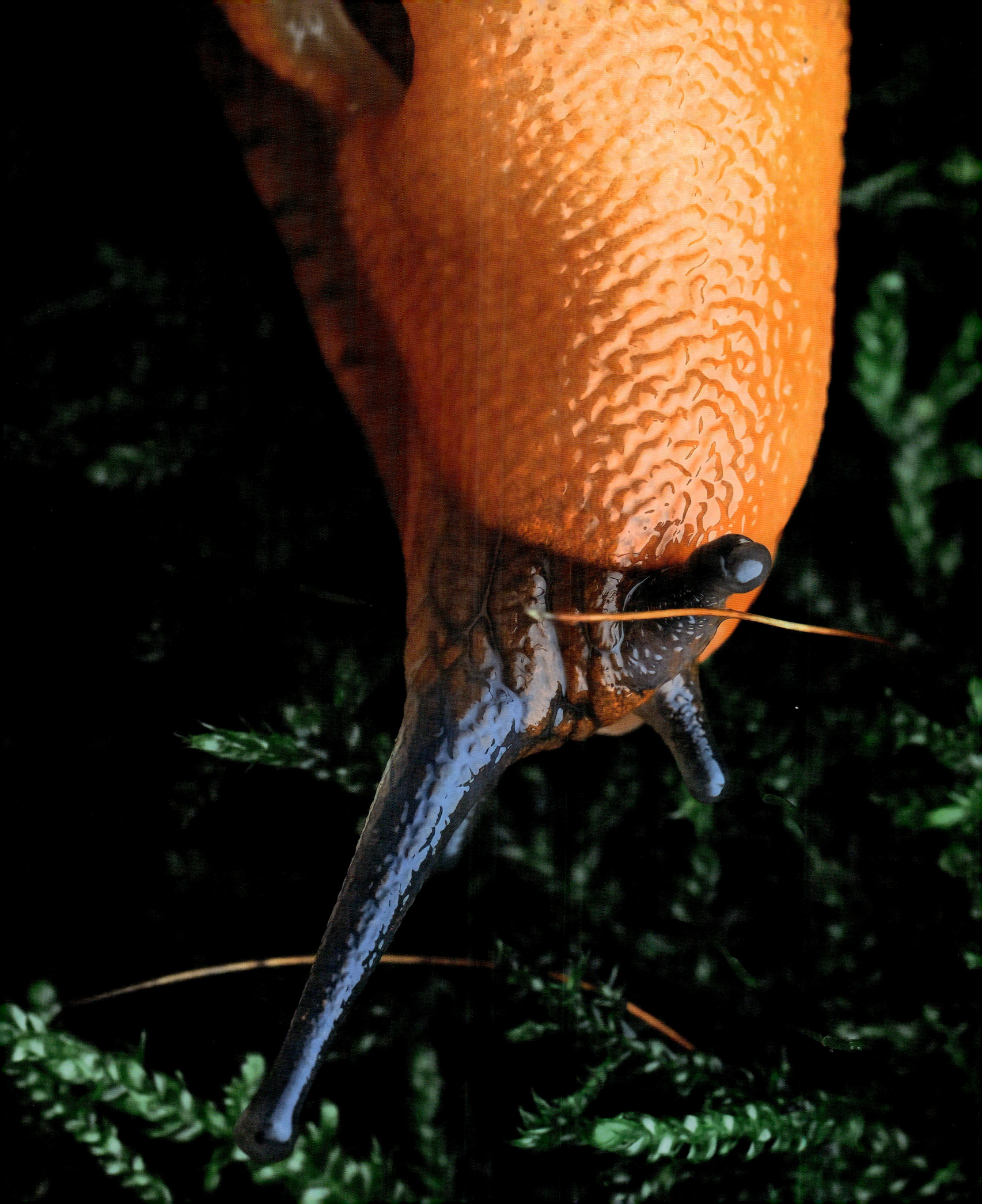

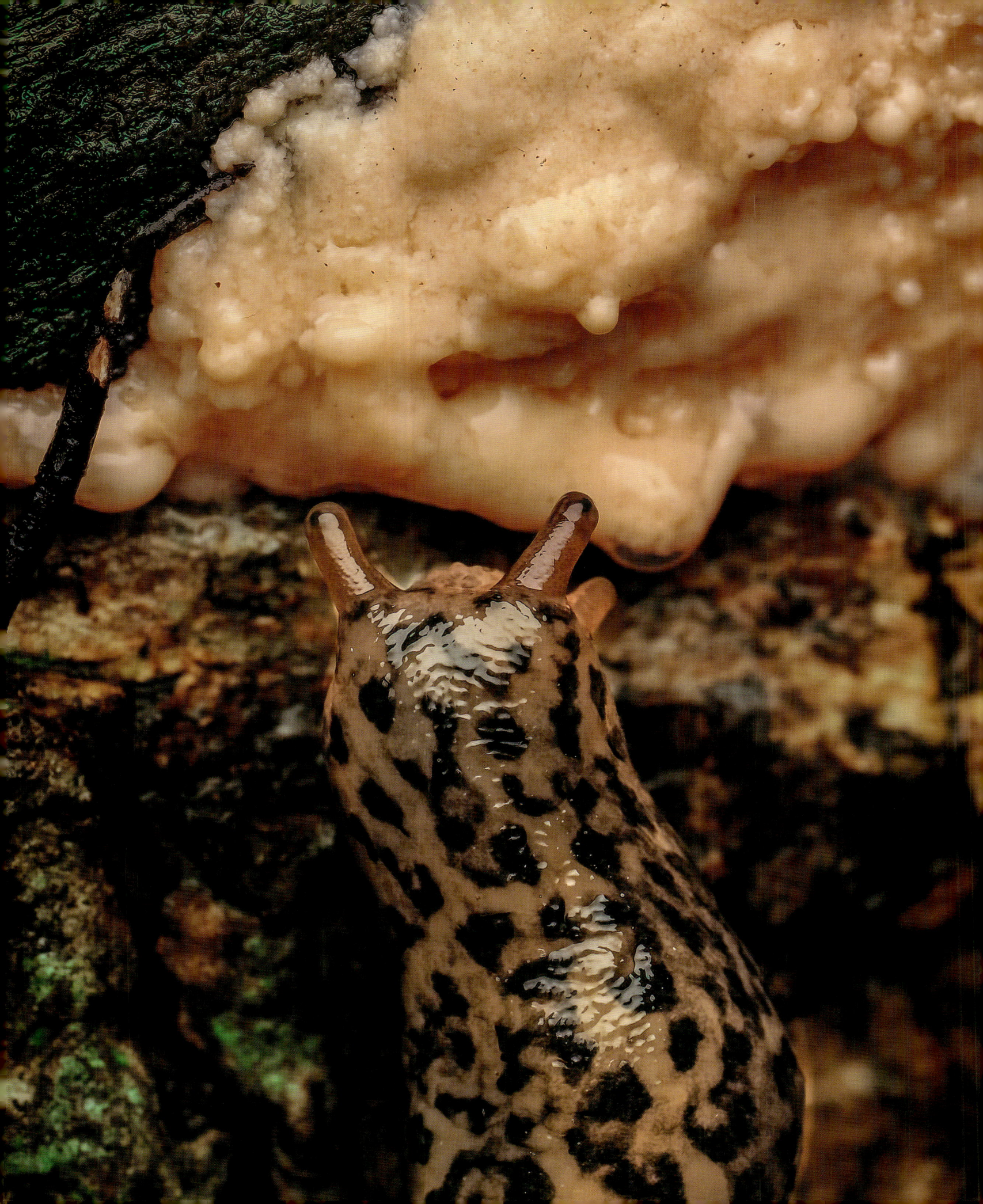

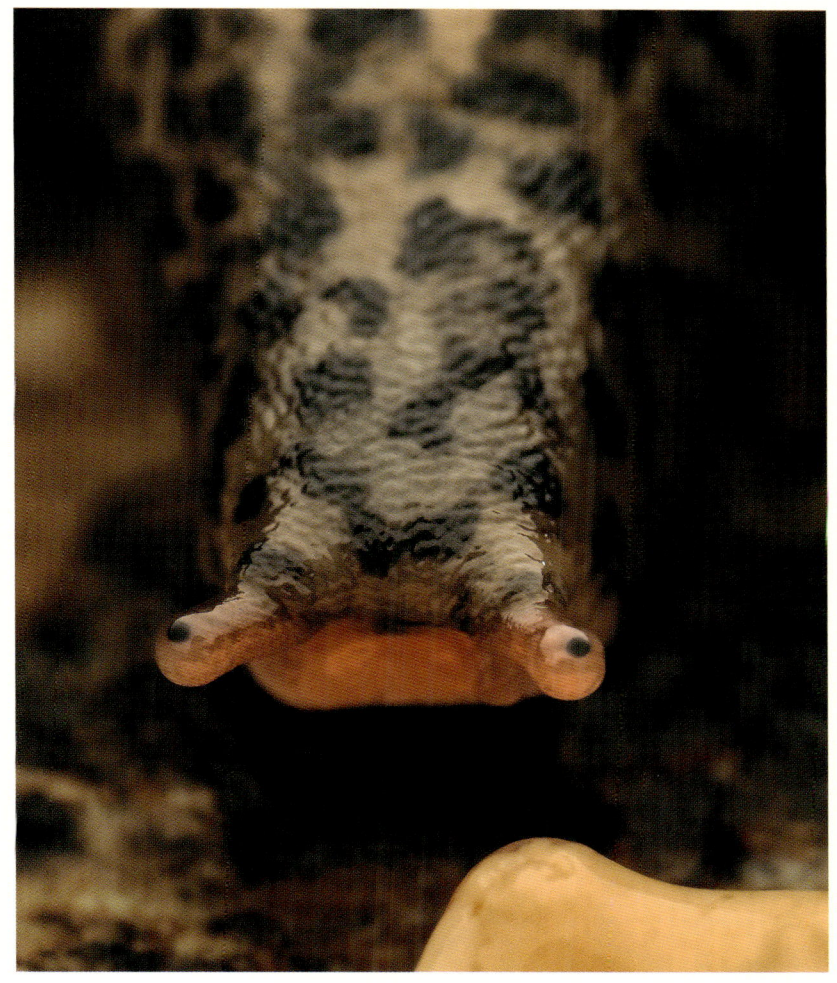

ABOVE AND OPPOSITE: LEOPARD SLUG (*LIMAX MAXIMUS*).

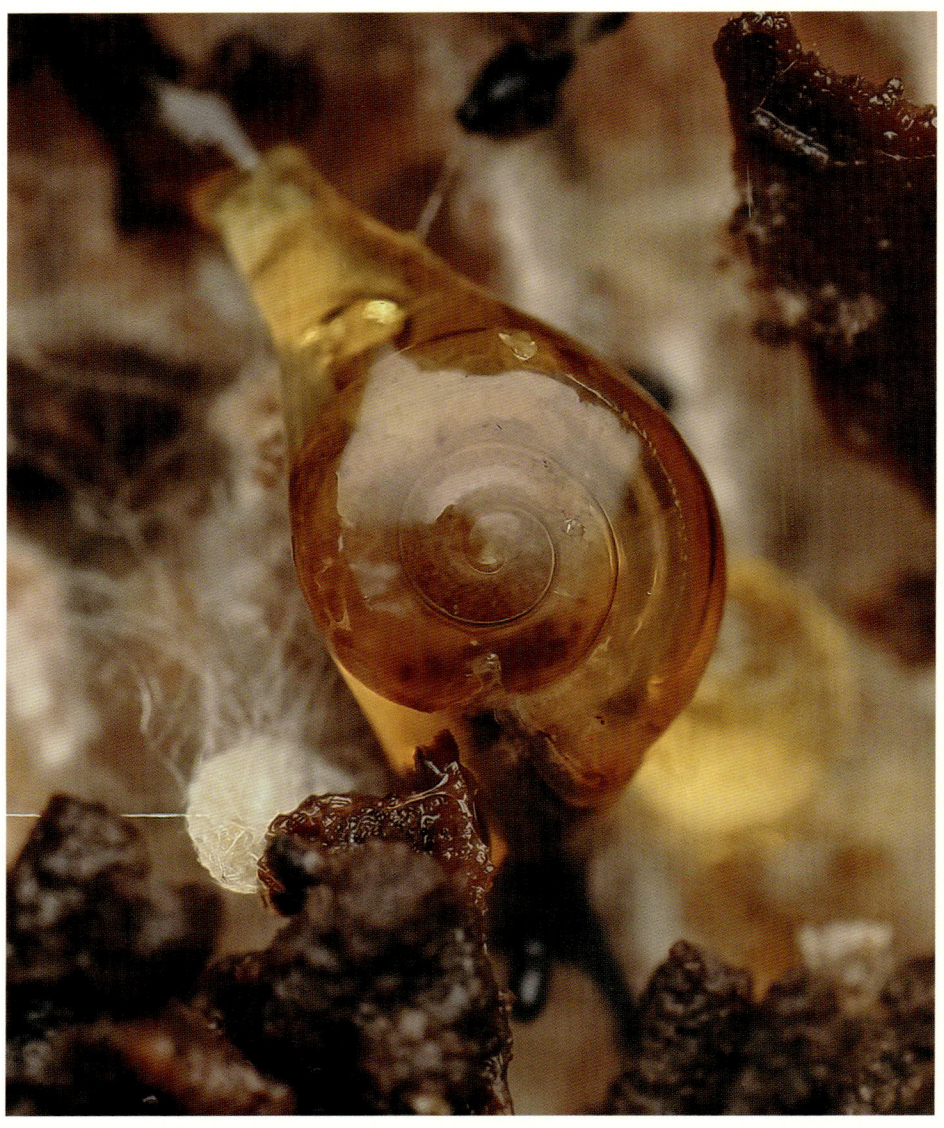

Snails are one of the tiny creatures that eat fungi and help spread the spores through their waste, aiding in the spread and reproduction efforts. They have even been used in folk medicine, believed to have healing properties when applied to wounds or consumed in certain concoctions. I really do love them, but not enough to rub their mucus all over my face. No thanks. I'll stick to photographing them and saving them from being crushed on busy sidewalks.

Gastropods also consume slime molds, further supporting the larger decomposition cycle. They are primarily herbivores, preferring plants and fungi, but they really aren't that picky. I've personally witnessed them munching on worms and dog poop and even a half-eaten hamburger! However, birds, hedgehogs, beetles, insects, frogs, and toads prey on gastropods, so though gastropods eat everything in sight, they also end up as lunch. Such is the circle of life.

ABOVE AND OPPOSITE: GLASS SNAIL (*VITRINIDAE* SP.).

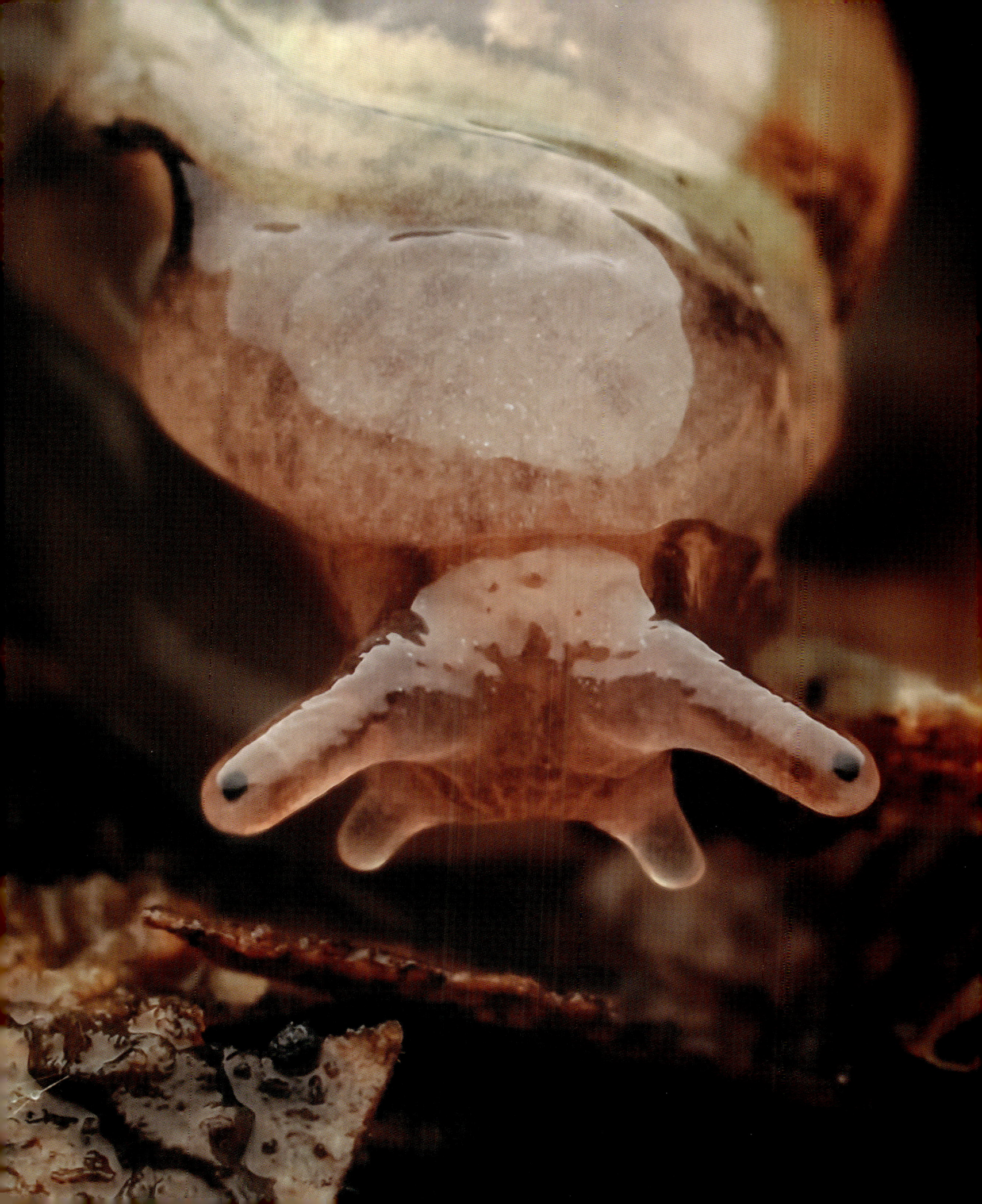

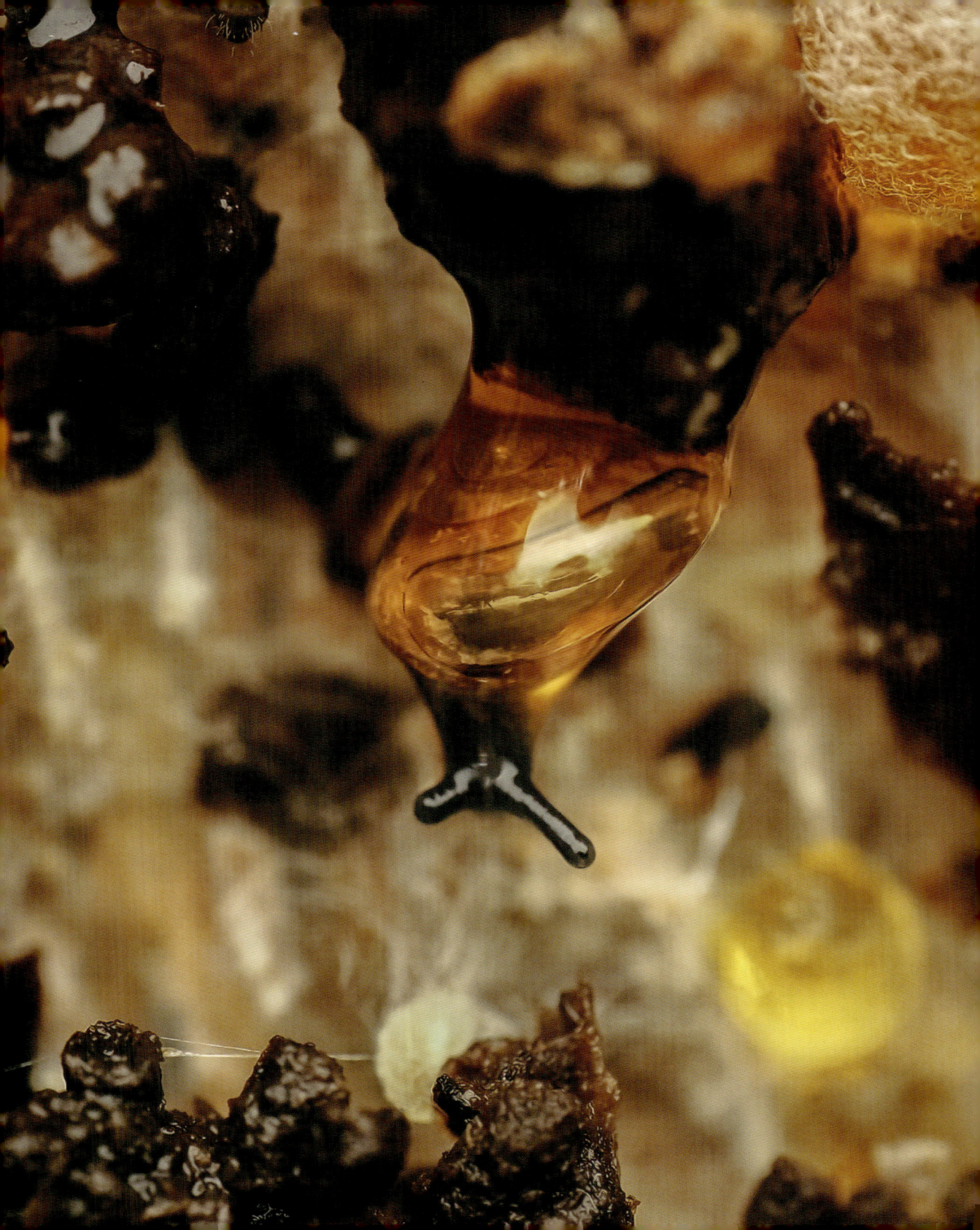

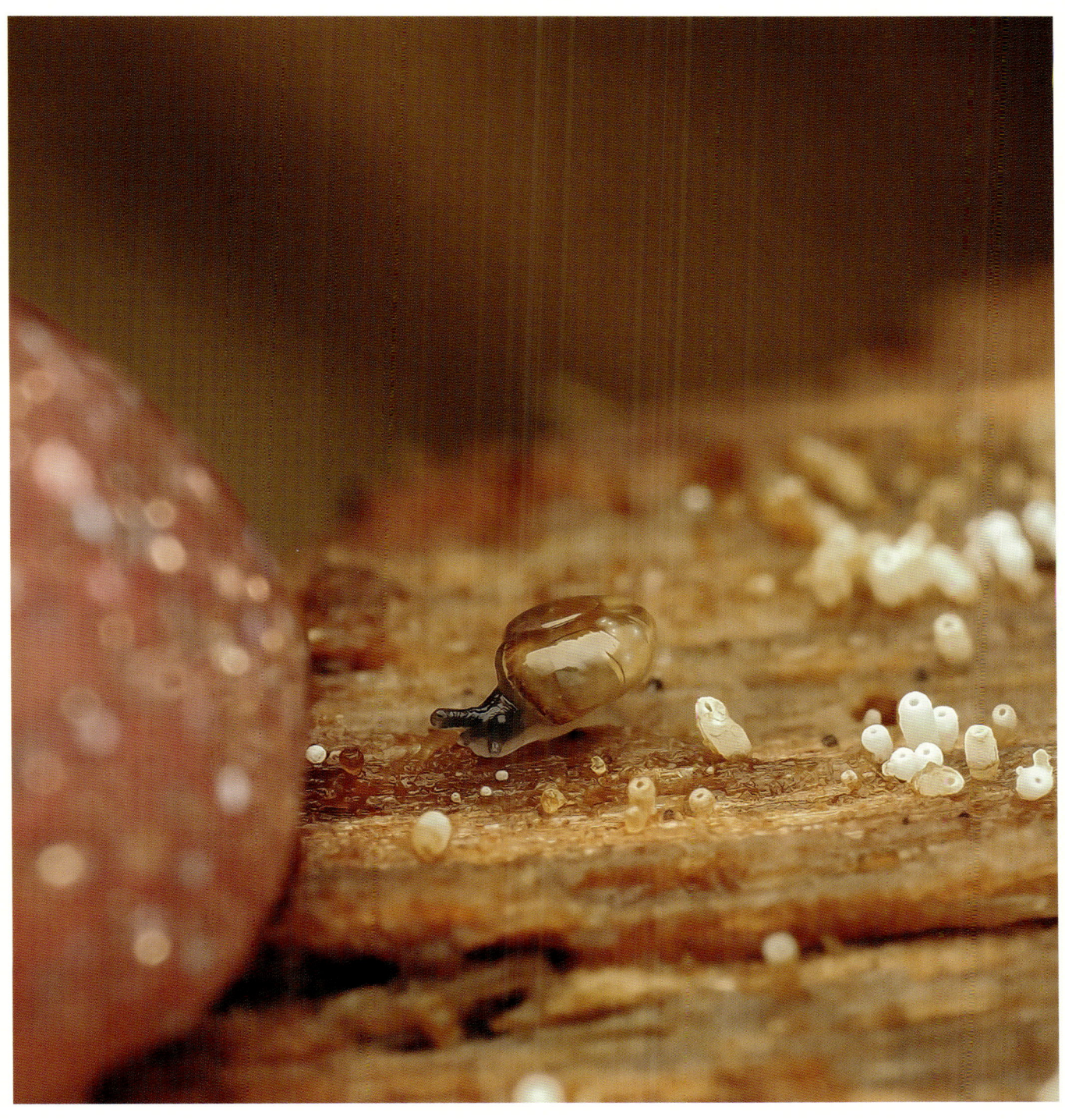

ABOVE: GLASS SNAIL (*OXYCHILIDAE* SP.). OPPOSITE: GLASS SNAIL IN MUSHROOM GUTTATION DROPLETS.

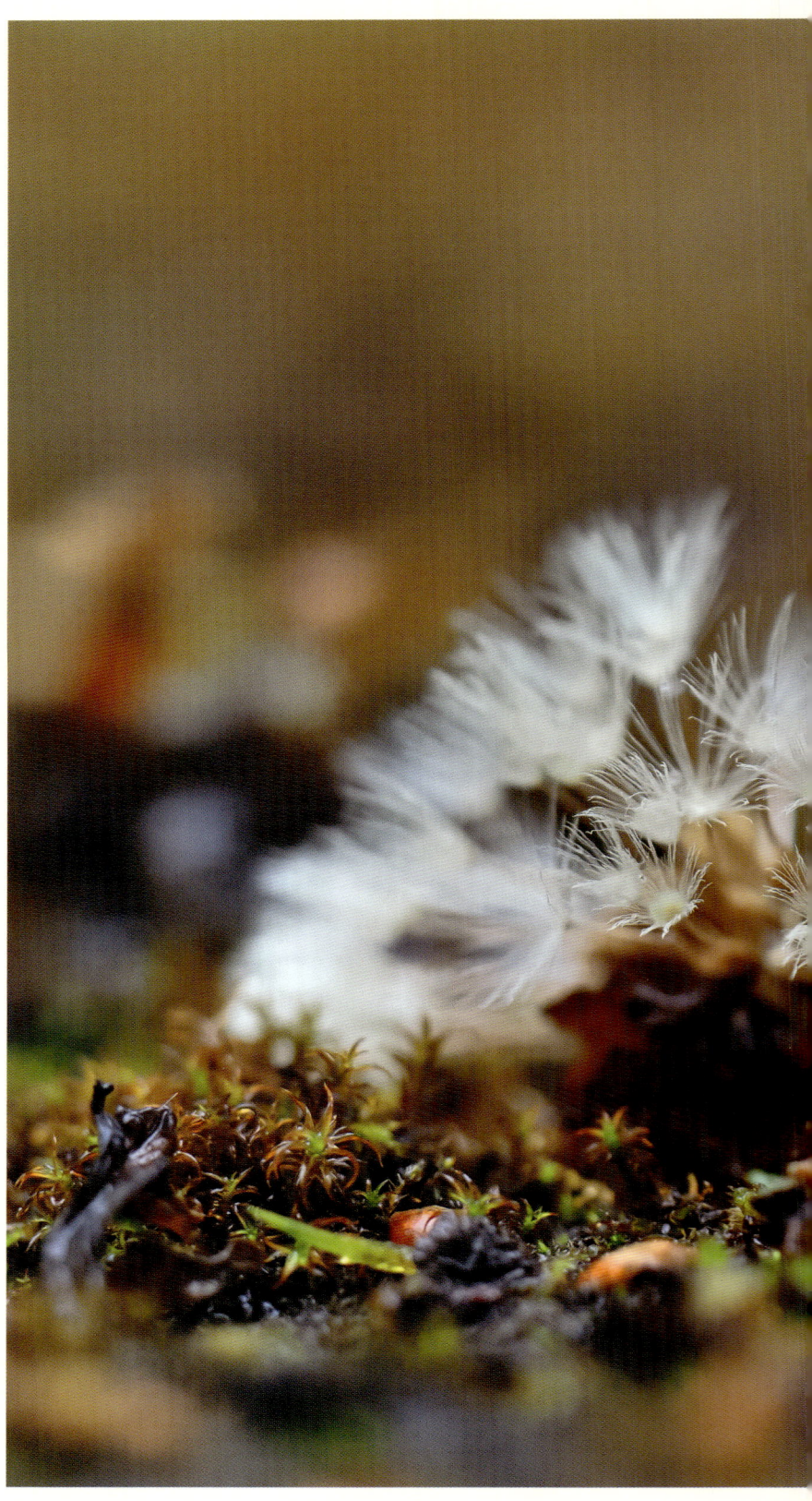

ABOVE: GARDEN SNAIL (*CORNU ASPERSUM*) WITH DANDELIONS.

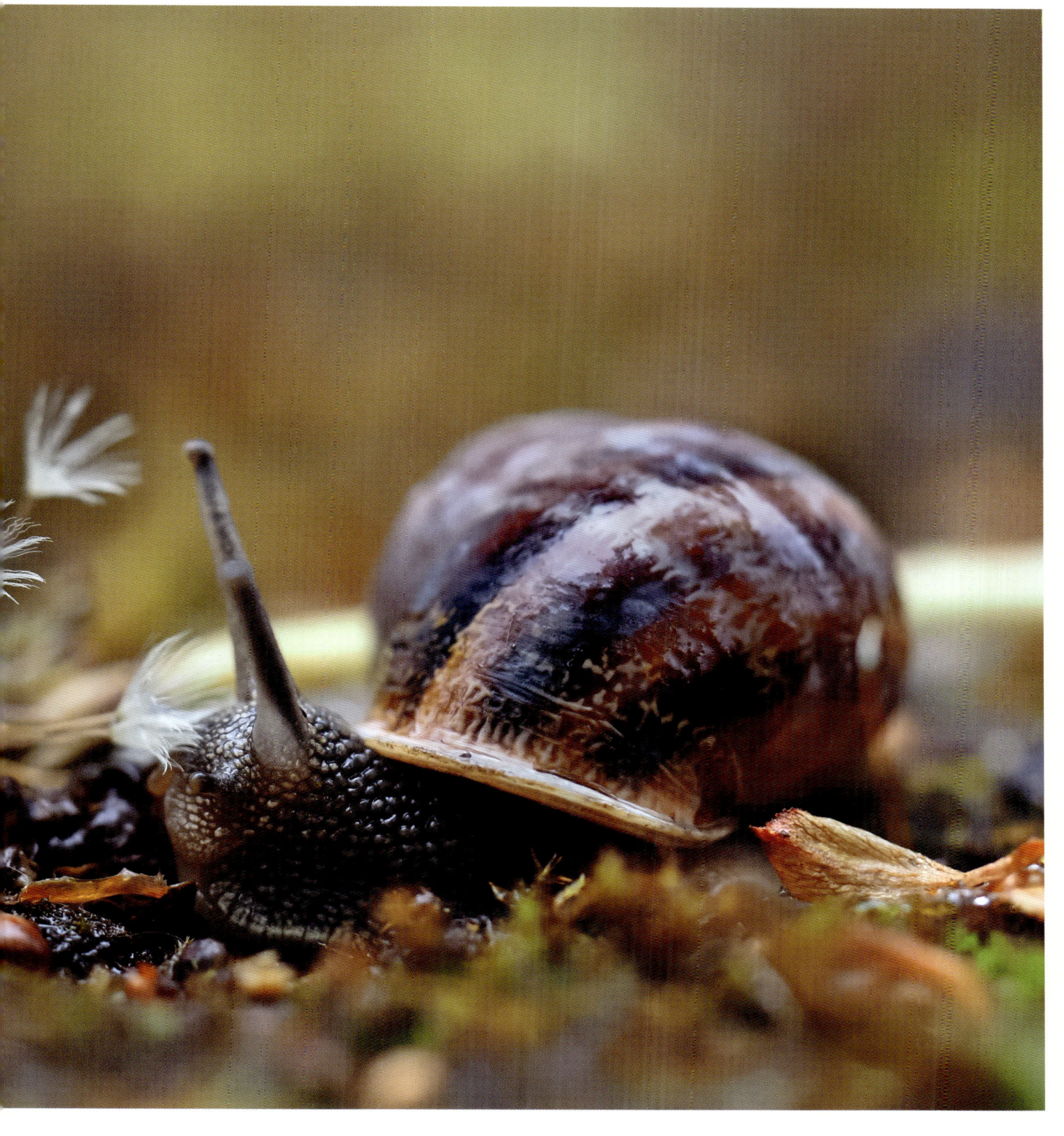

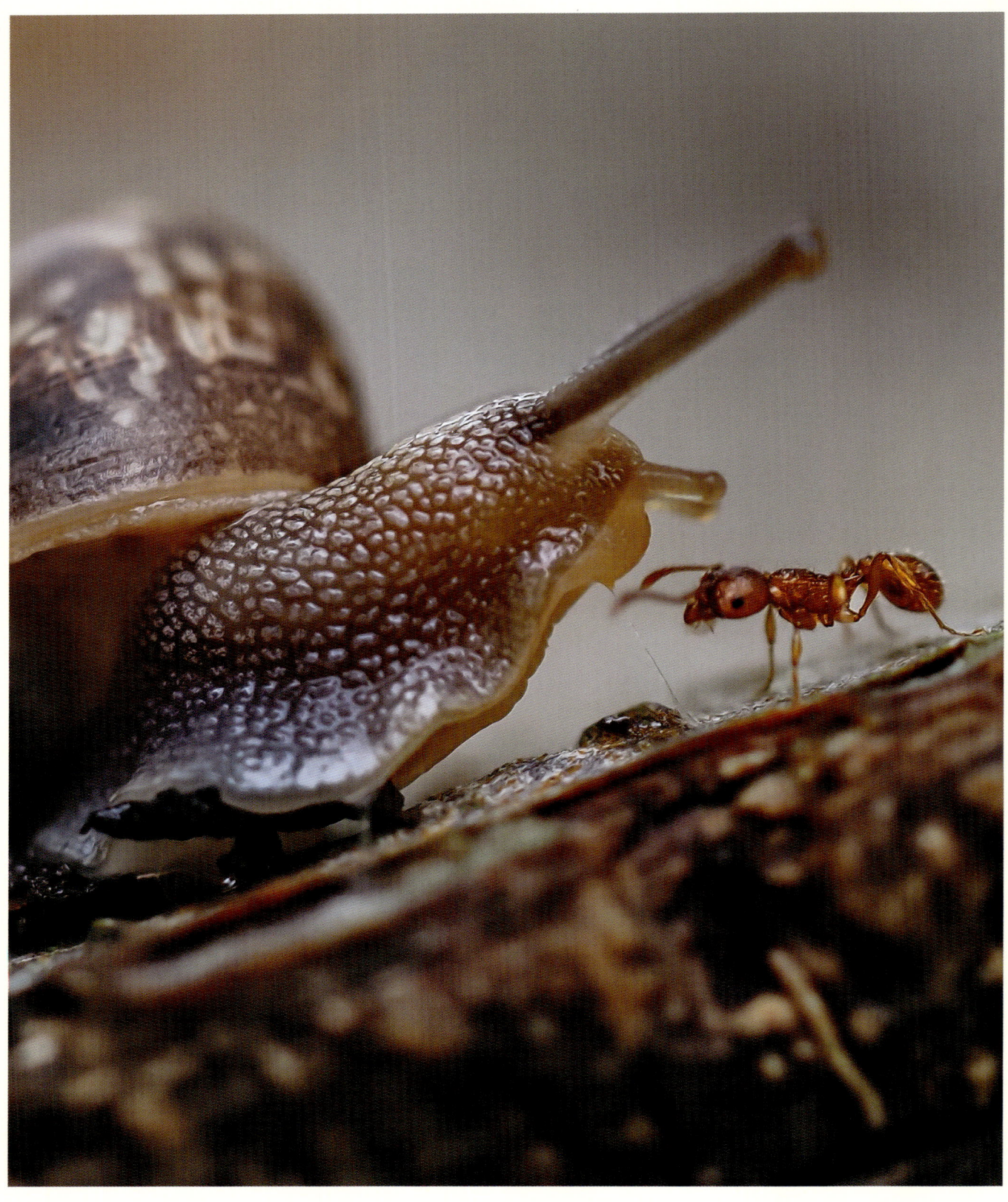

ABOVE: GARDEN SNAIL (*CORNU ASPERSUM*) AND ANT. OPPOSITE: RED SLUG (*ARION RUFUS* SPP.) EATING FLY AGARIC (*AMANITA MUSCARIA*).

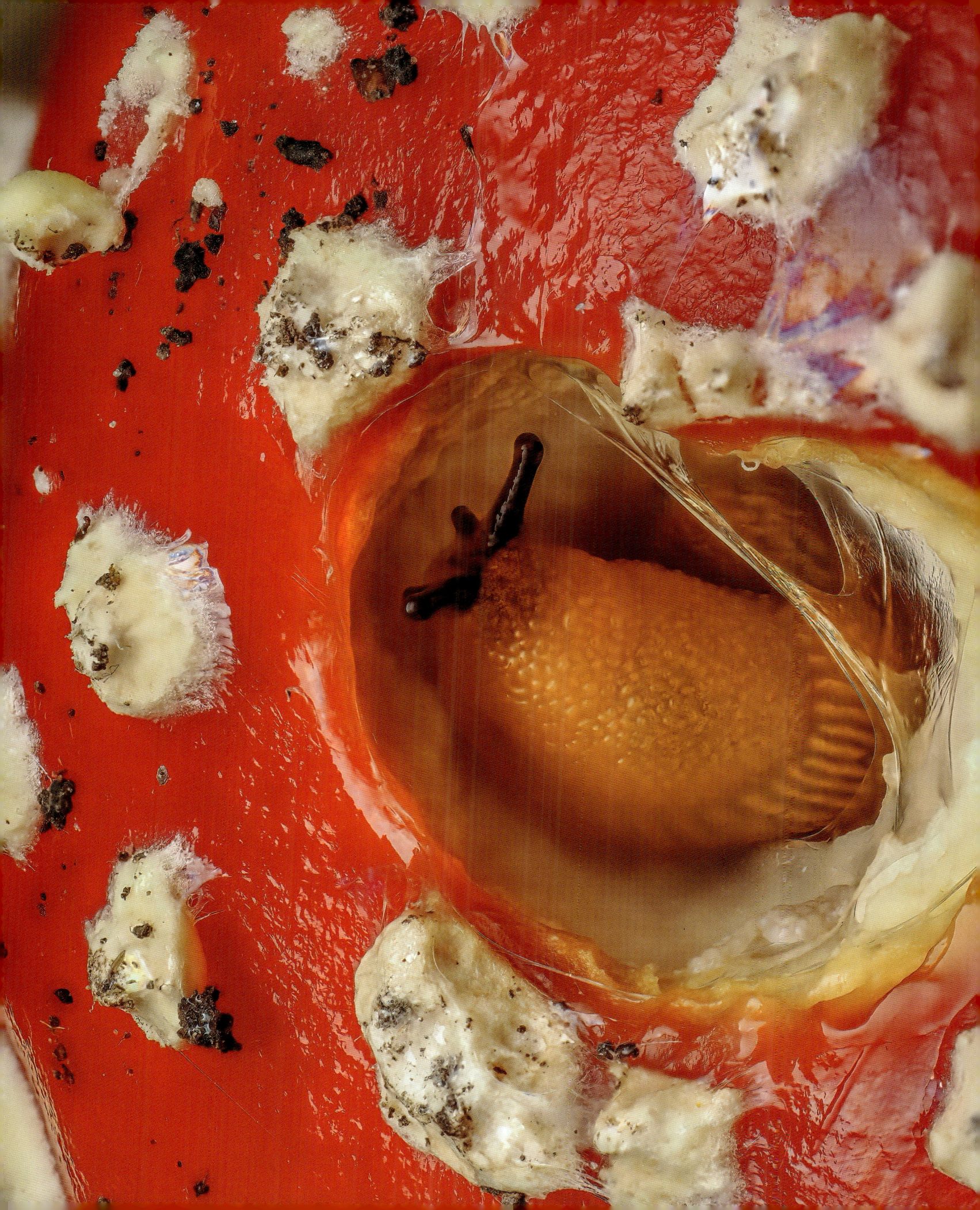

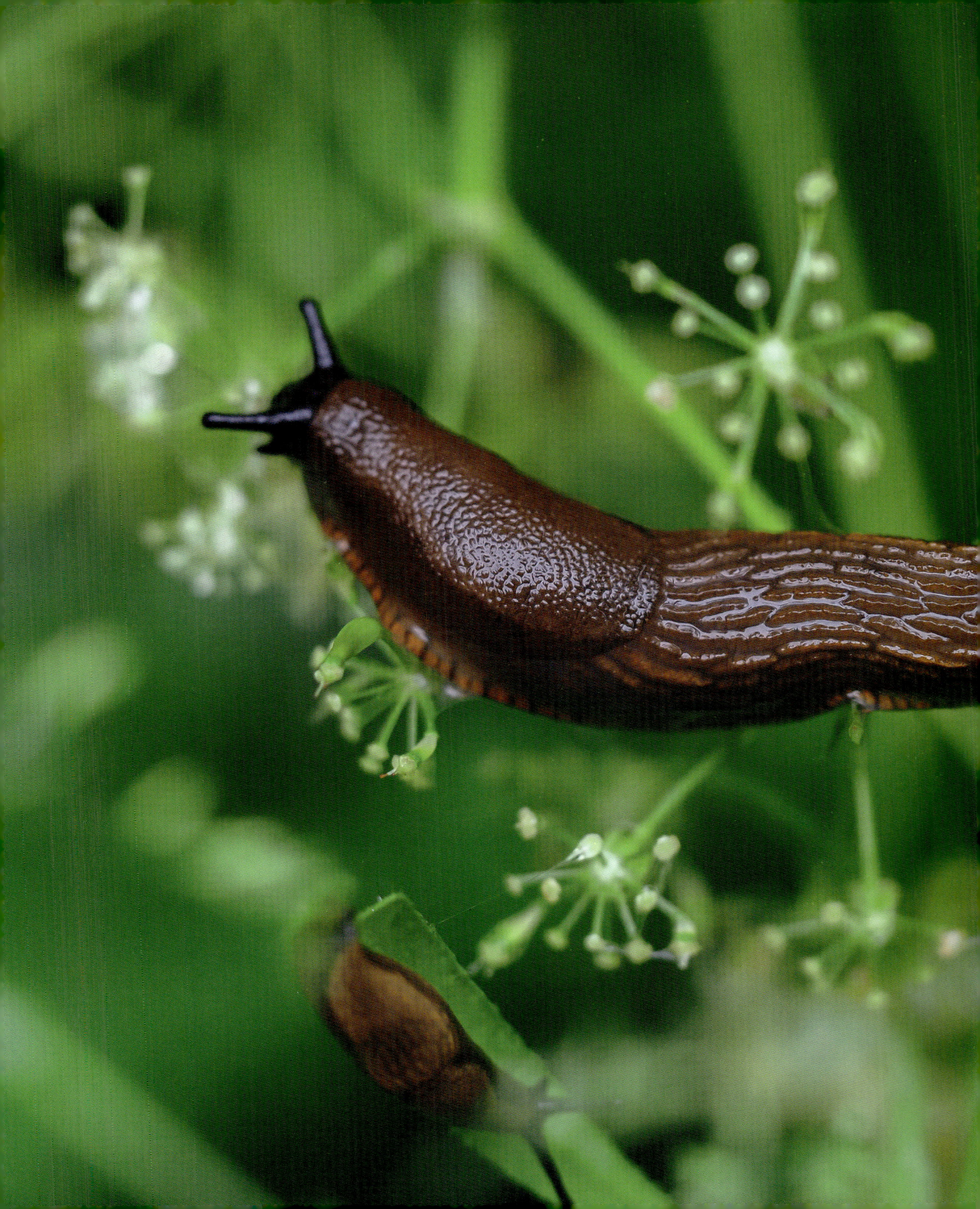

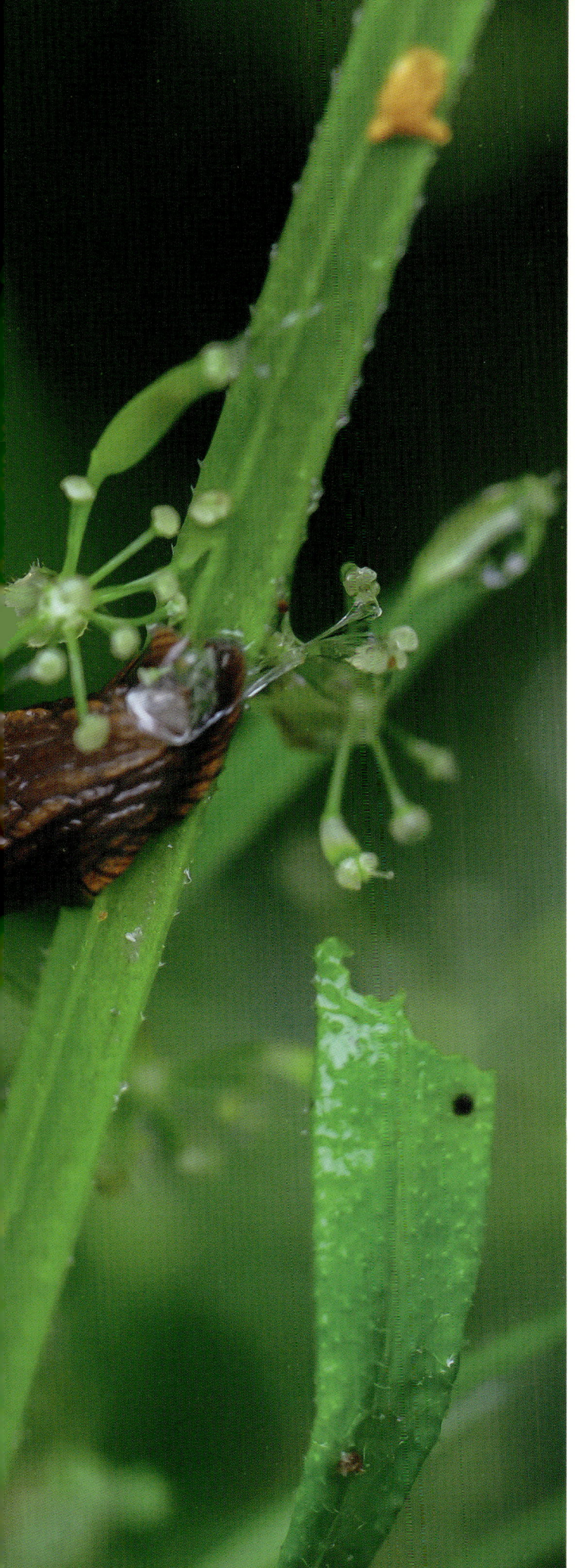

LEFT: *ARION RUFUS/VULGARIS.*

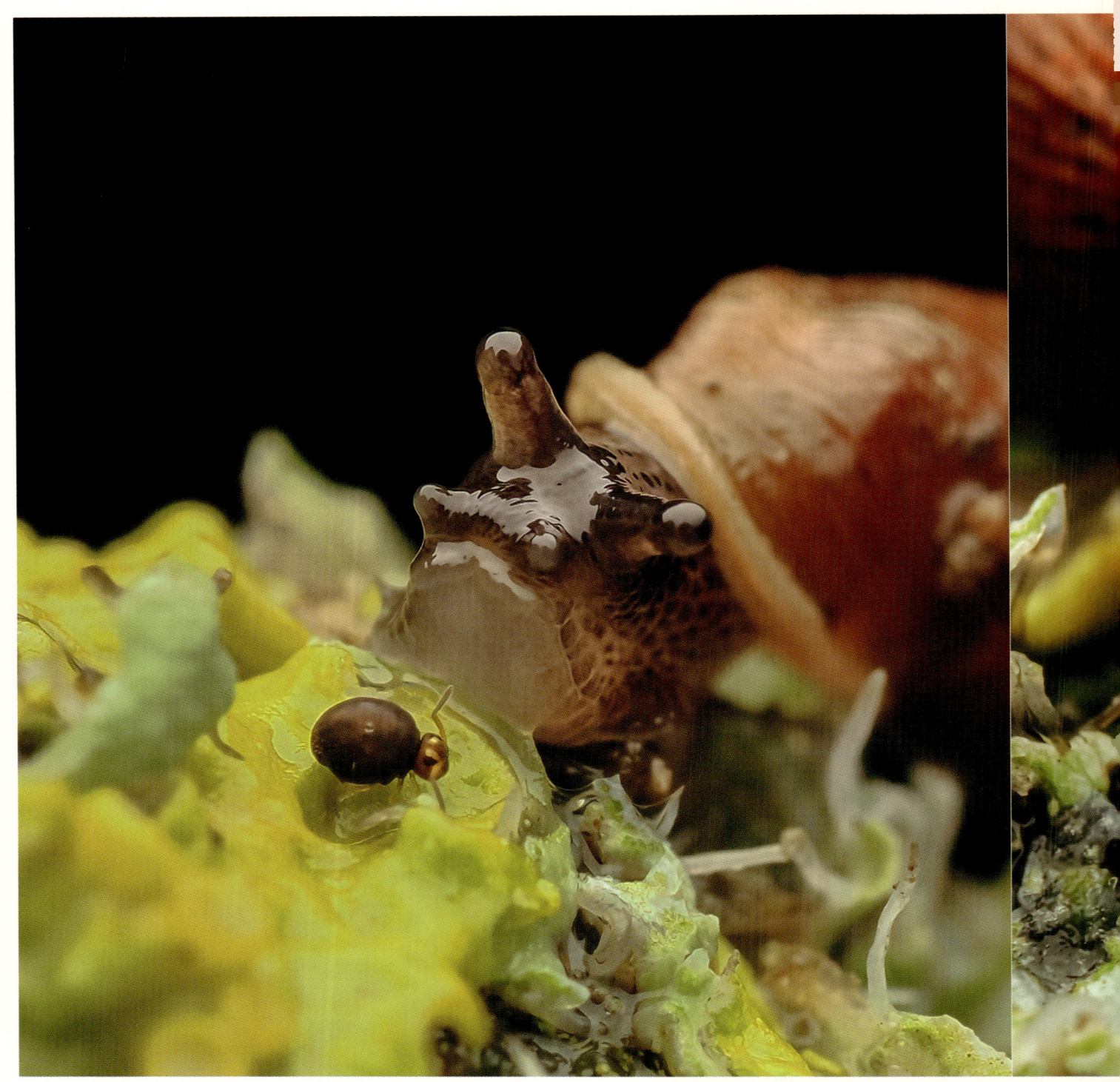

ABOVE AND OPPOSITE: DOOR SNAIL (*CLAUSILIIDAE* SP.) ON LICHEN.

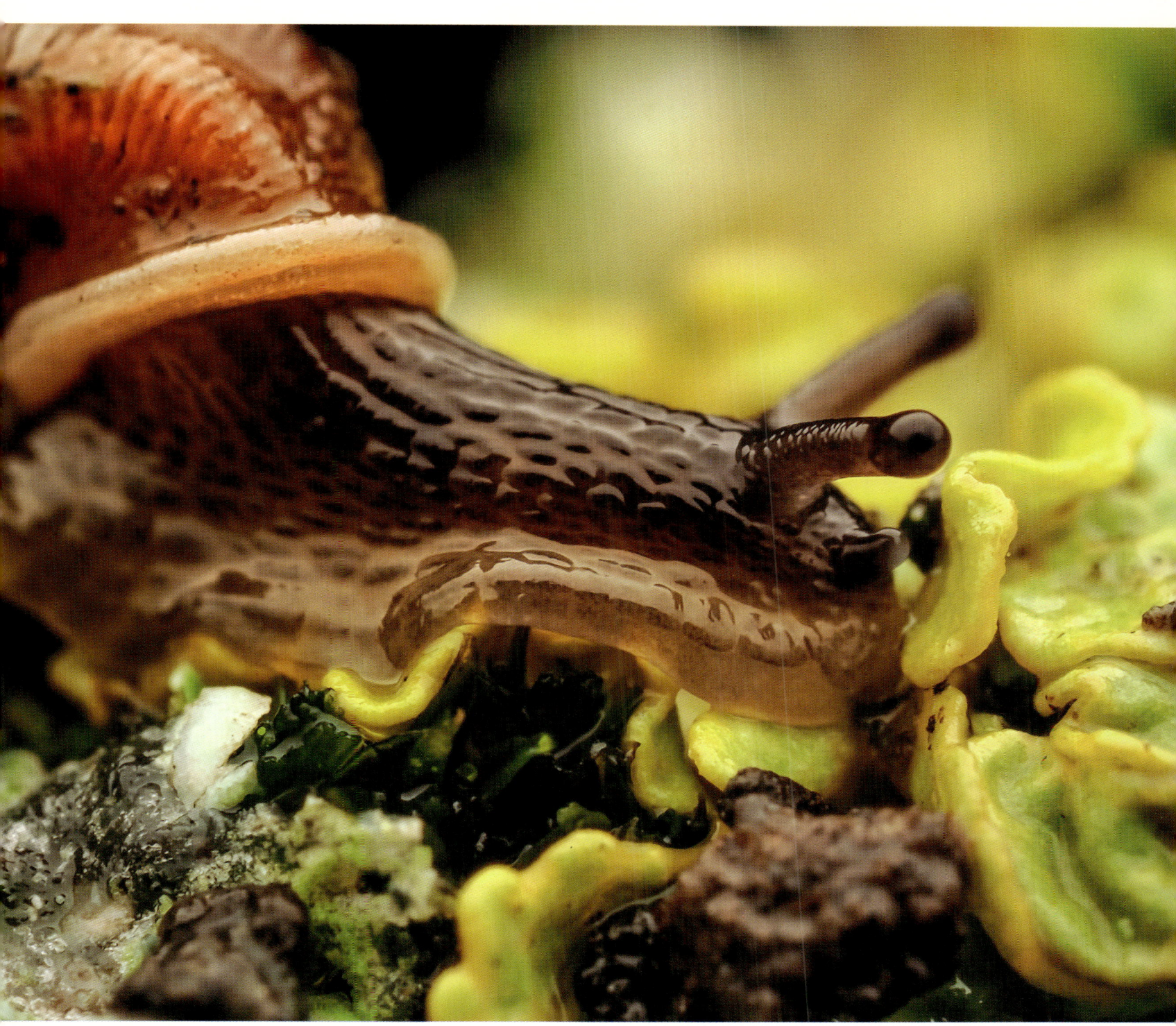

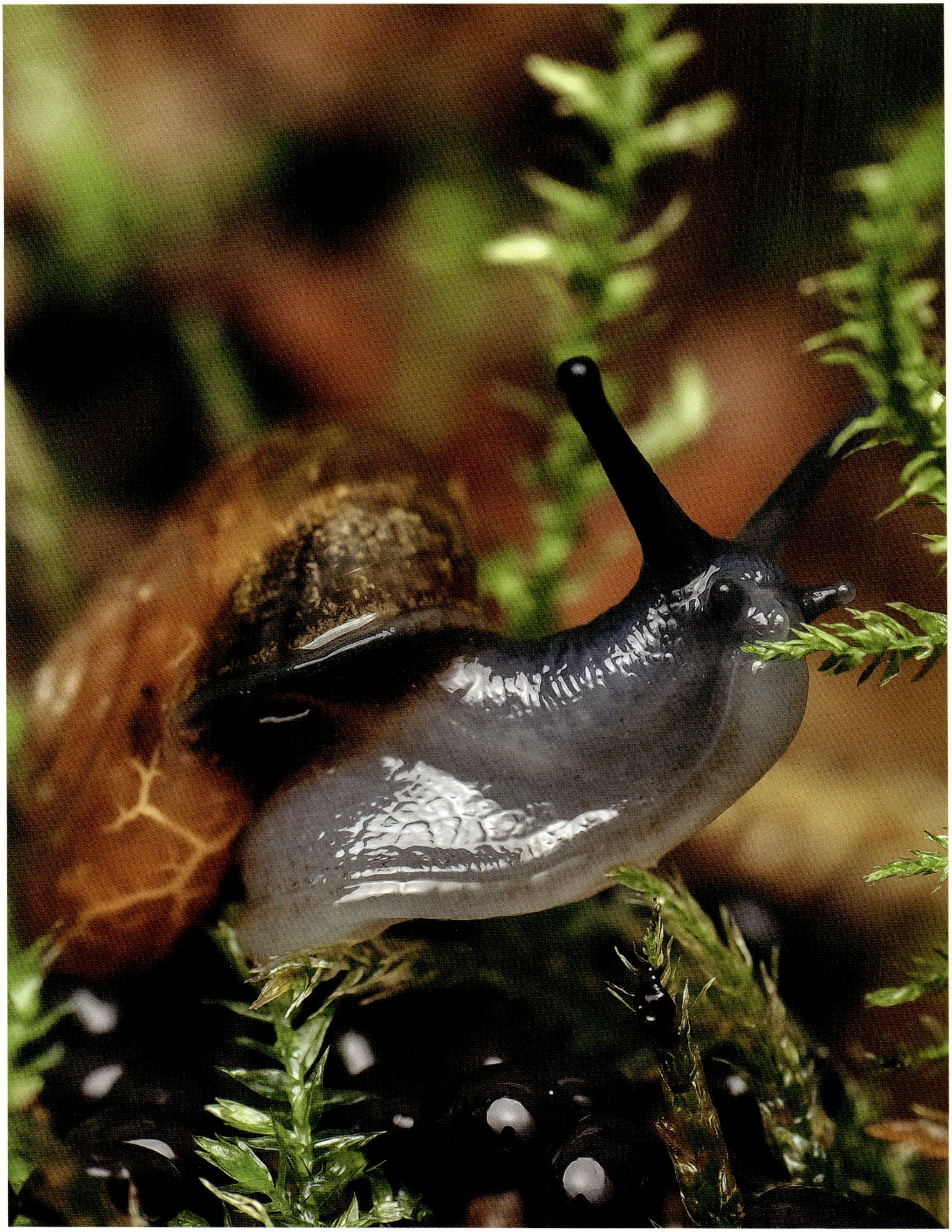

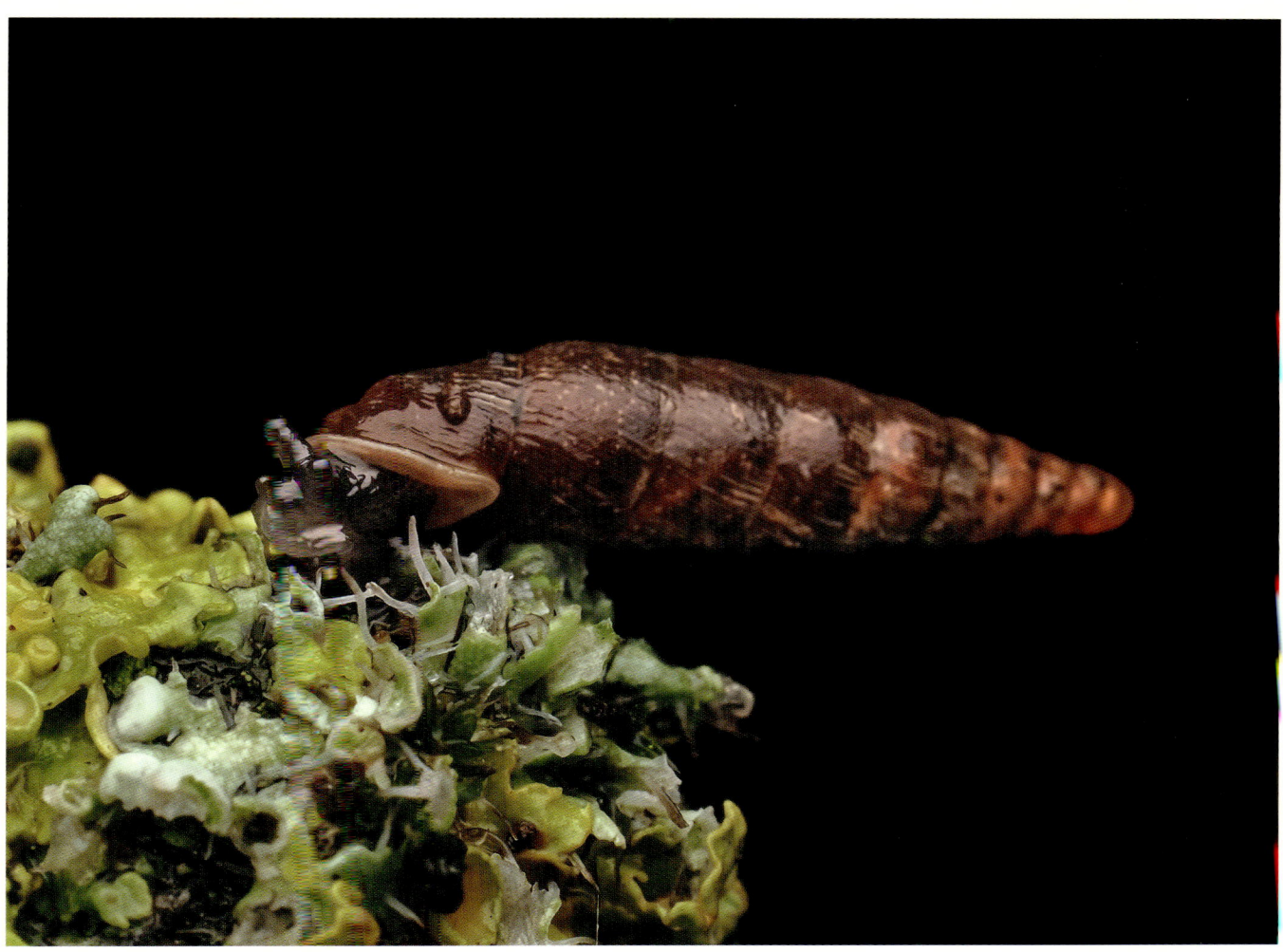

ABOVE: DOOR SNAIL (*CLAUSILIIDAE* SP.). OPPOSITE: UNKNOWN SNAIL EATING.

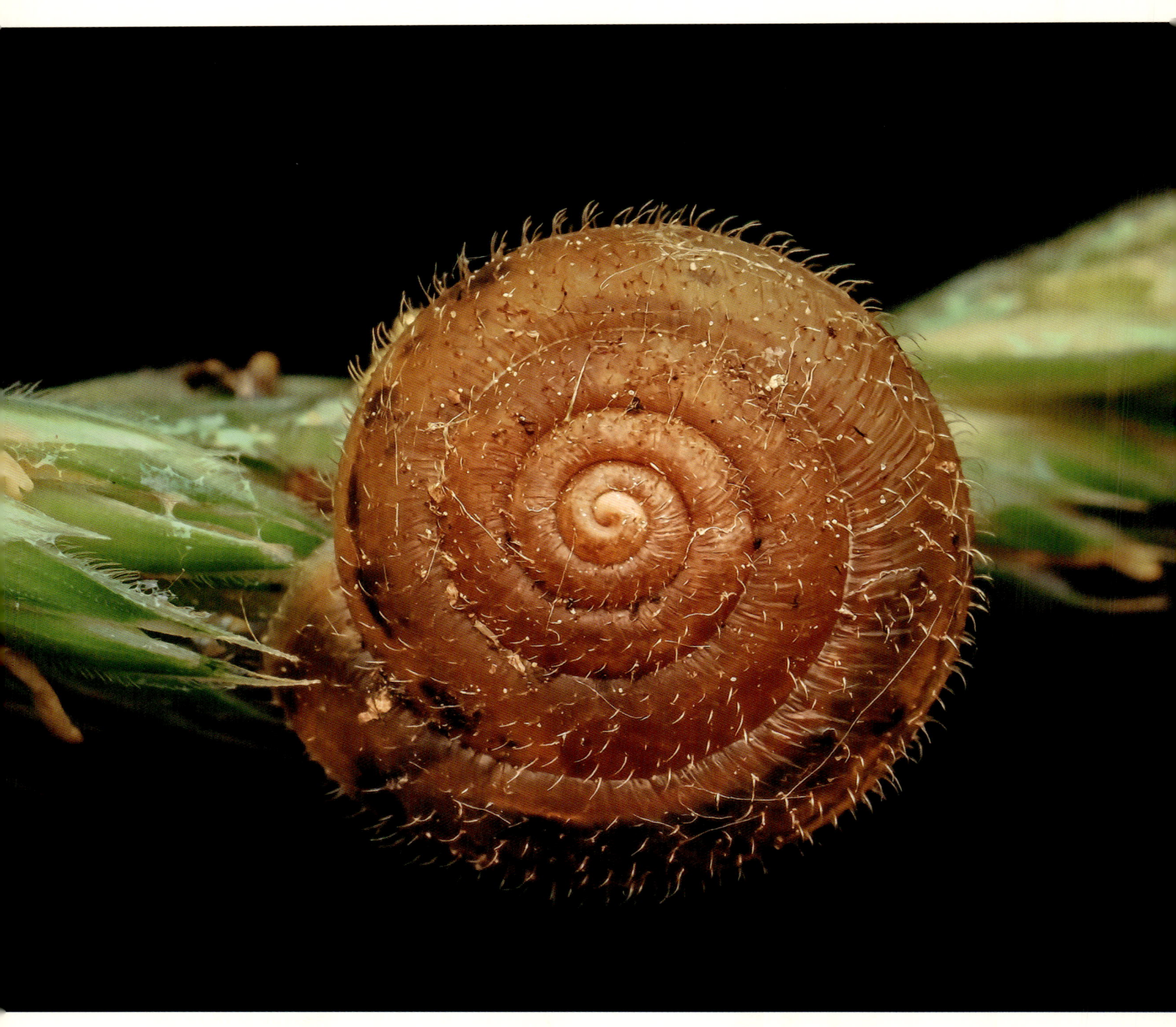

ABOVE AND OPPOSITE: HAIRY SNAIL (*TROCHULUS HISPIDUS*).

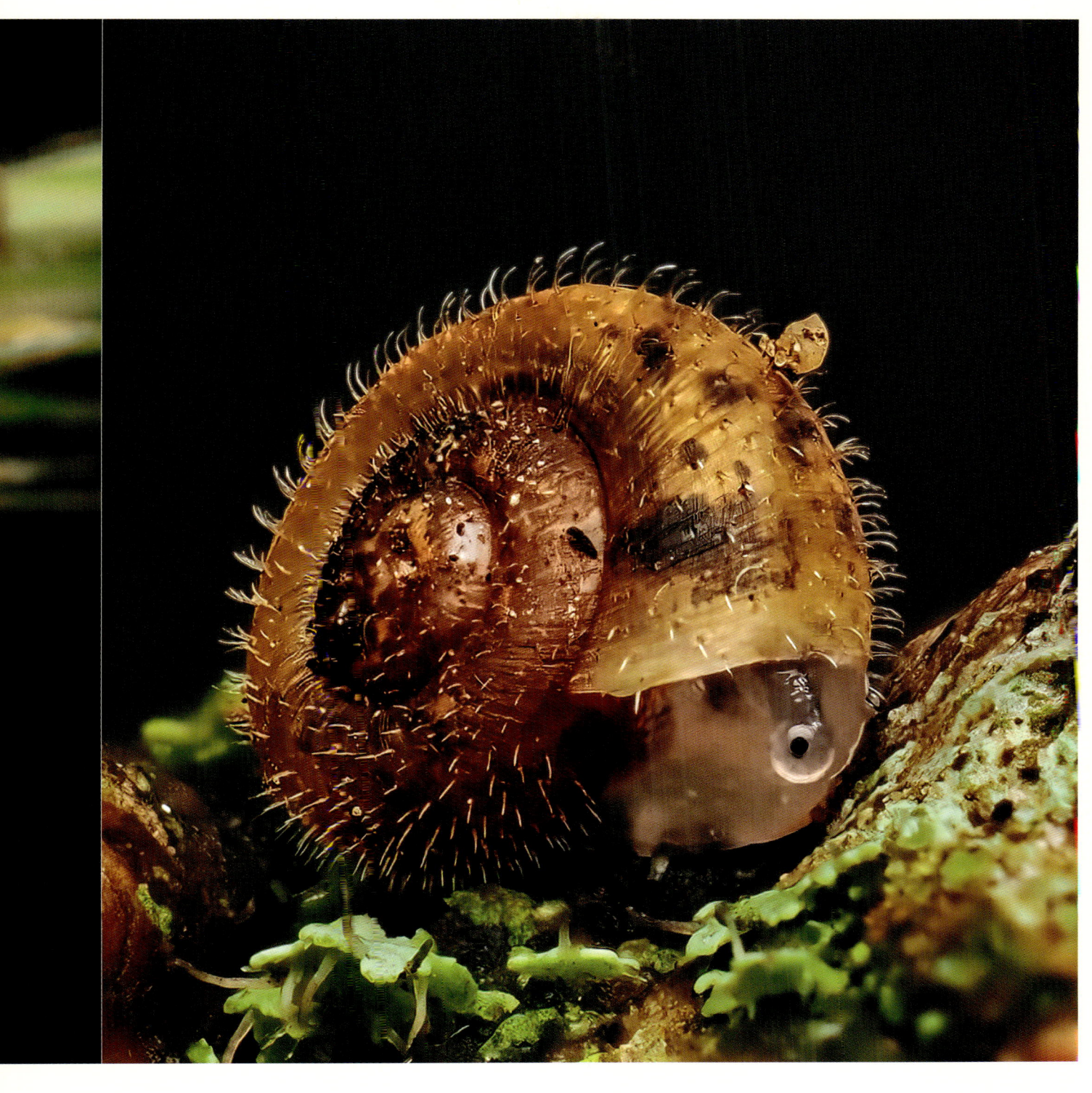

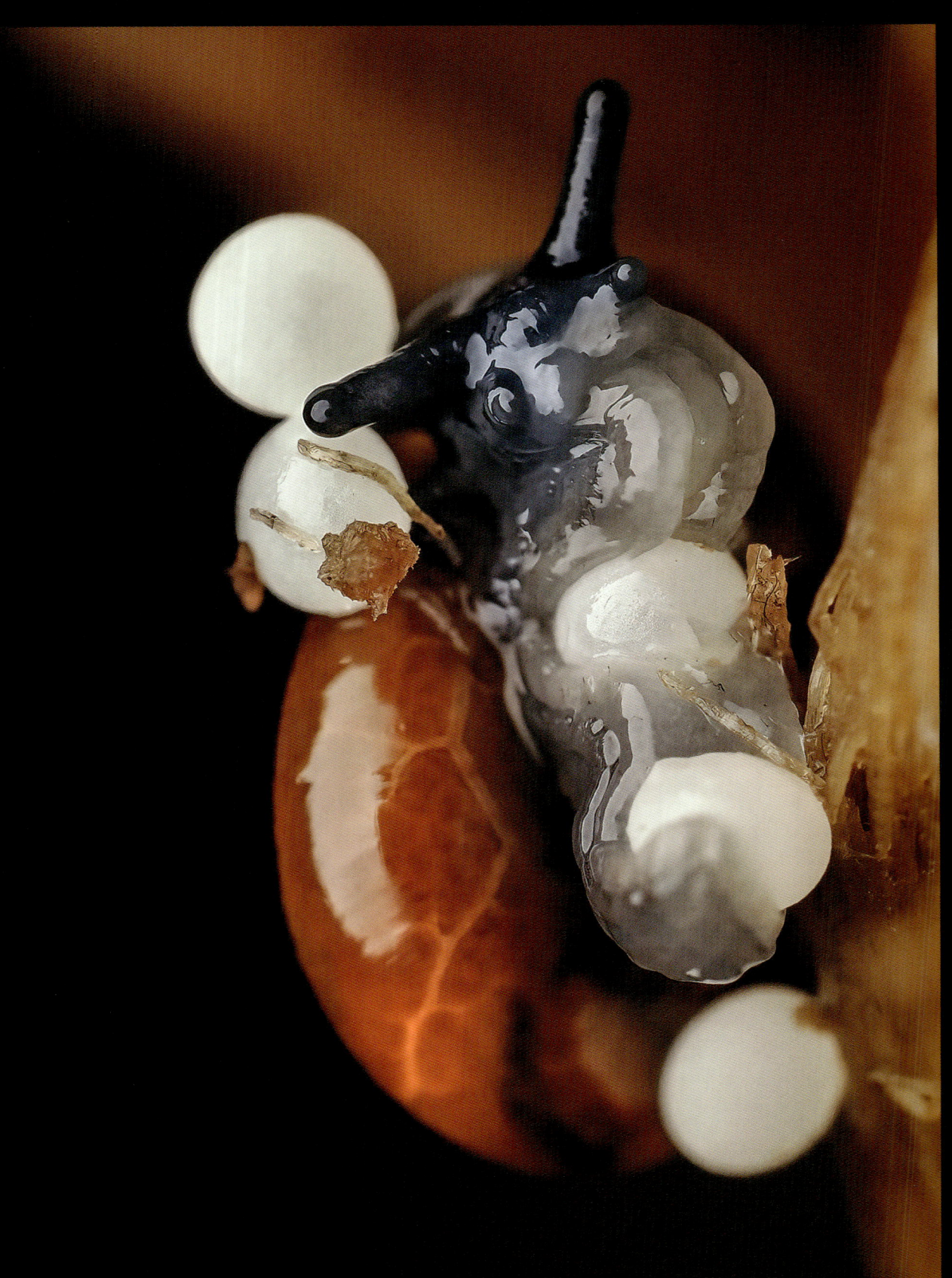

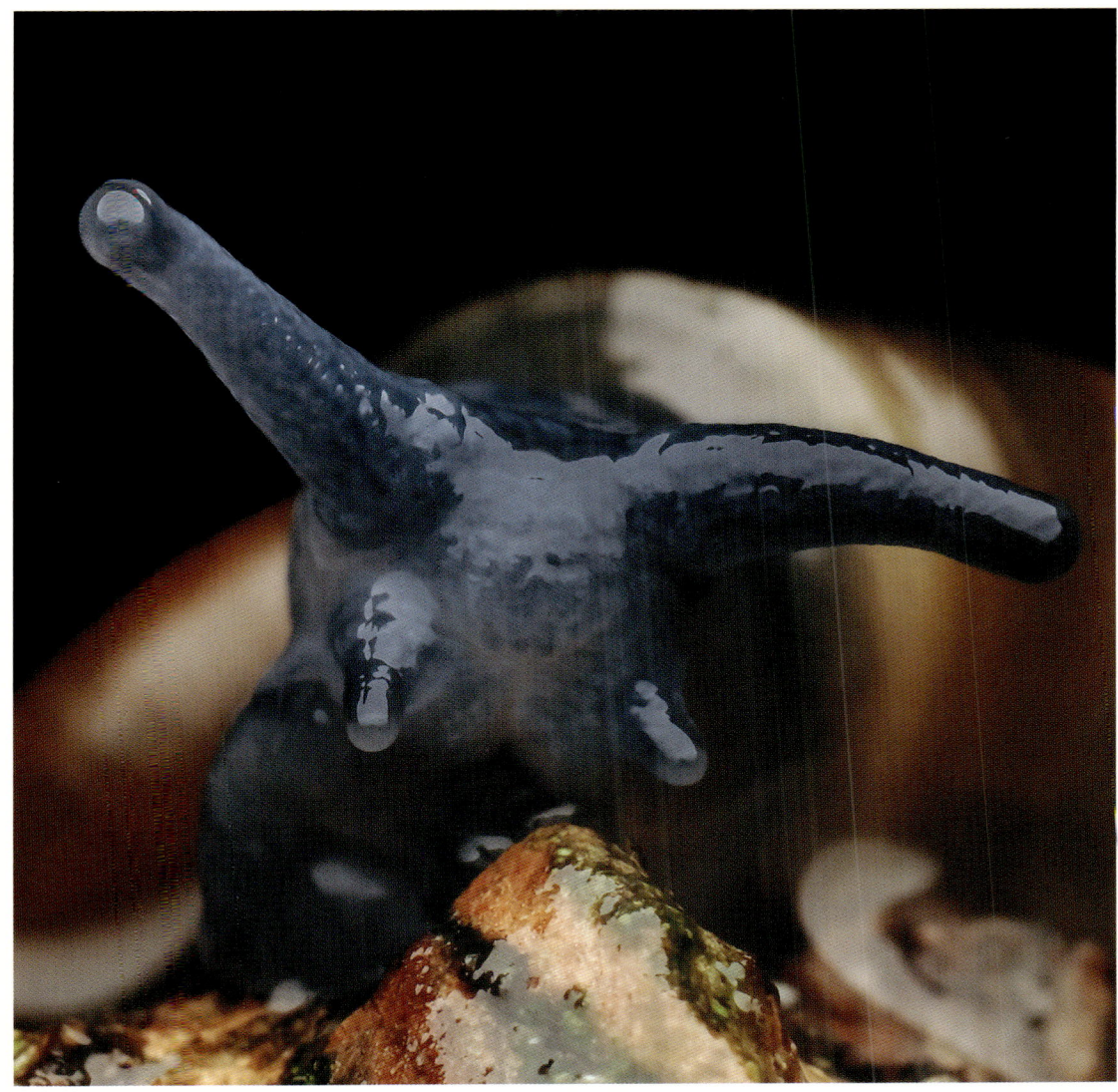

ABOVE: DRAPARNAUD'S GLASS SNAIL (*OXYCHILUS DRAPARNAUDI*). OPPOSITE: DRAPARNAUD'S GLASS SNAIL (*O. DRAPARNAUDI*) WITH EGGS.

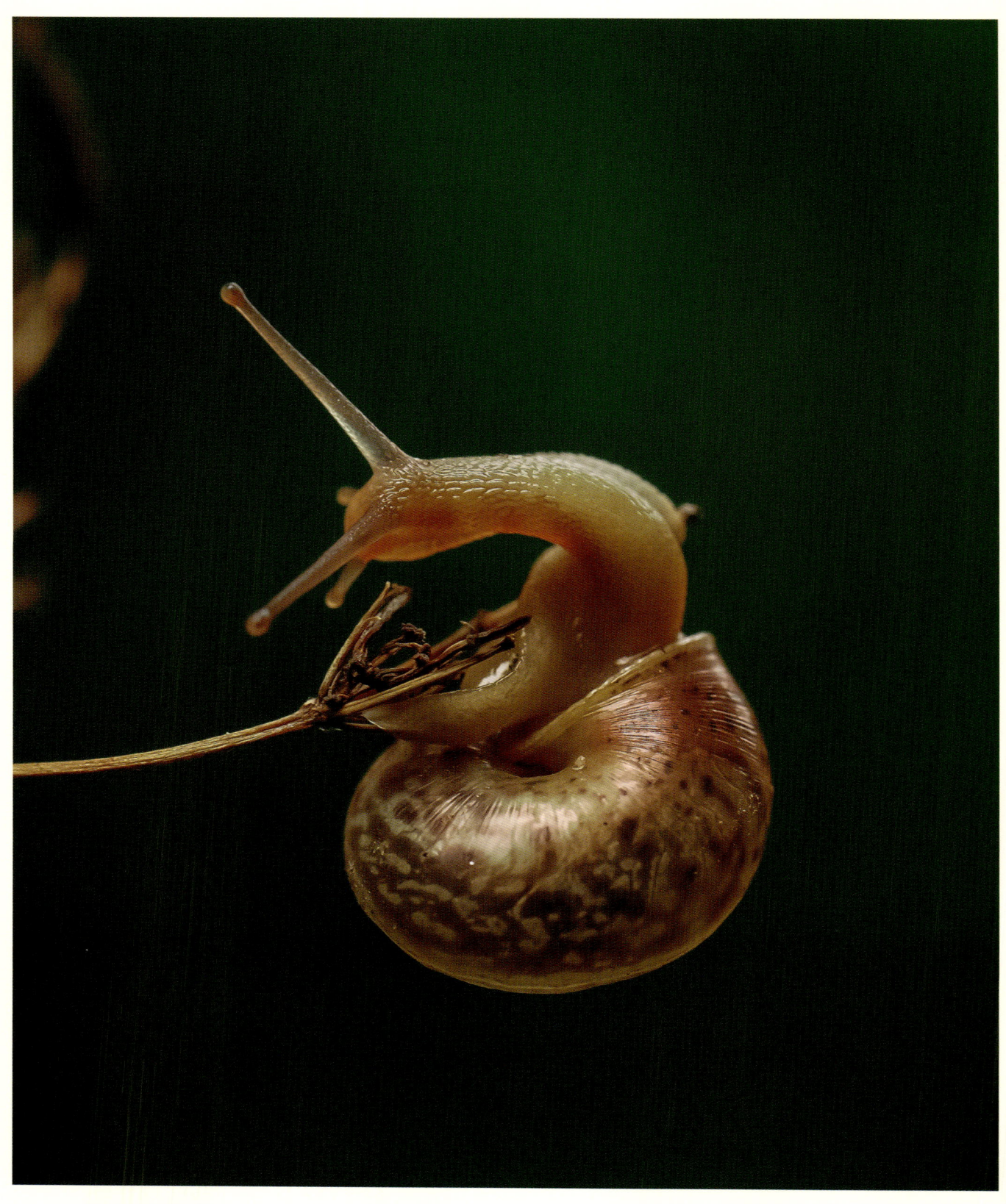

ABOVE: GARDEN SNAIL (*CORNU ASPERSUM*). OPPOSITE: ROUNDED SNAIL (*GONYODISCUS ROTUNDATUS*) EATING SLIME MOLD.

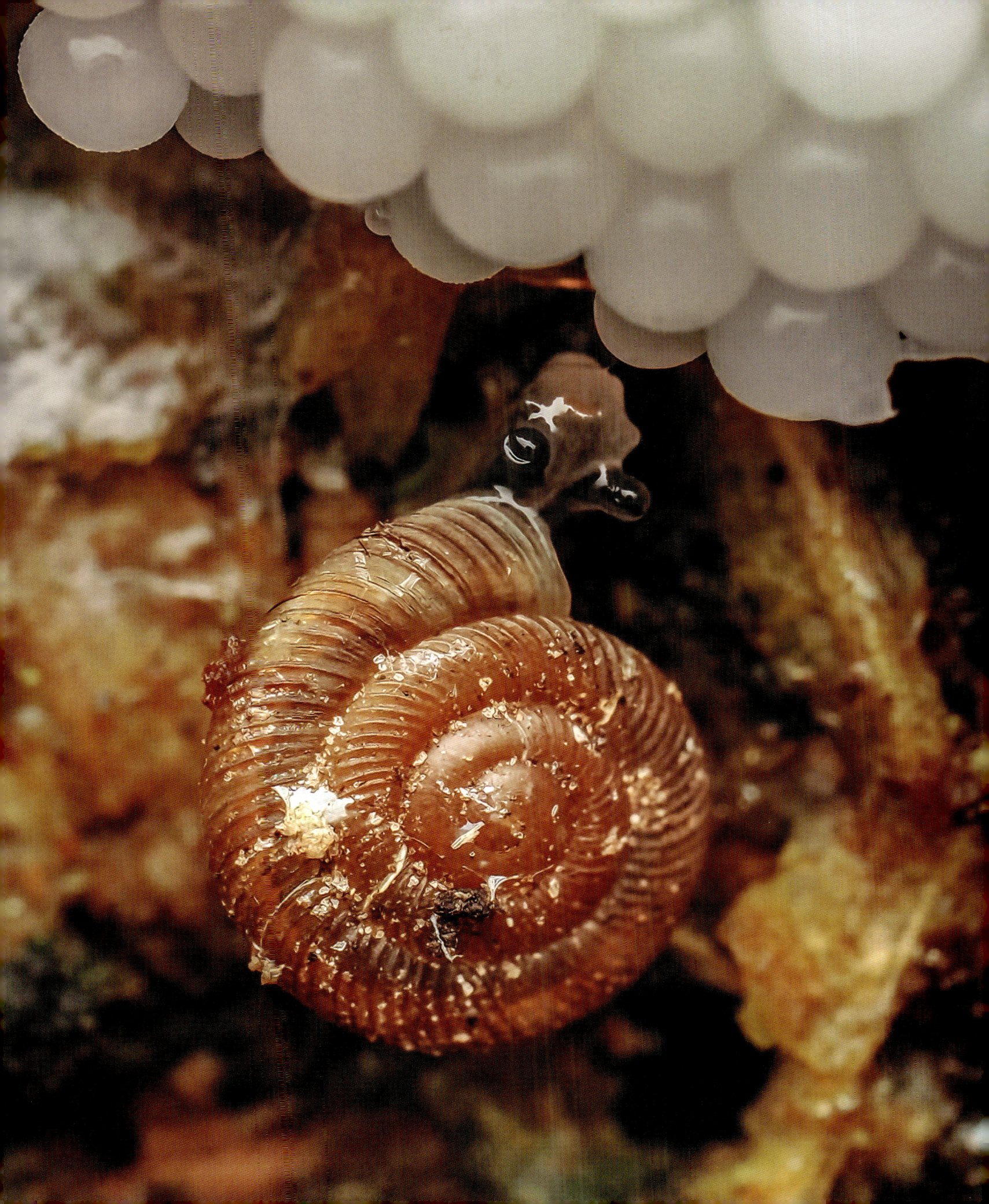

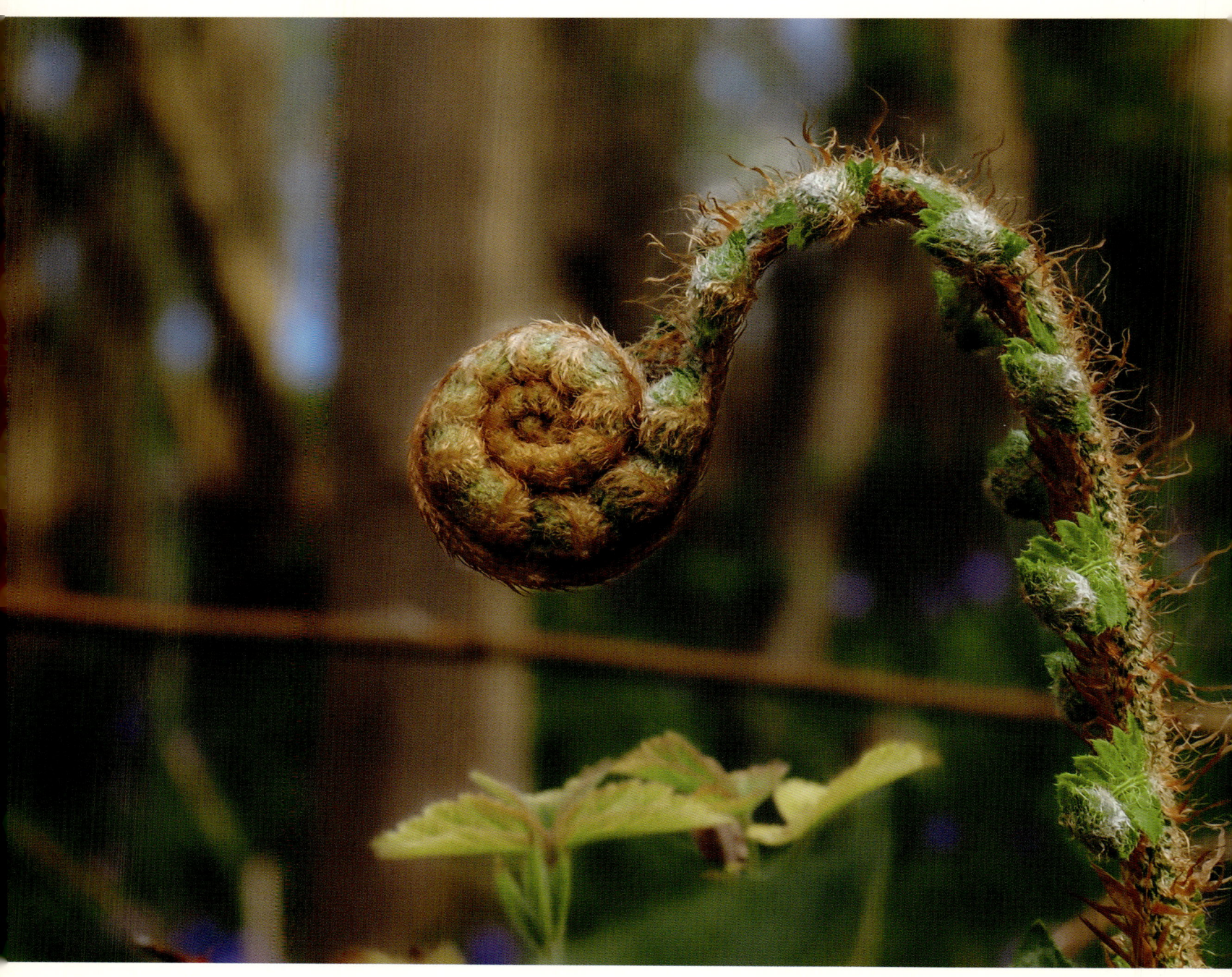

ABOVE: FIDDLEHEAD FERN UNFURLING. OPPOSITE: THREEBAND SLUGS (*AMBIGOLIMAX* SP.) ON LICHEN-COVERED STICK.

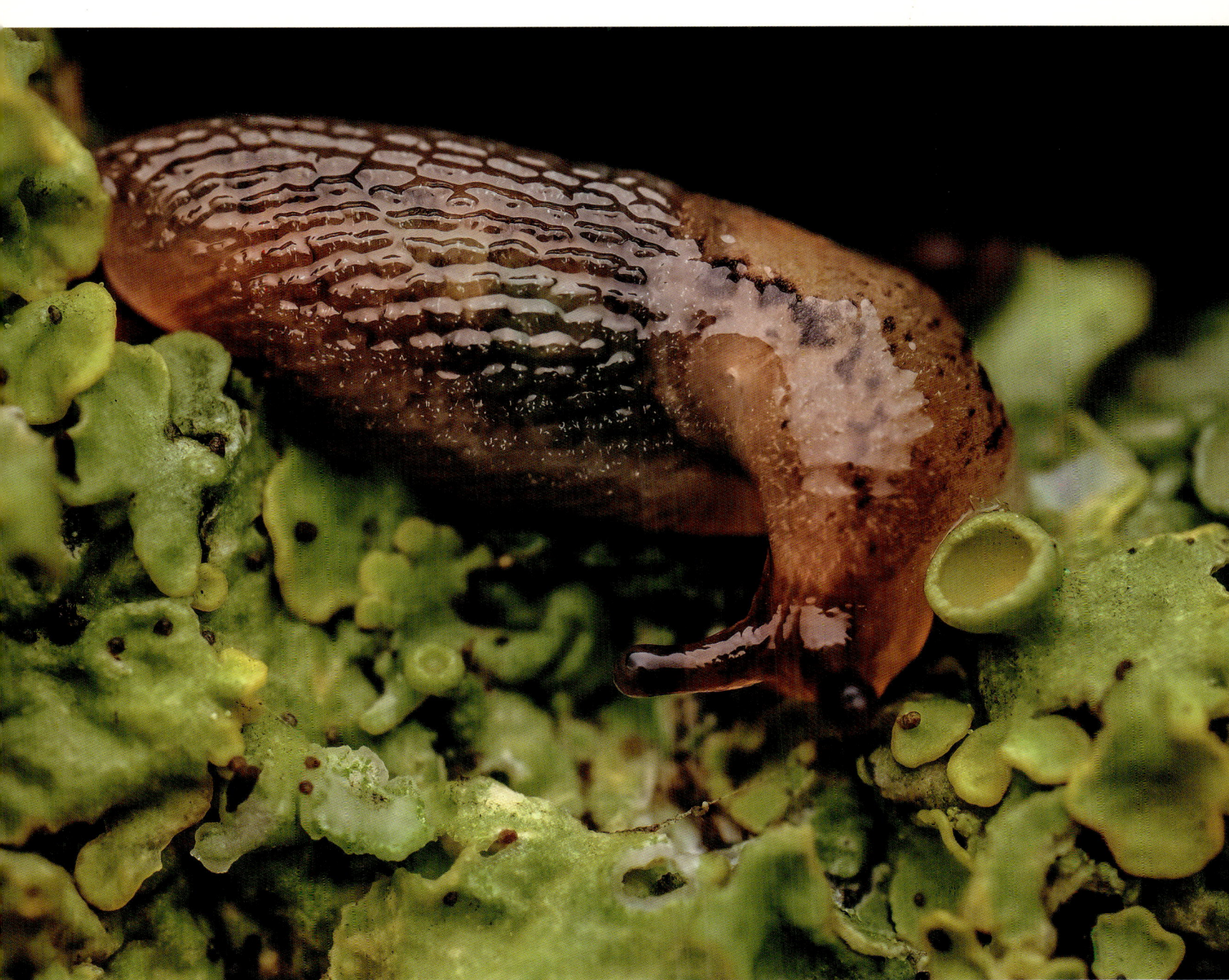

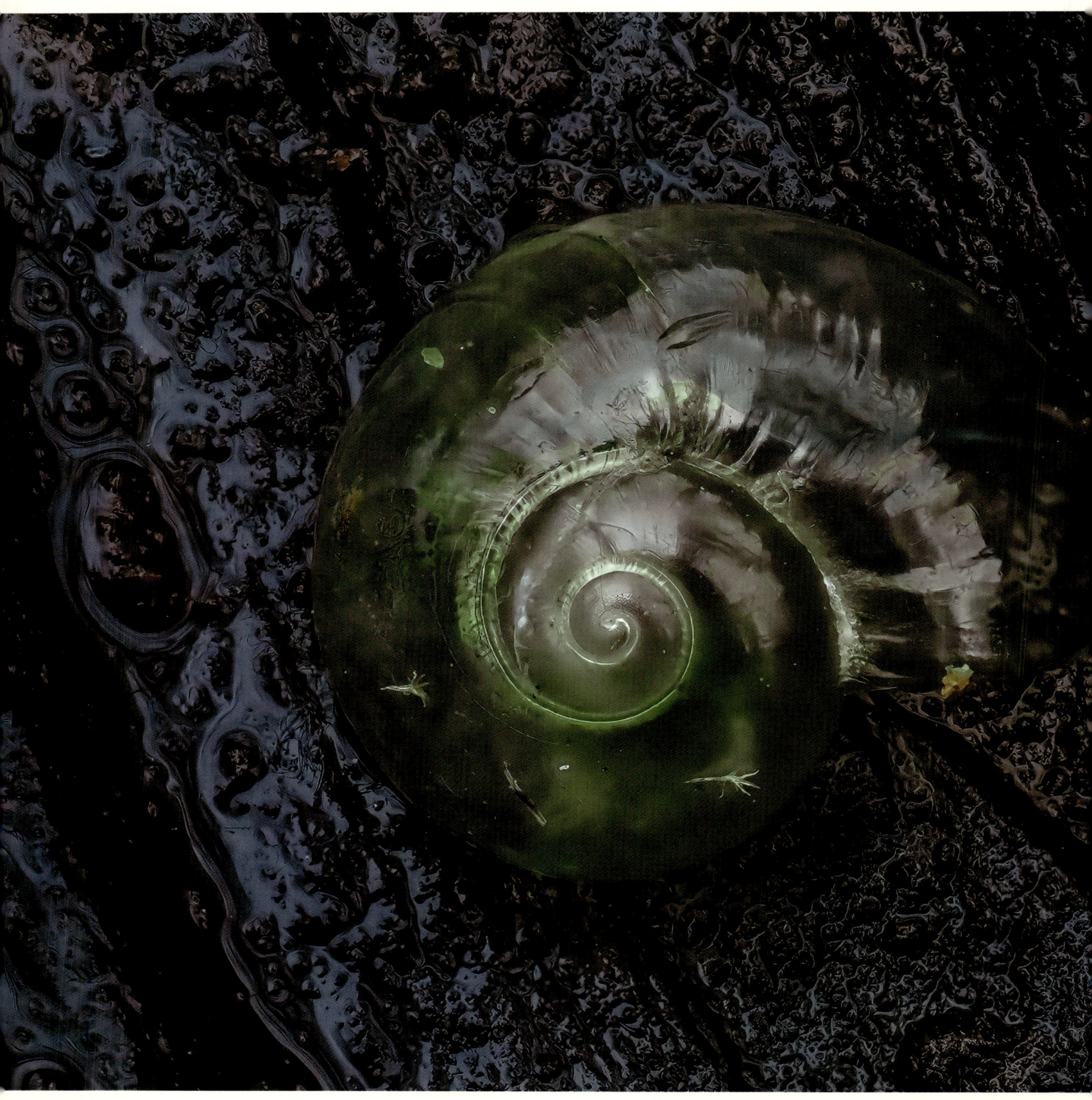

LEFT: SNAIL SHELL.

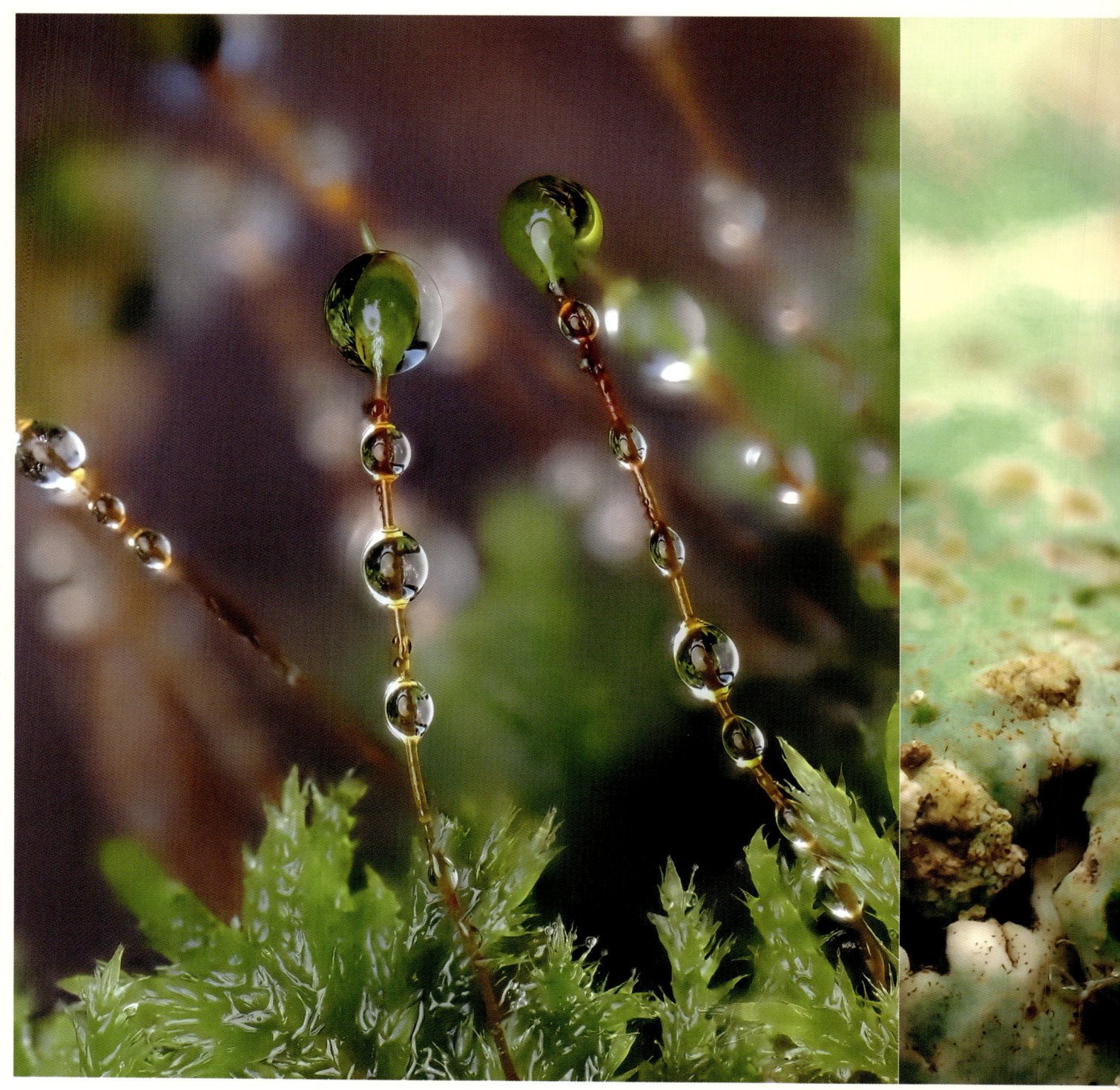

ABOVE: WATER DROPLETS ON MOSS. OPPOSITE: BABY SLUG IN NASHVILLE, TN, US.

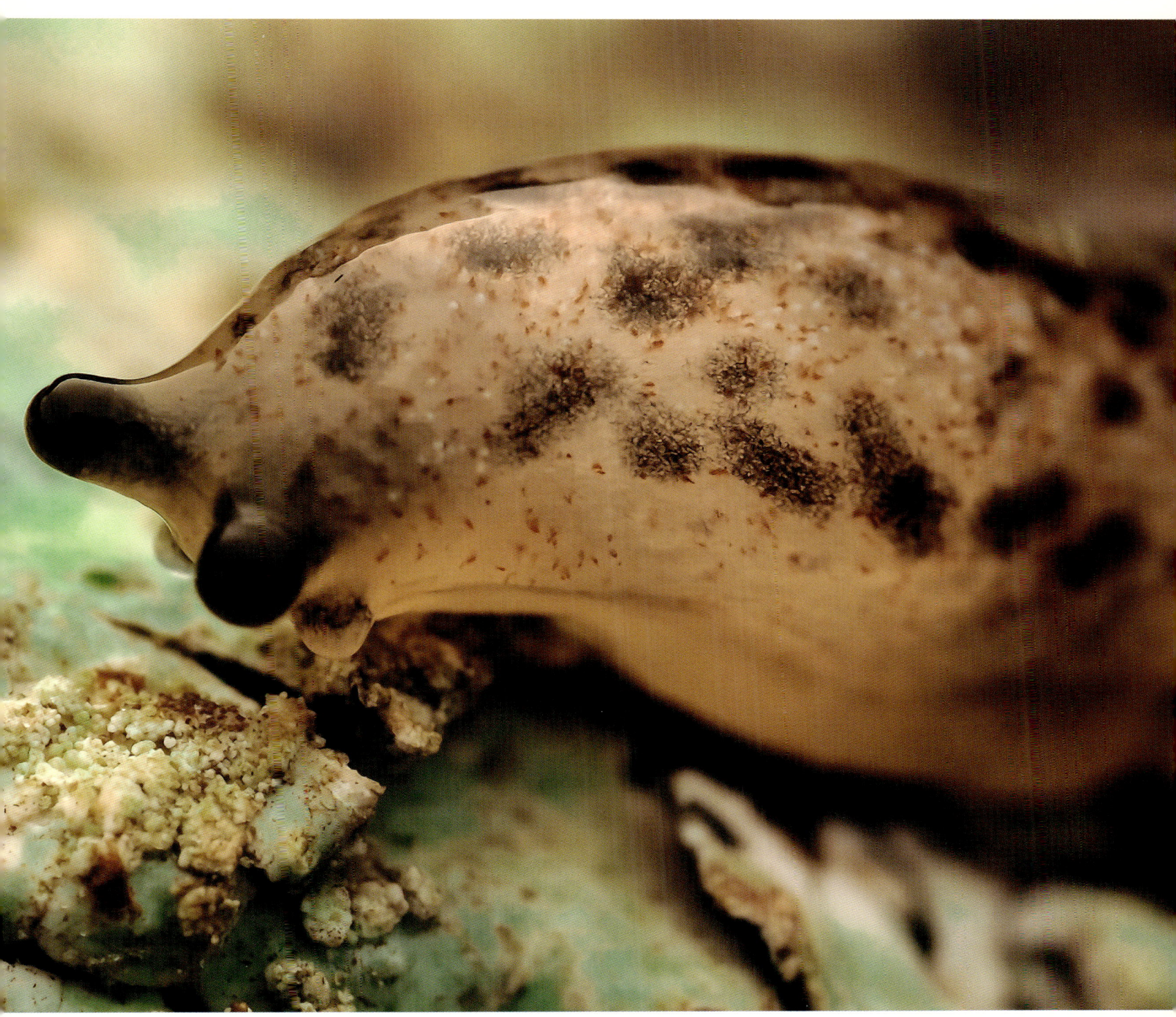

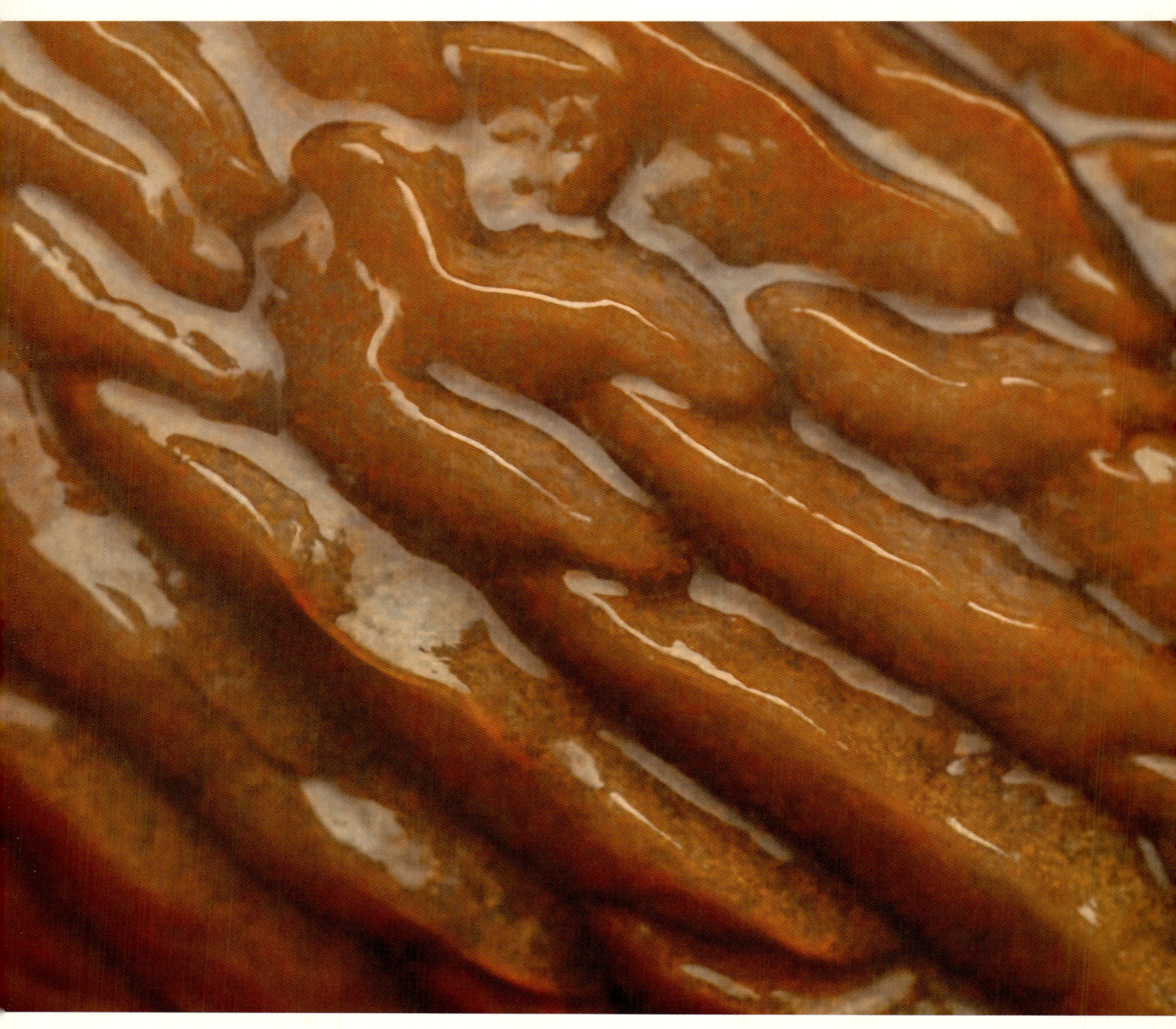

ABOVE: RED SLUG (*ARION RUFUS* SPP.). OPPOSITE: WATER DROPLETS AND SPIDERWEB.

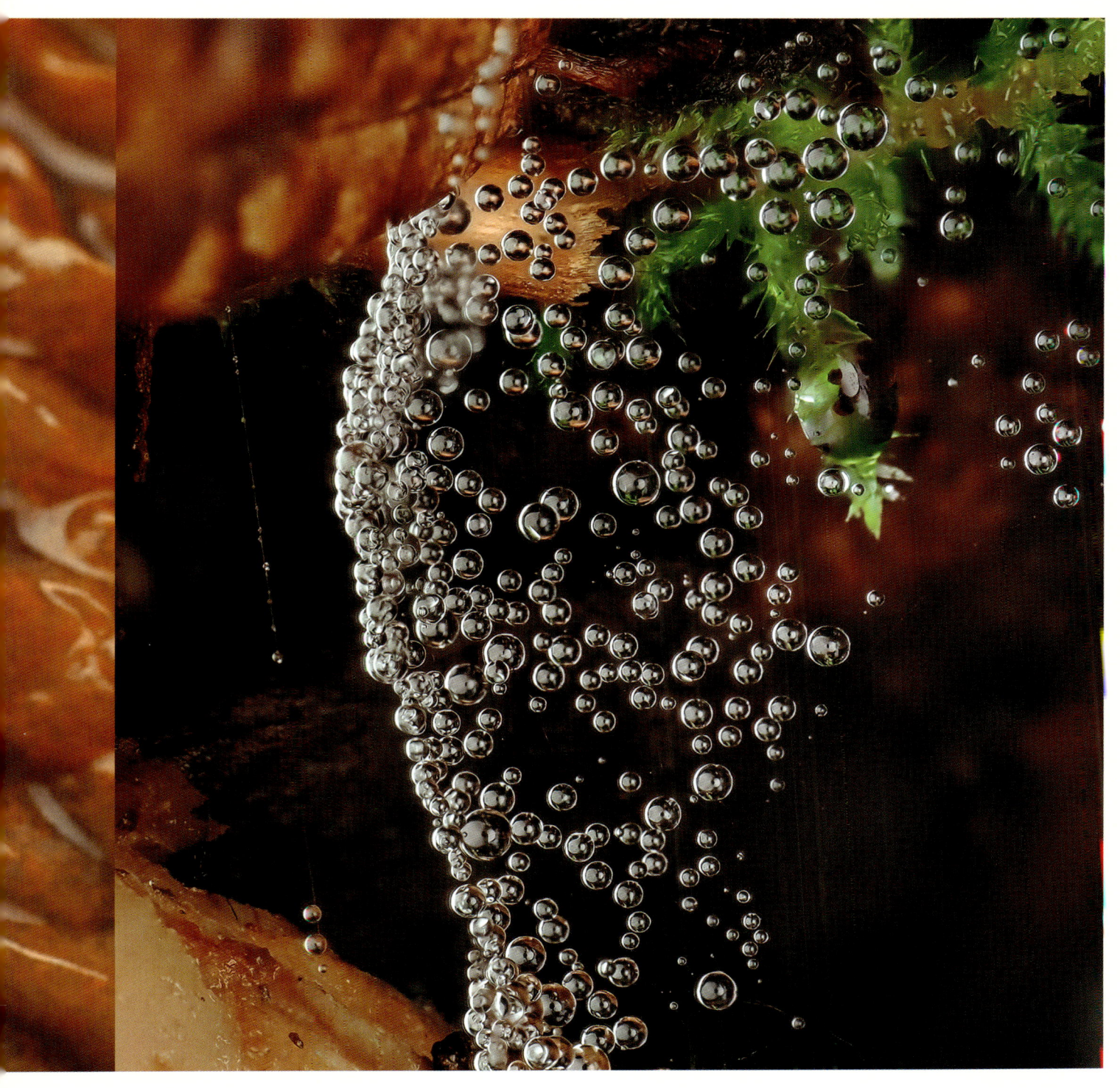

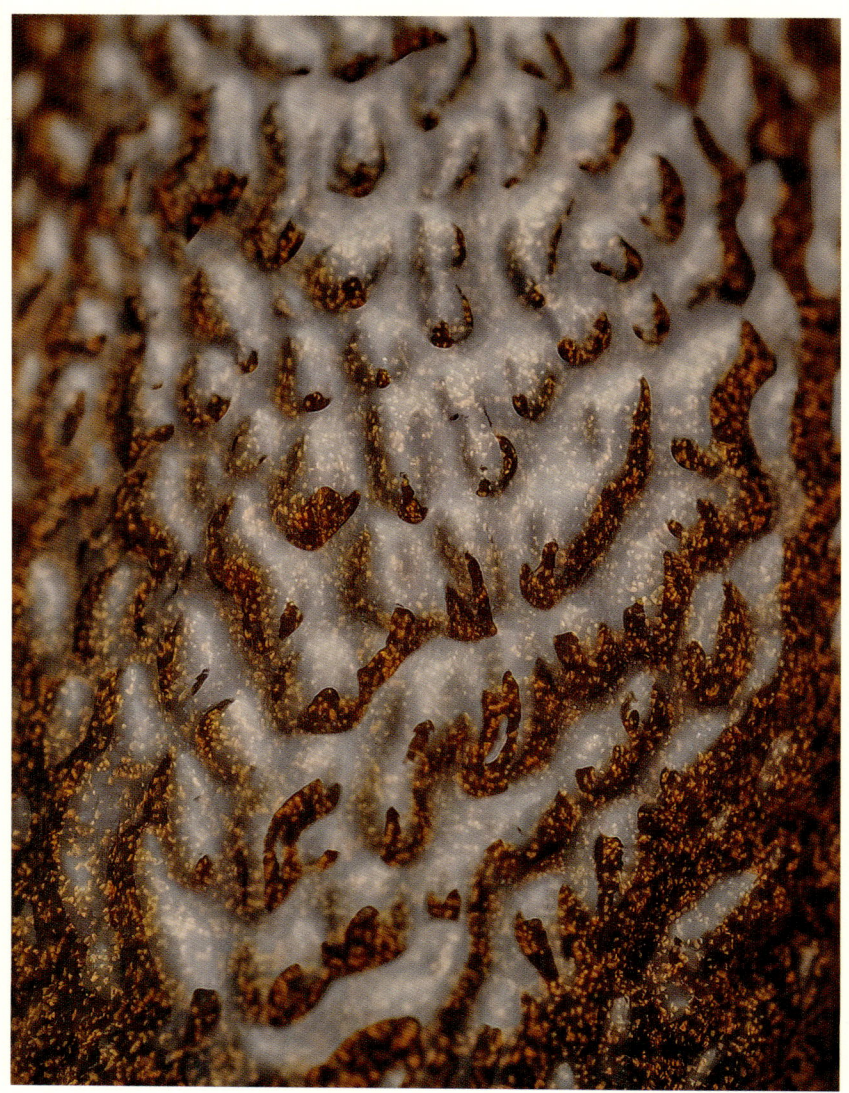

ABOVE: UNKNOWN SLUG. OPPOSITE: UNKNOWN SLUG WITH TINY SPRINGTAILS.

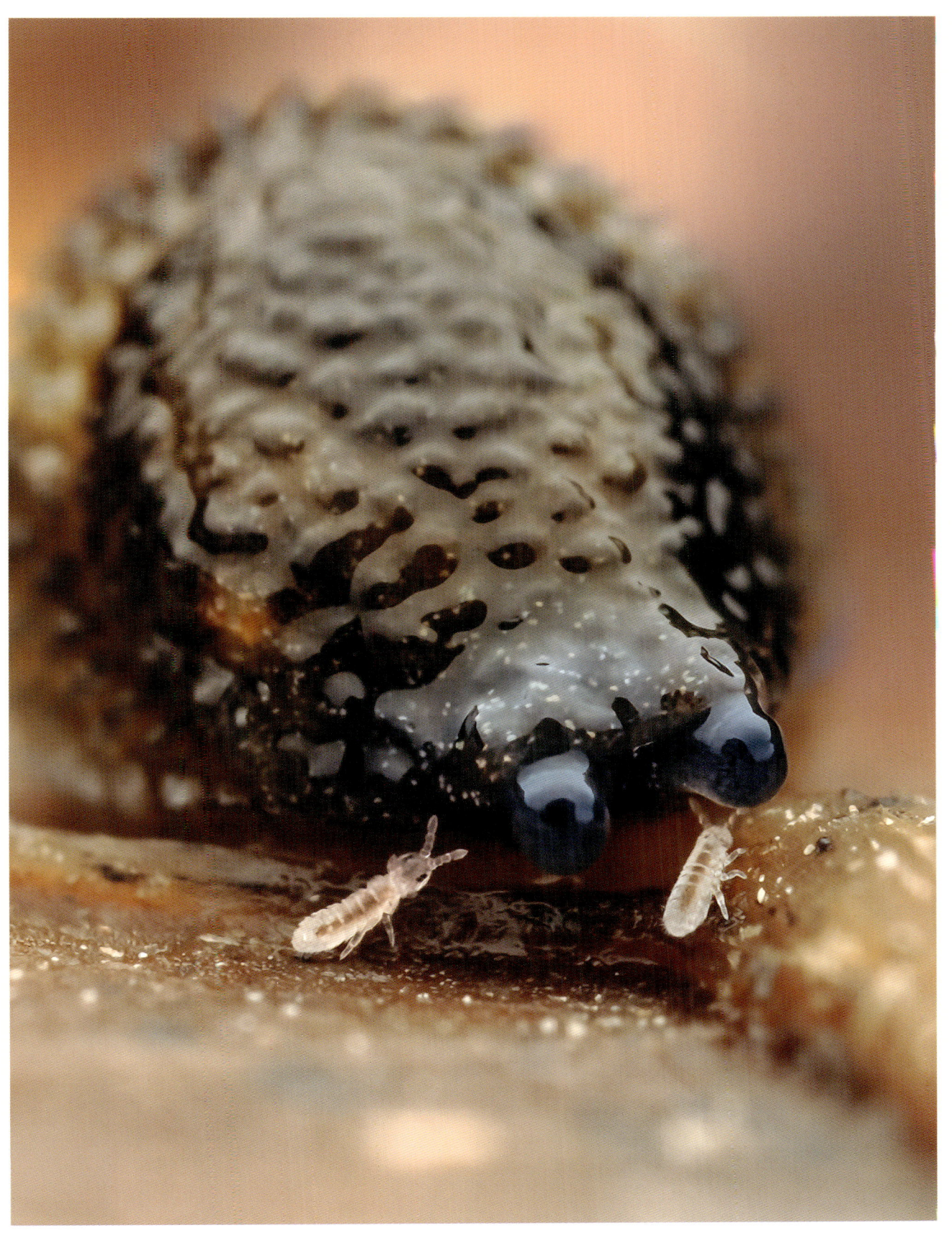

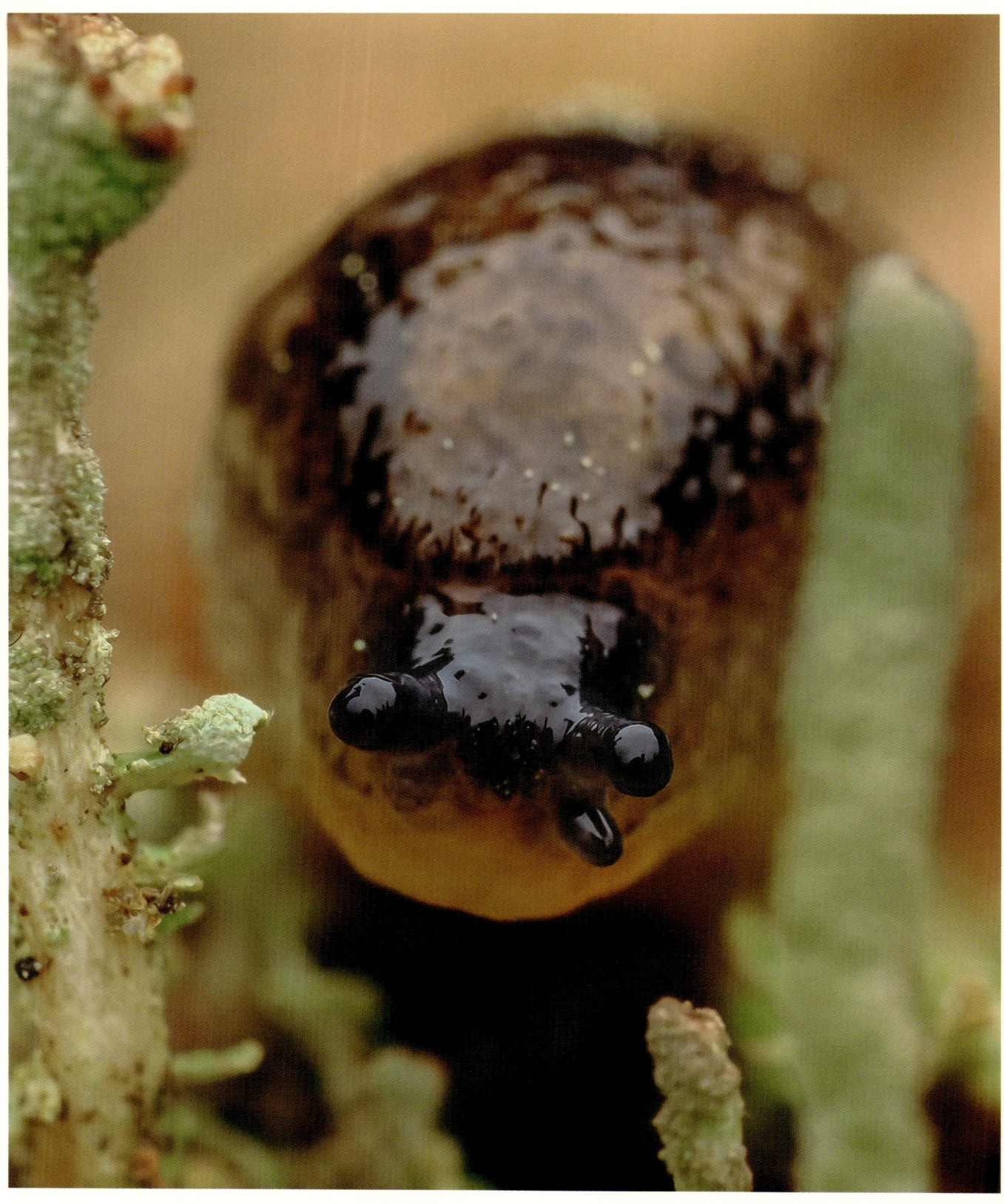

ABOVE: BABY SLUG. OPPOSITE: ROSETTE LICHENS (*PHYSCIA* SPP.).

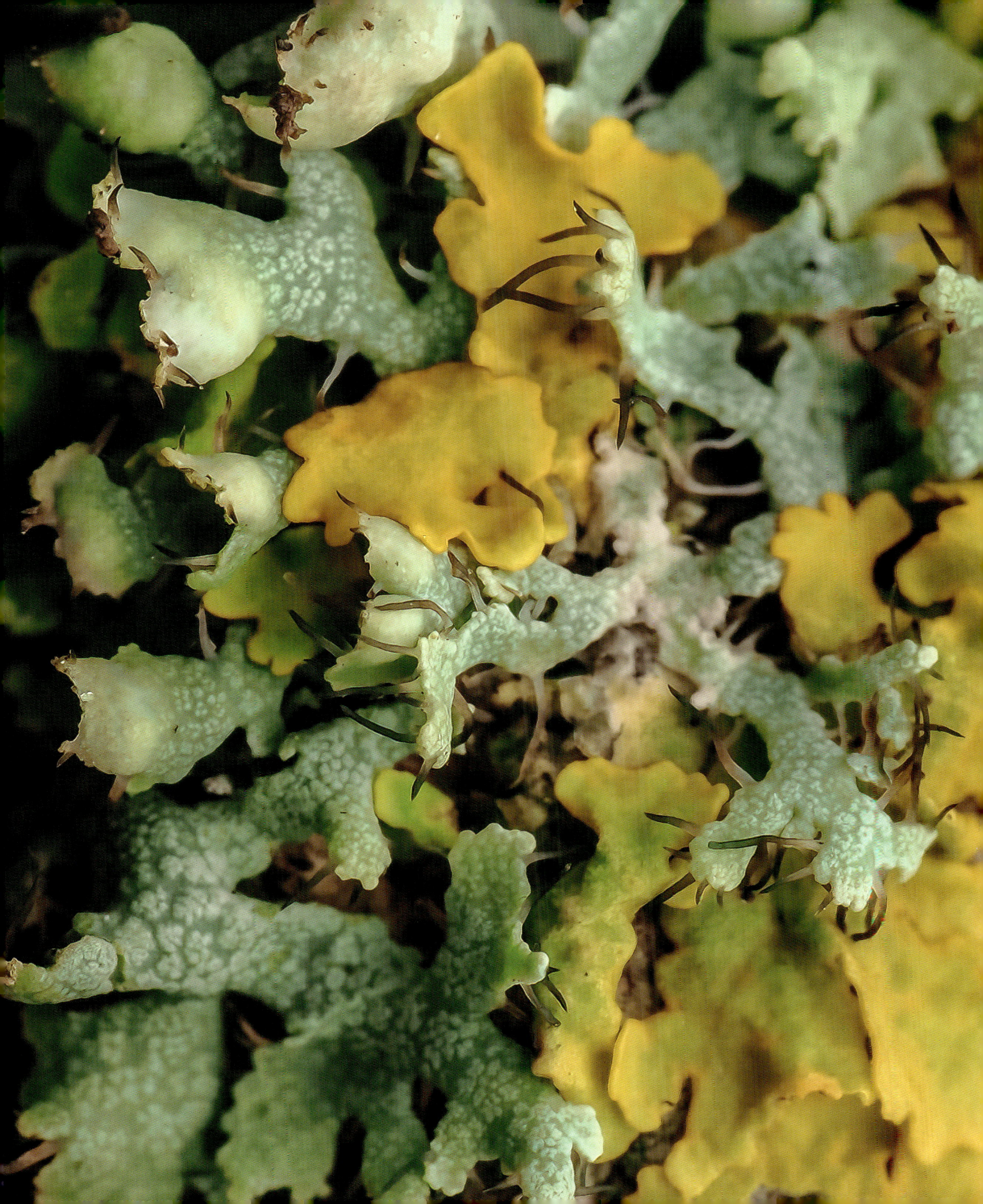

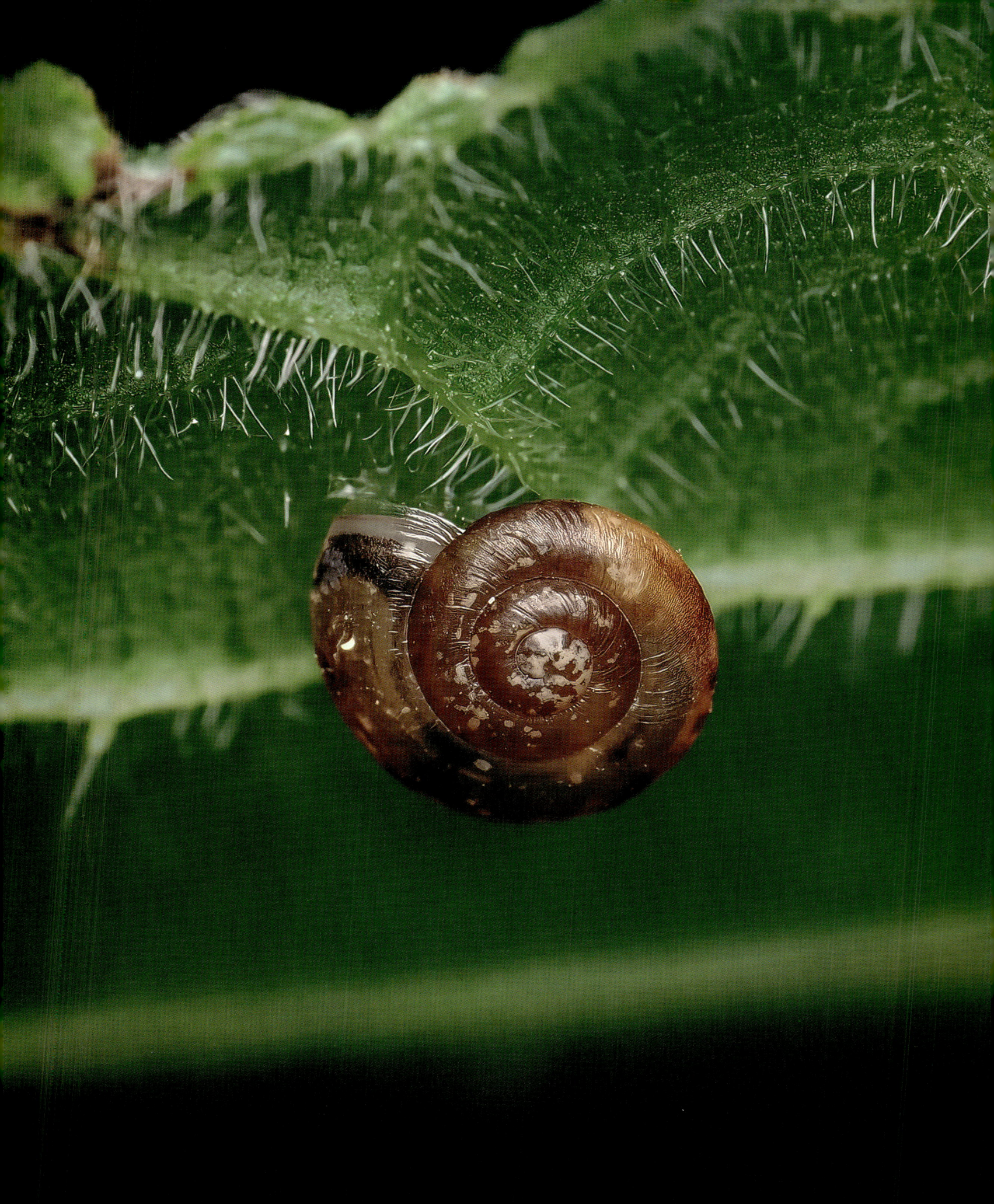

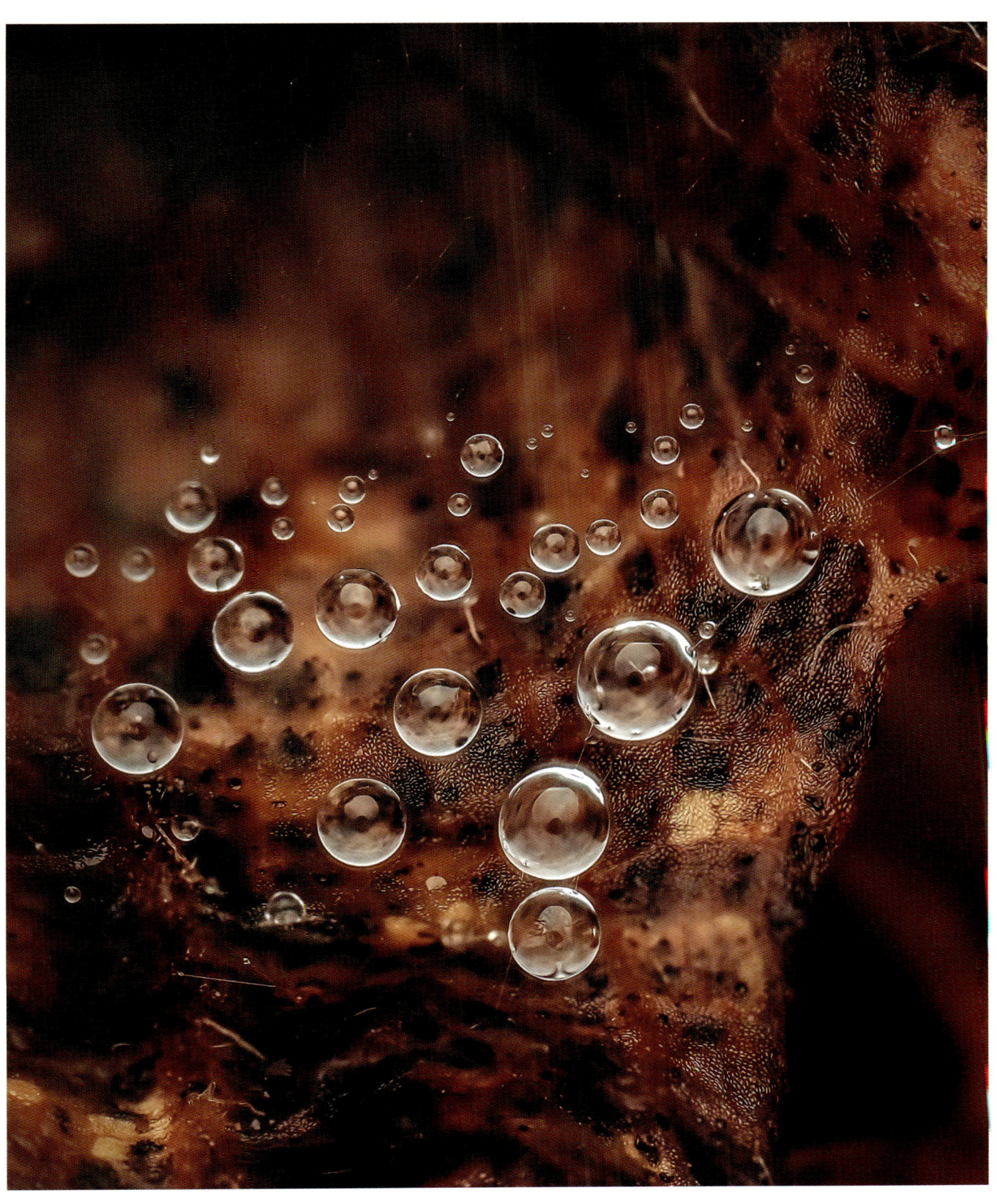

ABOVE: LEAF SNAIL. OPPOSITE: COPSE SNAIL (*ARIANTA ARBUSTORUM*).

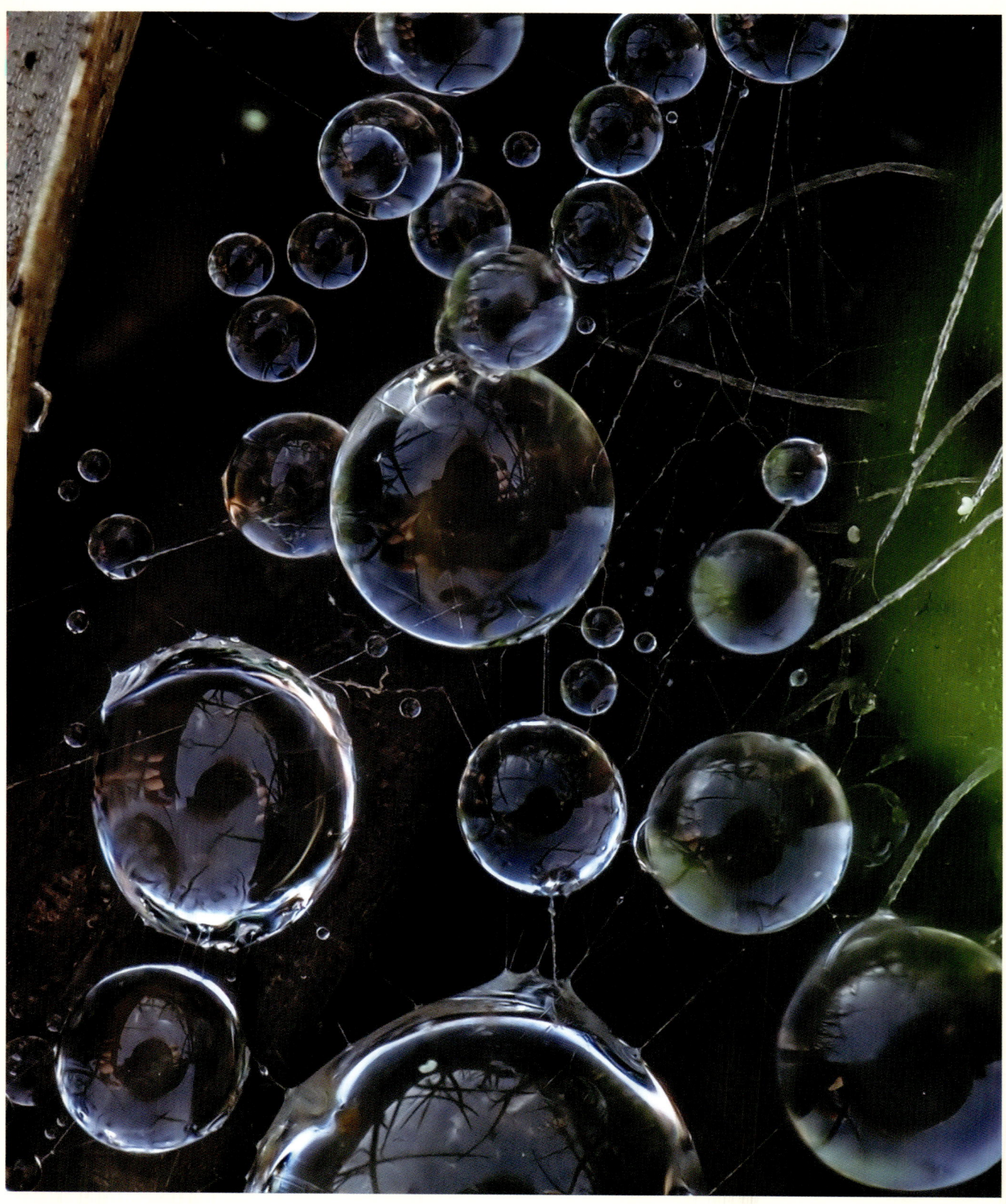

198

ABOVE: WATER DROPLETS AND SPIDERWEB. OPPOSITE: SLUG EGGS.

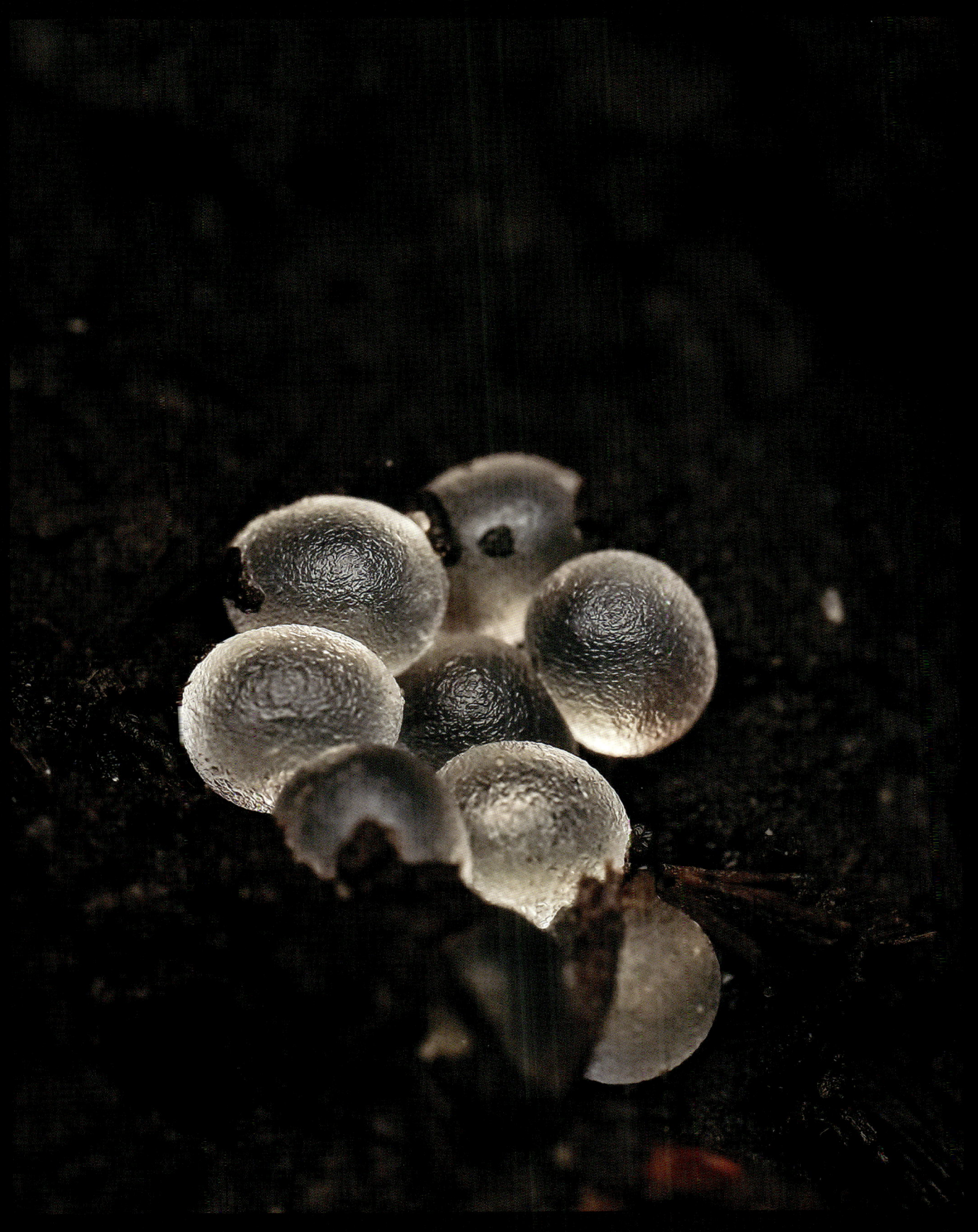

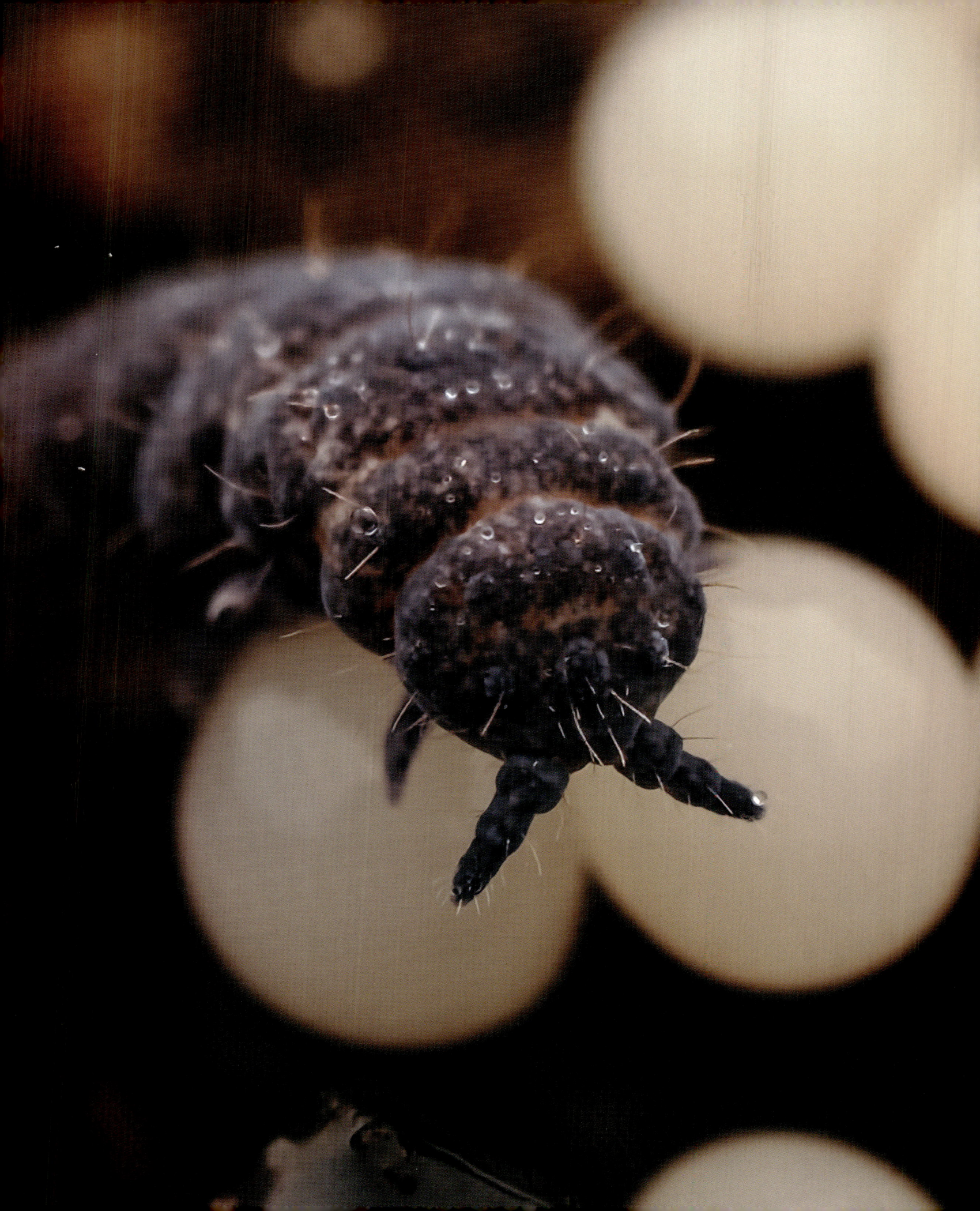

Soil Animal

By now you've probably gathered that there's a hidden world full of remarkable tiny things often overlooked. And I've saved the most hidden of them all for last: soil animals. Creatures like springtails, earthworms, millipedes, centipedes, ground beetles, ants, terrestrial isopods, mites, and other tiny arachnids. Soil animals feed on decaying plant material, contributing to decomposition and nutrient recycling. Some are also predators that help control the populations of other small soil animals.

I used to be in the "not so into bugs" camp. I had a "safe bugs" list and a "scary bugs" list. Macrophotography has been the ultimate exposure therapy, providing me with the opportunity to educate myself and reframe bugs as they truly are: vital to the ecological balance and resilience of the environment, but also beautiful. Wild, strangely beautiful. Up close, they are mesmerizing and alien-esque. Through my constant exposure to them, I was able to fight past my lifelong fear in an effort to capture their incredible colors, textures, and patterns. ⟿

OPPOSITE AND FOLLOWING SPREAD: MOSS SPRINGTAIL (*NEANURA MUSCORUM*).

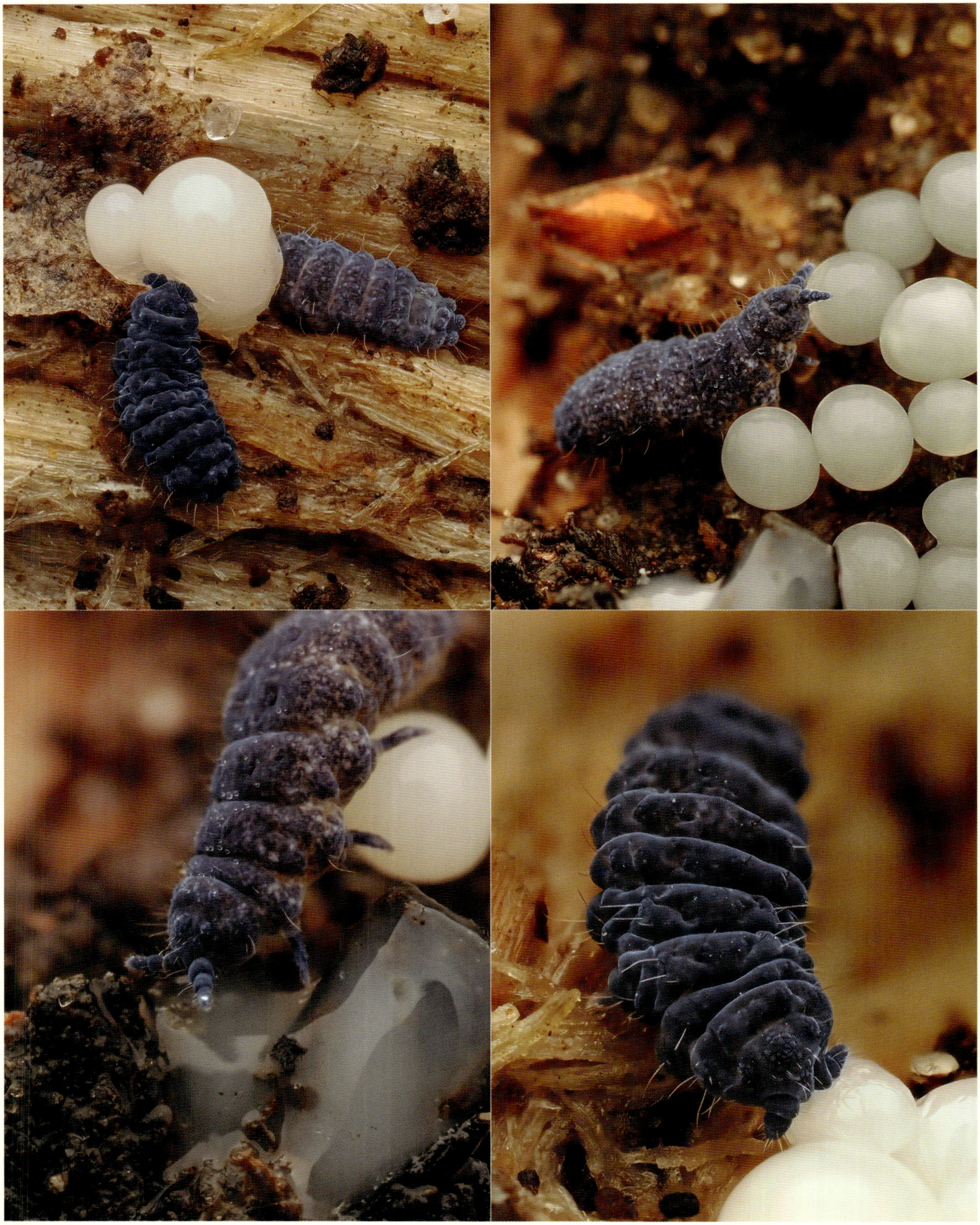

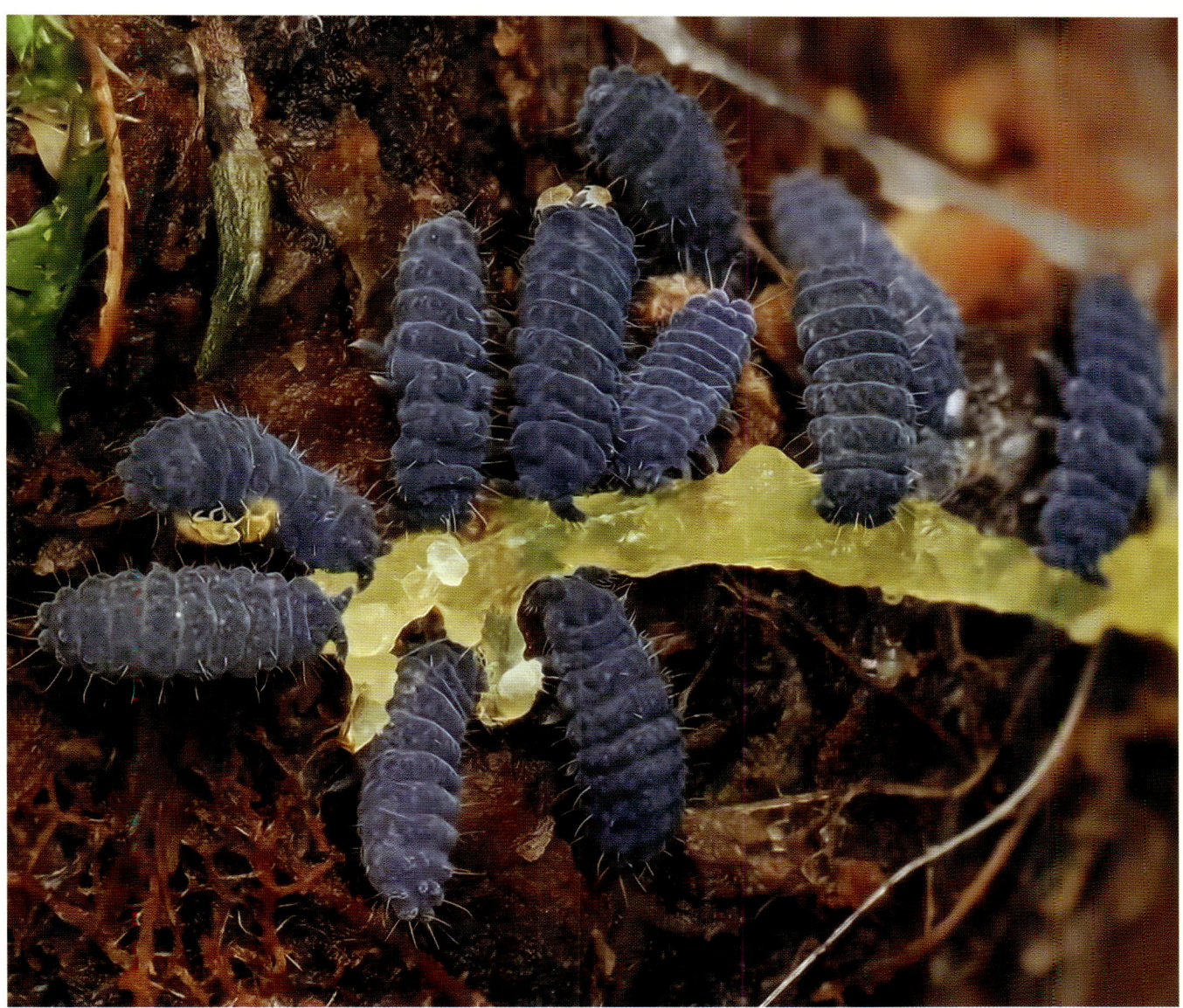

Macrophotography allowed me to see the world from an insect's perspective and thereby repositioned my view of them. When I look at insects now, it's one of sheer reverence, appreciation, and wonderment. Macrophotography highlights their unique adaptations and intricate features in a way that demystifies them.

Through my efforts "to get the shot," I've spent countless hours hovering over nettles and flowers with things buzzing all around me, but never harming me, and I gained a new viewpoint: bugs are friends.

Springtails, or *Collembola*, might be at the top of my list of soil animals so I'll spend most of my time here. With around 250 species in the UK—which is incredibly abundant and diverse—soil animals are small, wingless, six-legged hexapods. Dating as far back as 412 million years, they are known for their unique ability to appear spring-loaded by expertly launching themselves into the air thanks to a unique appendage called a furcula. This special tail helps them escape from danger, making them look like they've been shot out of a tiny cannon.

Although a useful ability, they are hilarious to watch but frustrating to photograph. By the time I've got them in my sights, they've usually got me in their sights and *sproing* off they go.

Like gastropods, fungi, and slime molds, springtails also play a crucial role in forest decomposition by feeding on fungi, bacteria, and decaying organic matter. Over the last four years, I've photographed my fair share of springtails but there are just so many! I'm terrible at identifying them because the species often look so similar, and well . . . they are tiny.

Springtails can be broadly categorized into three main types based on their habitat and behavior: soil, surface, and tree-dwelling. They are insanely cute little beings, with very cartoonish bodies. Poduromorpha species always remind me of an inflatable pool raft with rabbit-like gummy ears. They have a shape that can only be described as "blorpy." Globular springtails, however, look like some sort of farcical rainbow-splattered children's drawing of a space creature. Some are hairy, some translucent, and some can come in a truly candy-colored variety of hues.

The easiest way to get springtail photographs is to know where they are likely to be in abundance. Fungi, slime molds, and moist lichen-covered branches littering the understory are likely springtail haunts. Often, I can sit among wet leaves, gingerly turning them over one by one, and find springtails slumbering. This makes them a bit easier to capture a photo of; however, despite your best efforts not to disturb them, they immediately catapult themselves into the stratosphere. You win some, you lose some.

204

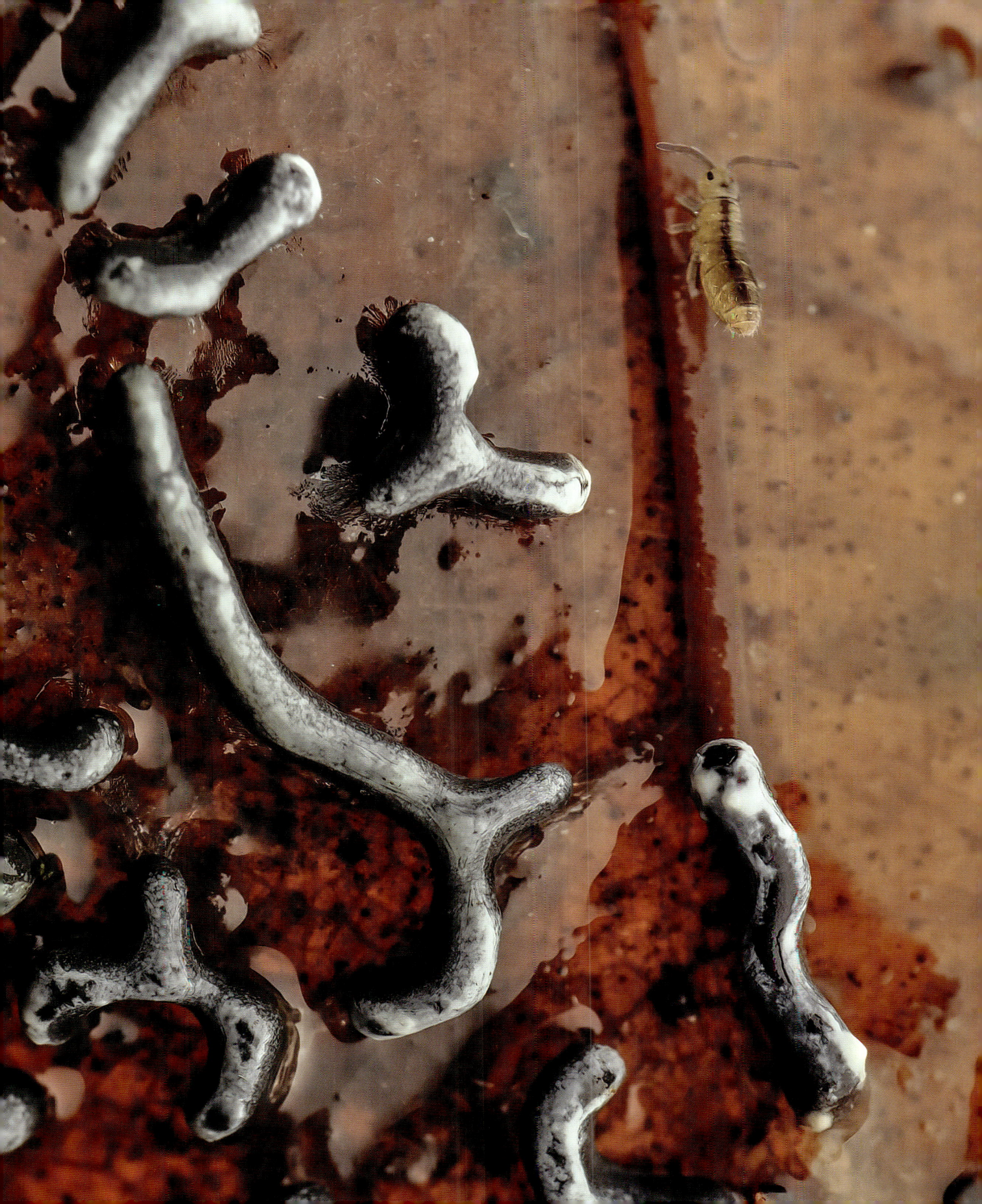

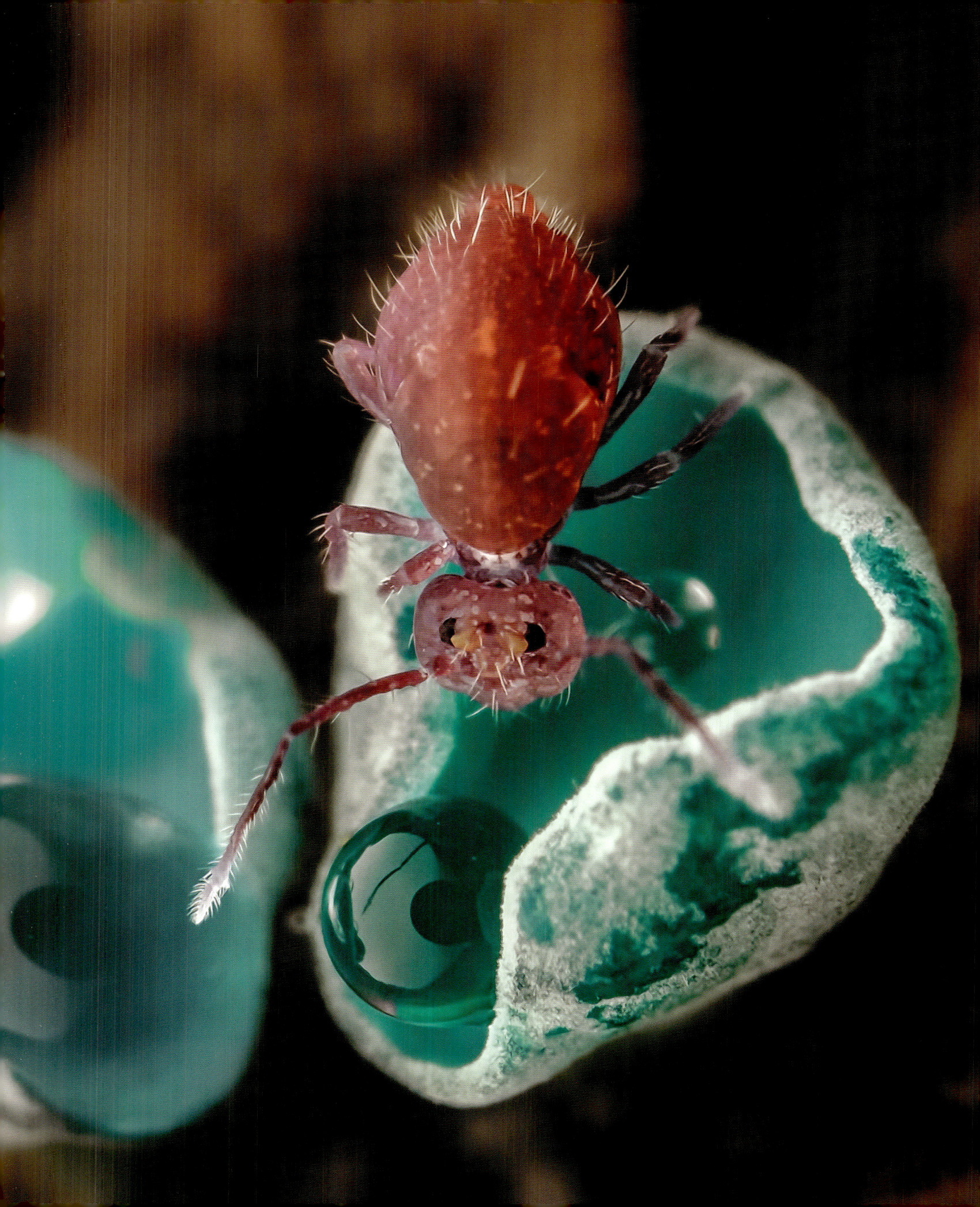

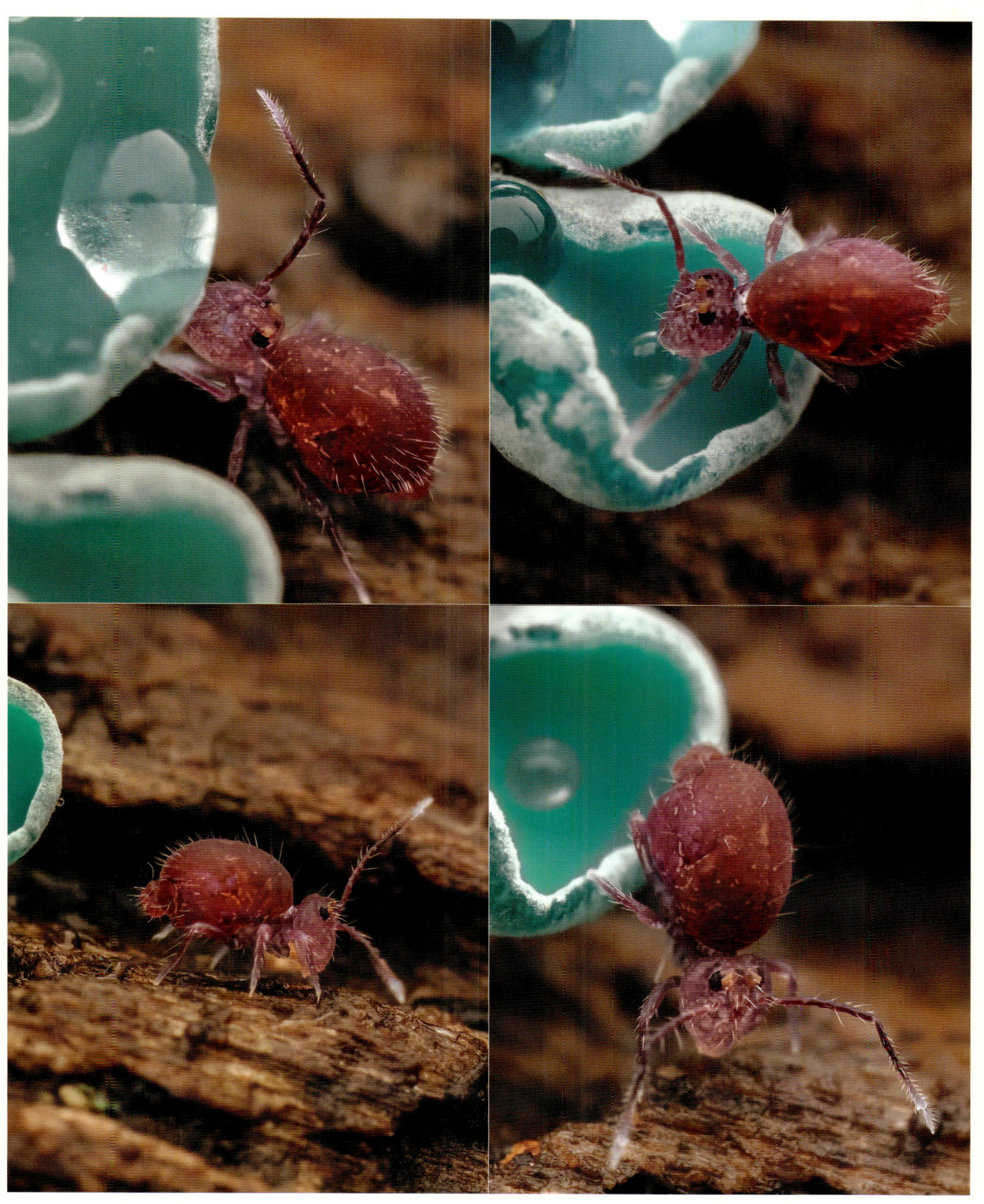

ABOVE AND OPPOSITE: GLOBULAR SPRINGTAIL (*DICYRTOMA* SP.) AND GREEN ELFCUPS (*CHLOROCIBORIA* SPP.) FUNGI.

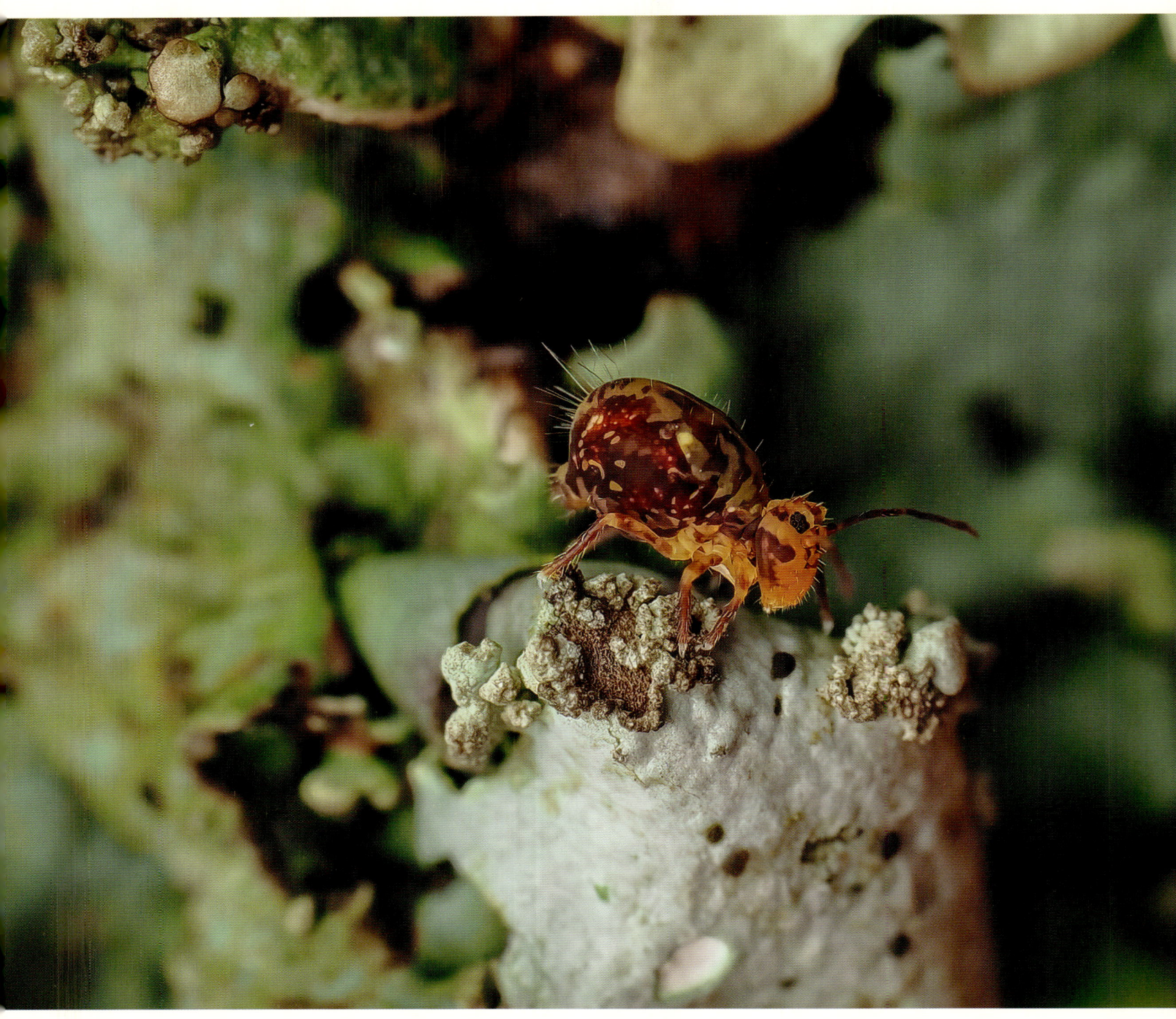

ABOVE AND OPPOSITE: GLOBULAR SPRINGTAIL (*DICYRTOMINA ORNATA*) ON LICHEN.

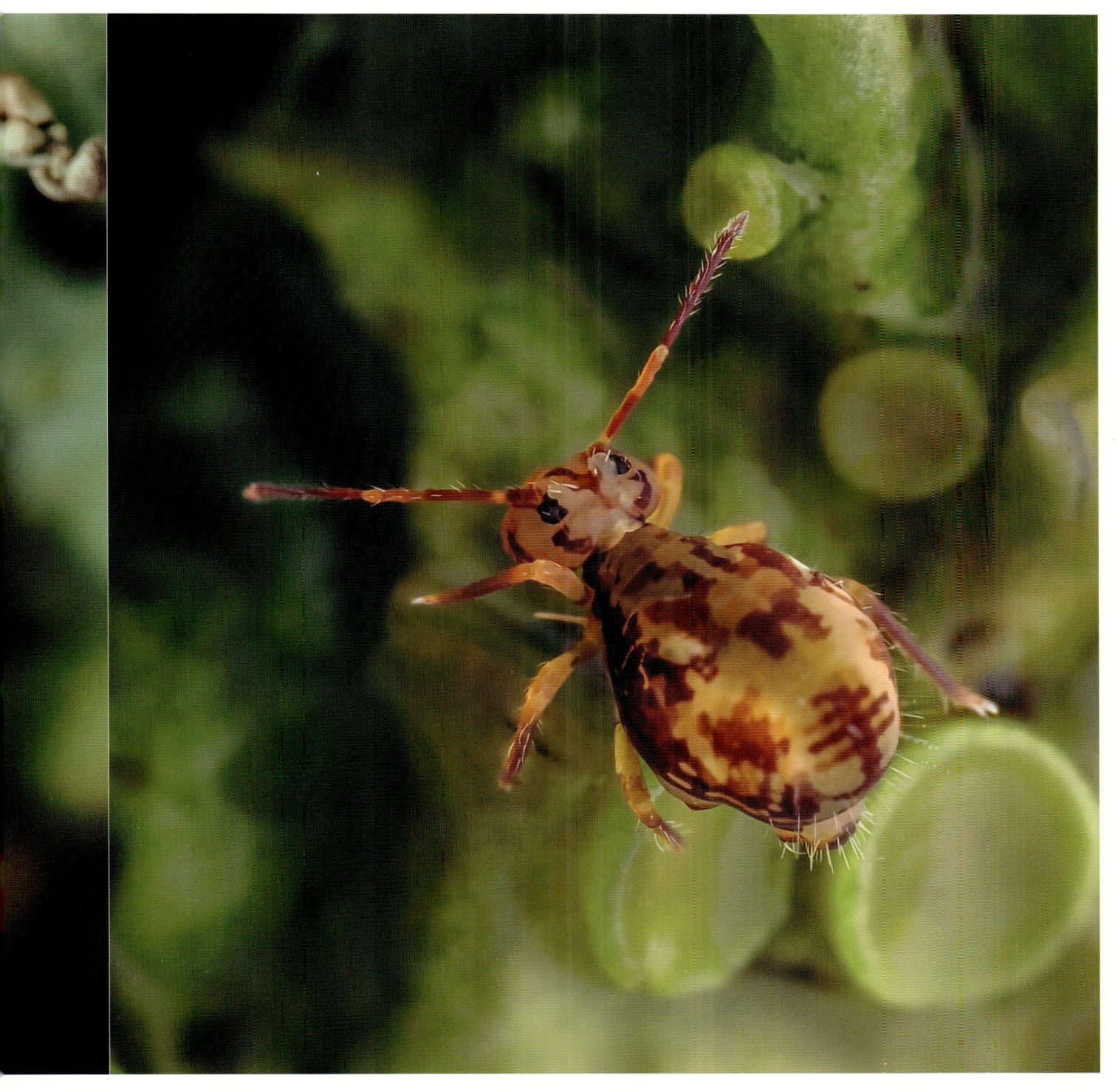

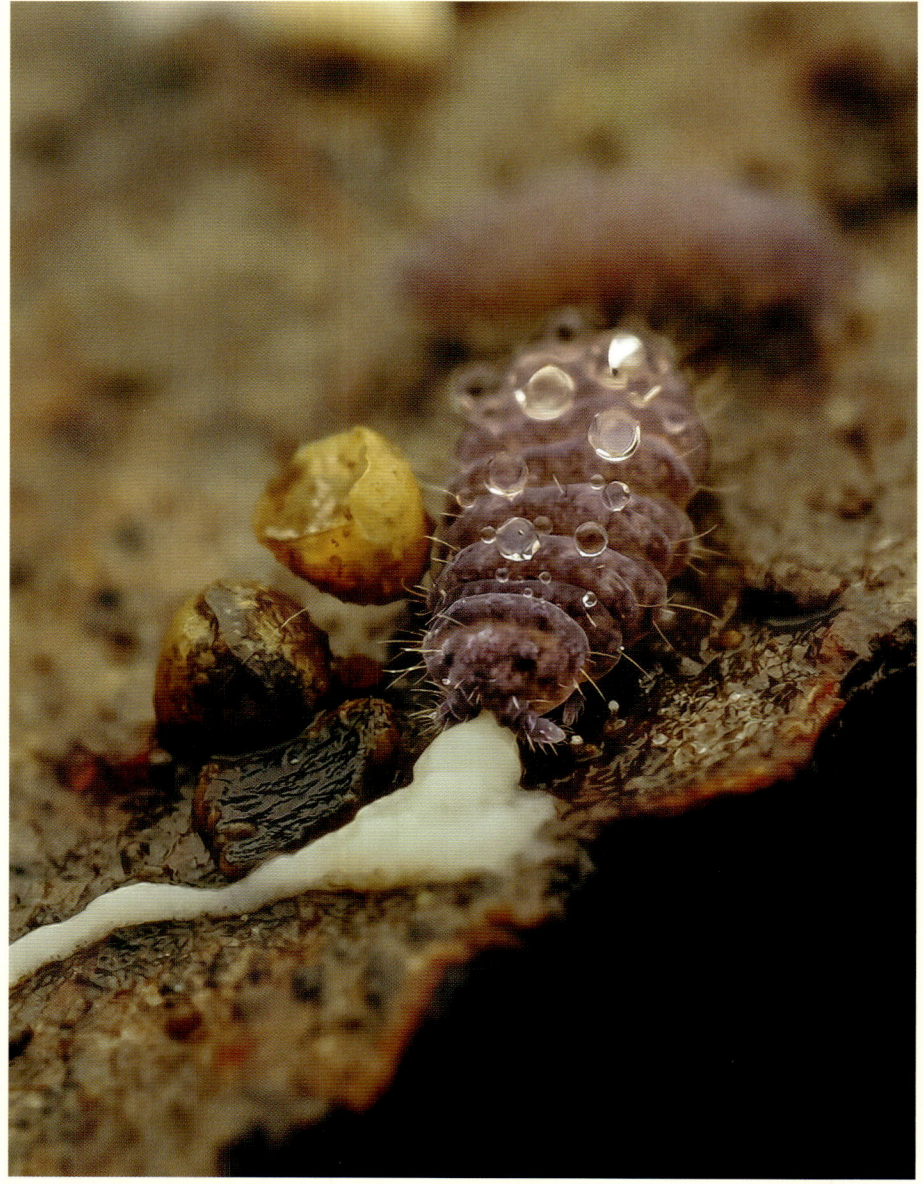

ABOVE: MOSS SPRINGTAIL (*NEANURA MUSCORUM*) EATING. OPPOSITE: UNKNOWN SPRINGTAIL ON CORAL FUNGI (*RAMARIA* SP.).

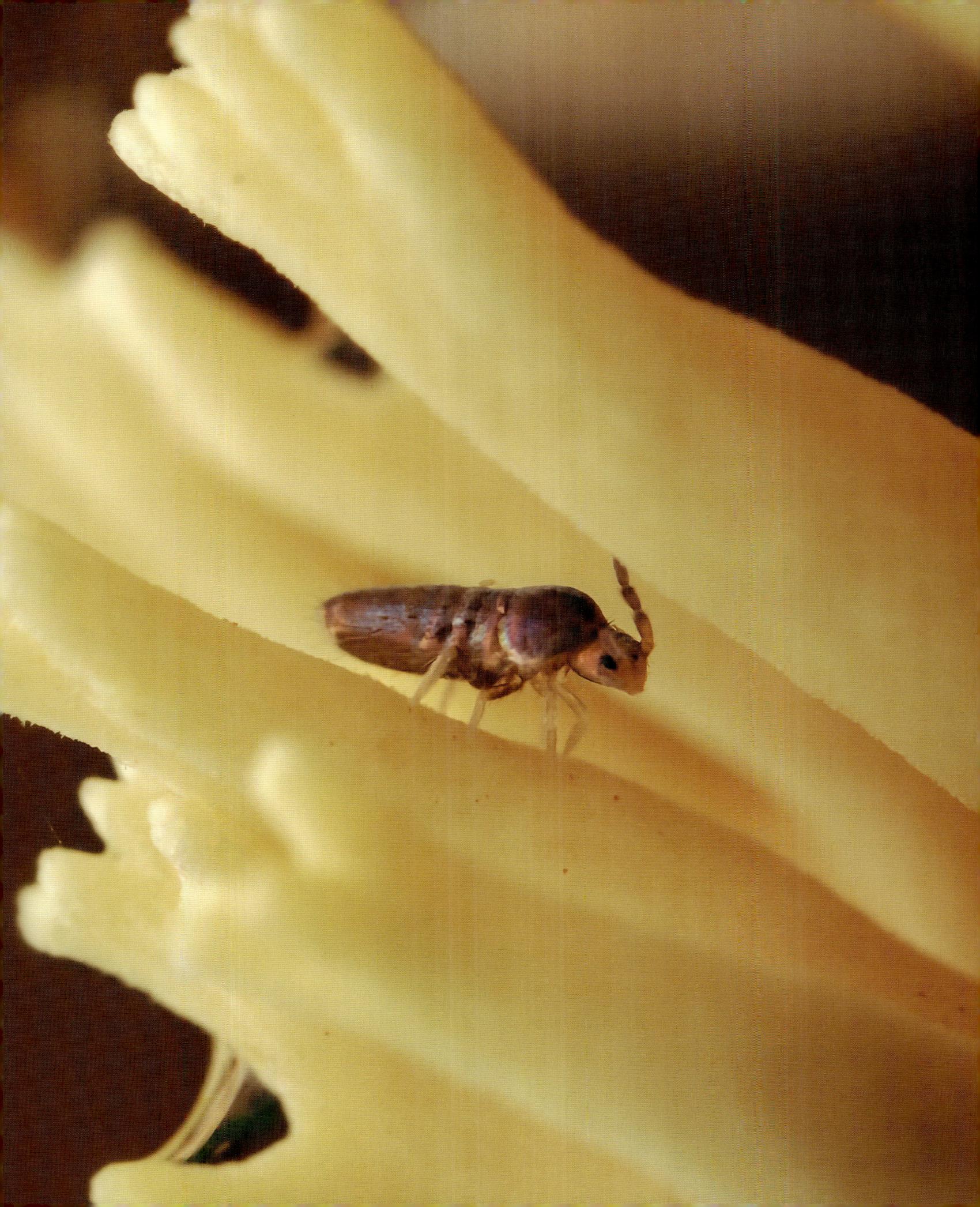

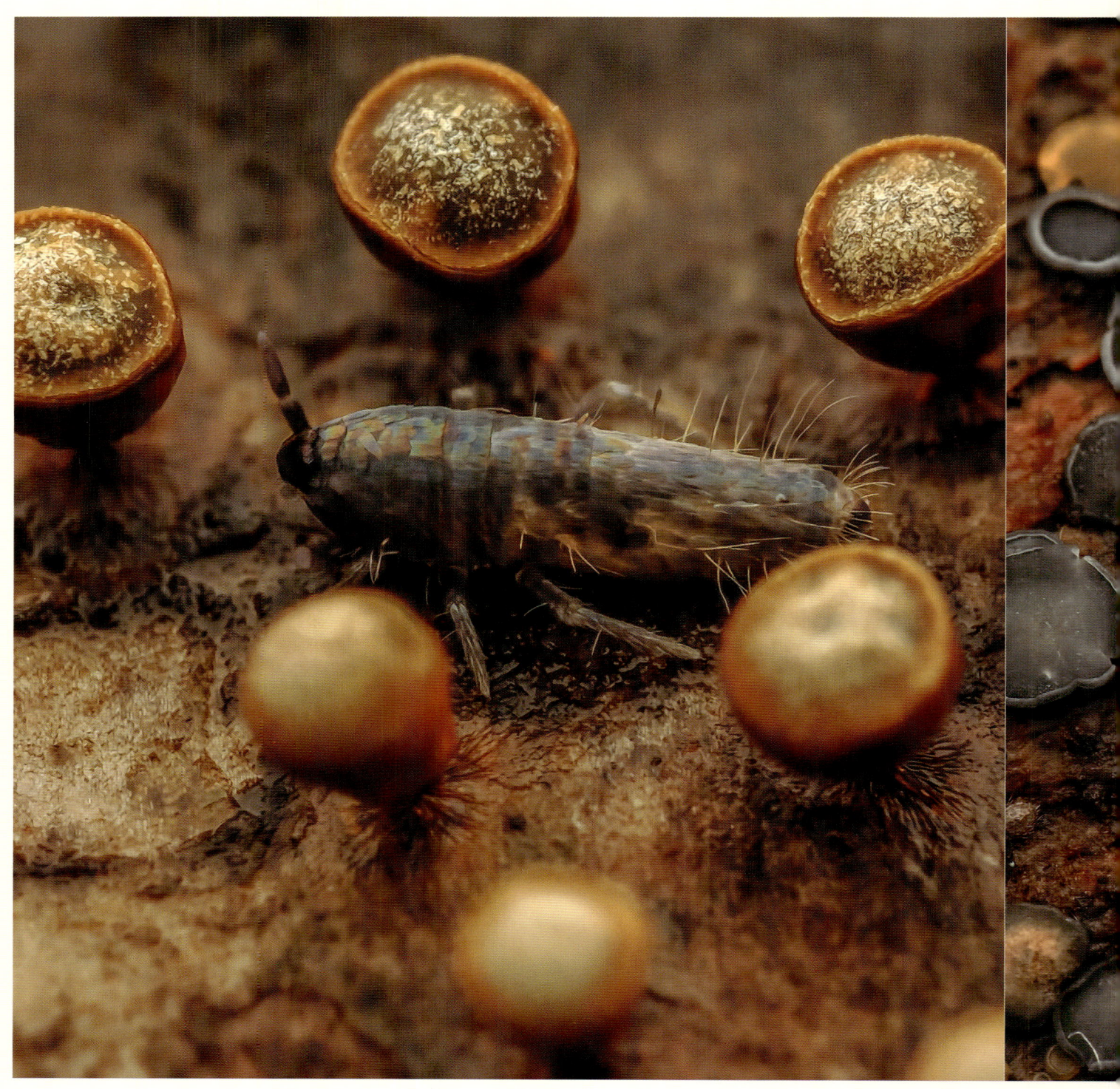

ABOVE: *LEPIDOCYRTUS* SP. WITH *CRATERIUM* SP. OPPOSITE: GLOBULAR SPRINGTAIL (*SYMPHYPLEONA* SP.) AND FUNGI.

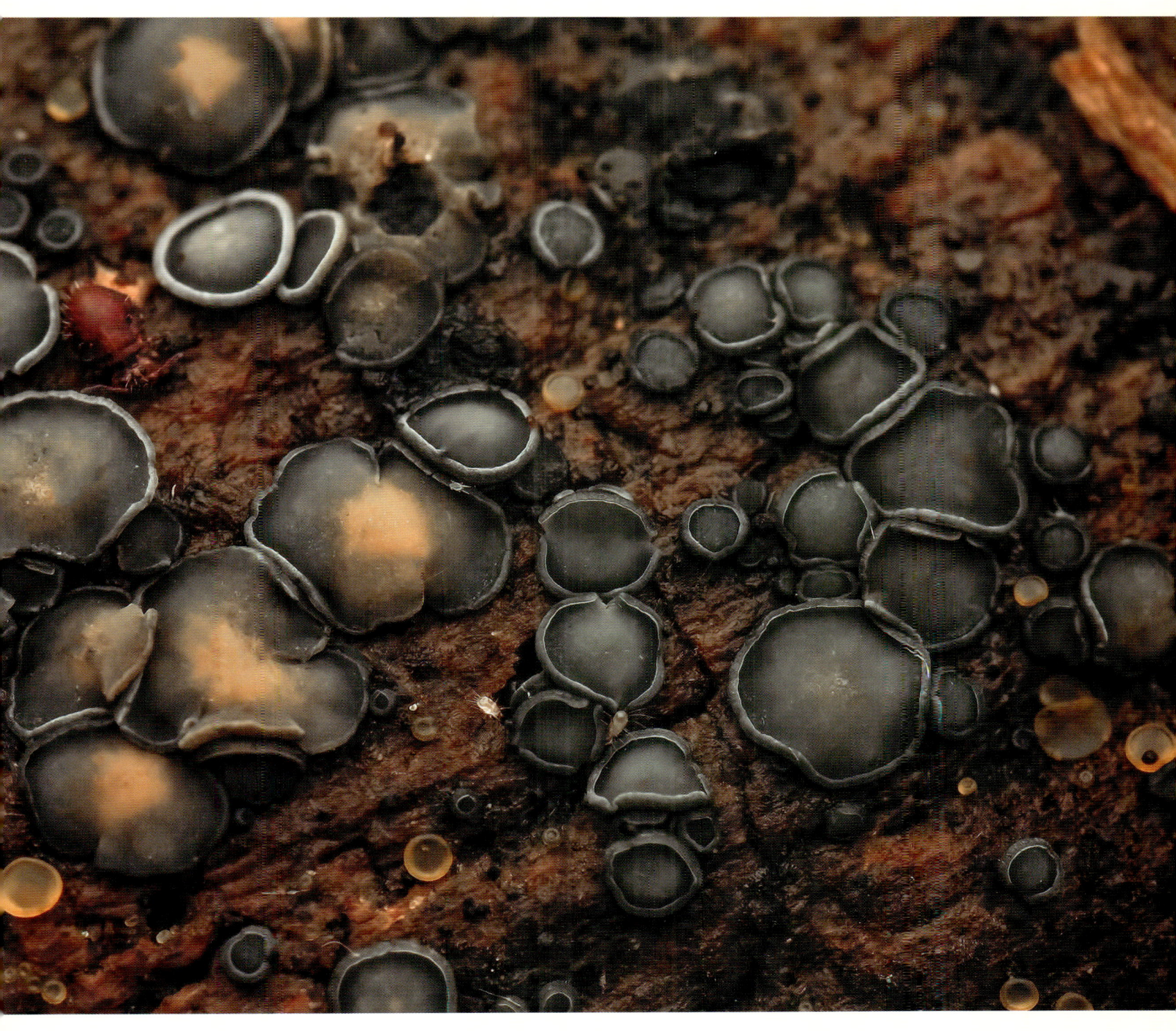

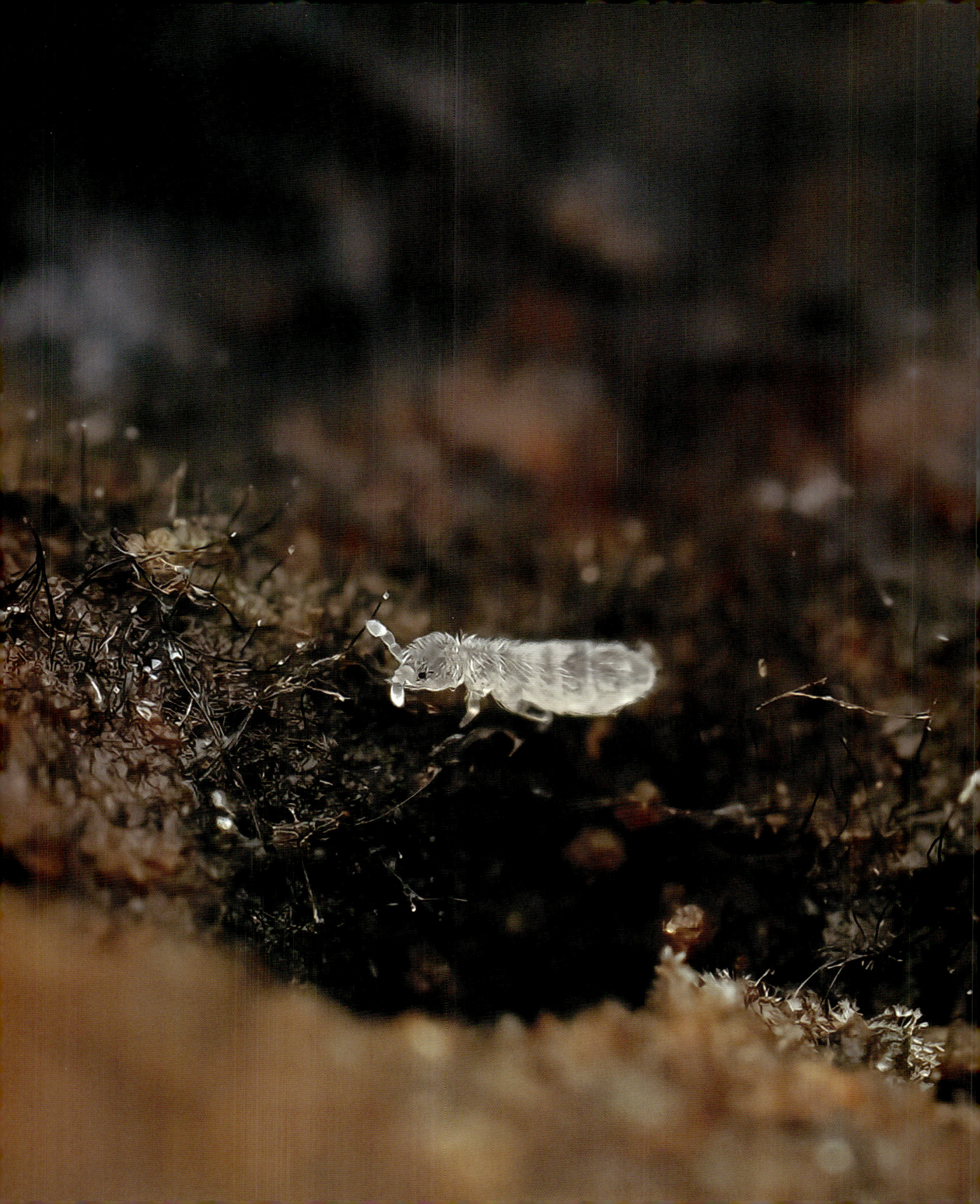

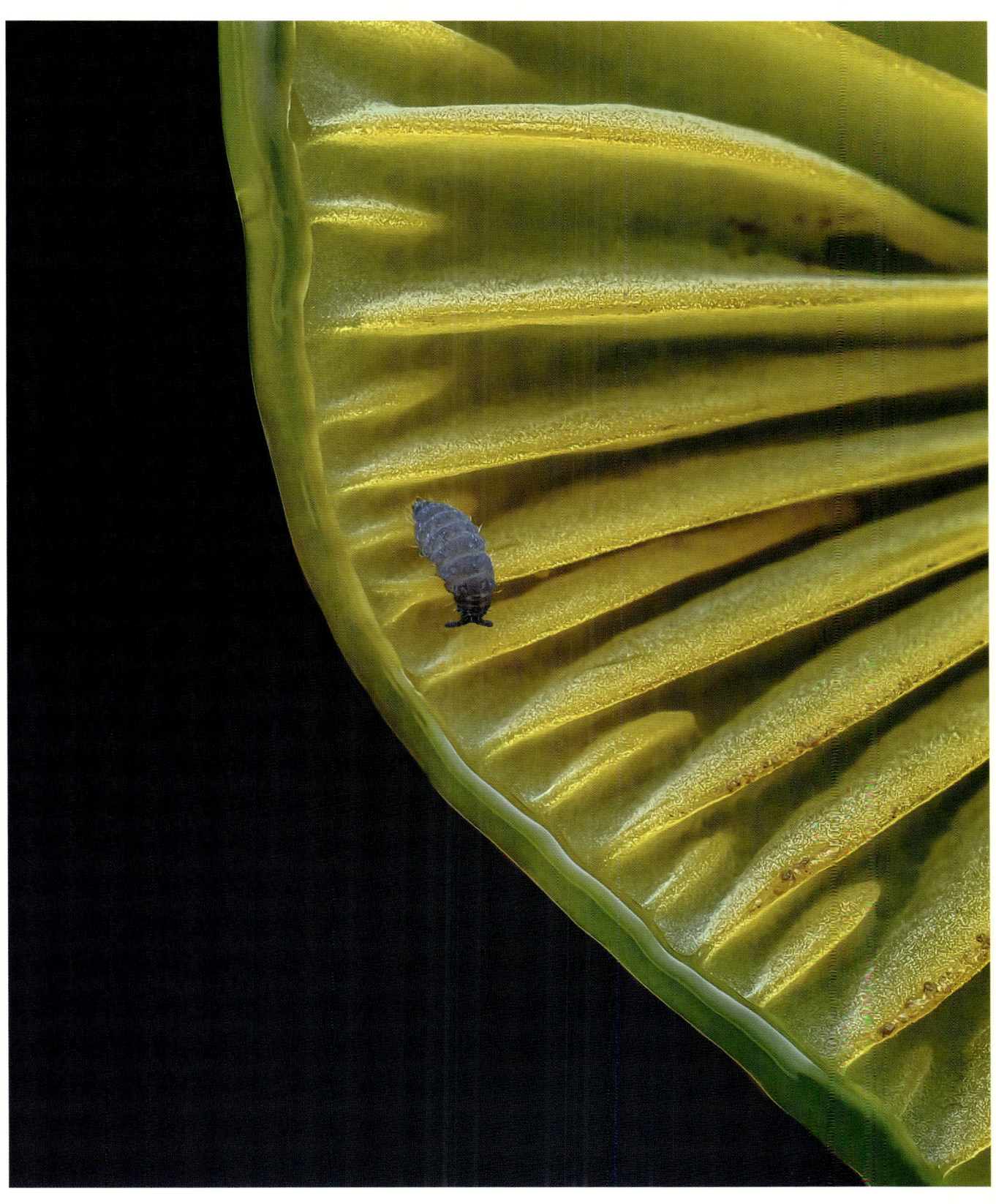

215

ABOVE: *PODURA* SP. AND PARROT WAXCAP (*GLIOPHORUS PSITTACINUS*). OPPOSITE: TRANSLUCENT SPRINGTAIL.

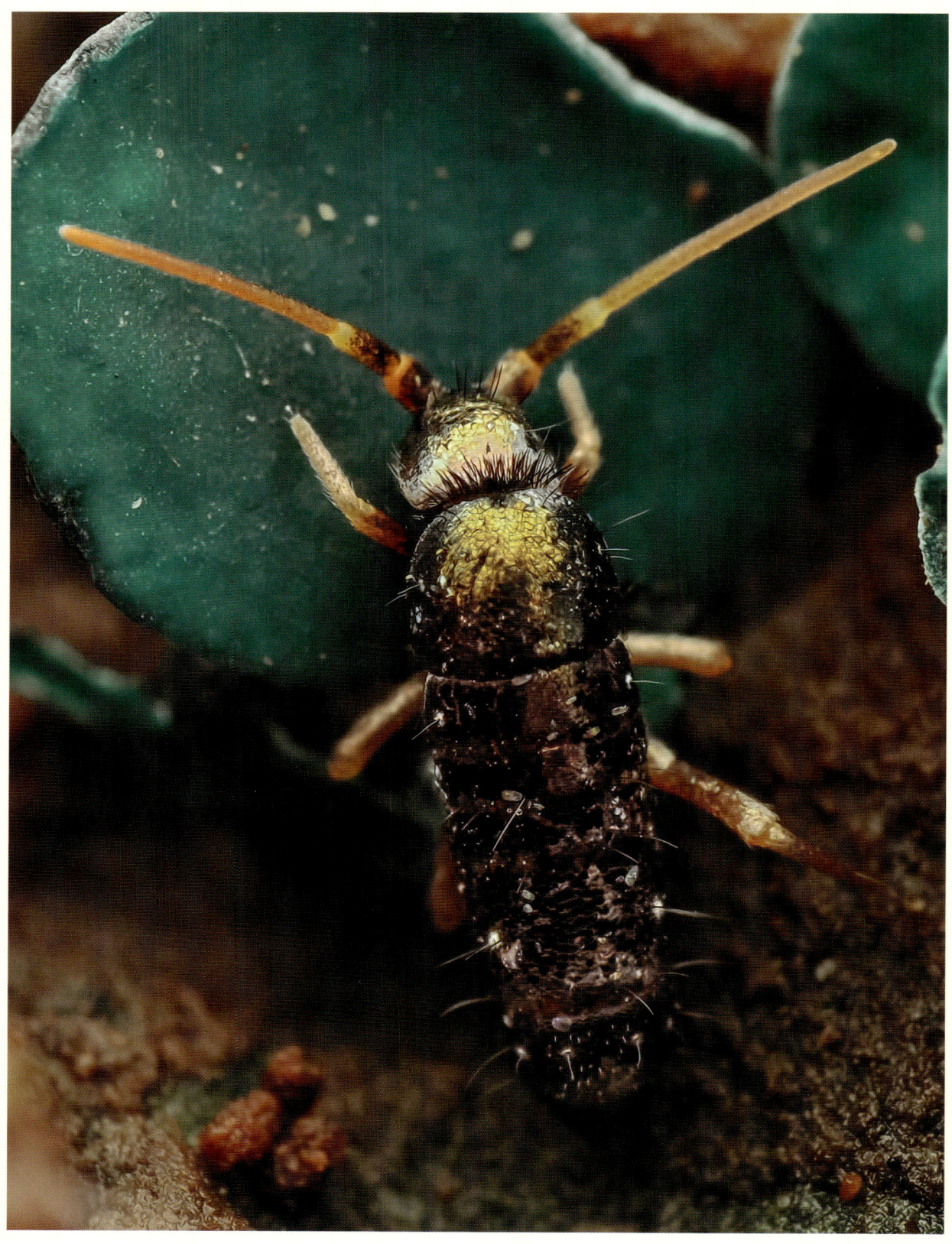

ABOVE: TOMOCERUS SP. AND GREEN ELFCUPS (*CHLOROCIBORIA* SP.) FUNGI. OPPOSITE: UNKNOWN SPRINGTAIL AND TEAL CONIFER-PIN (*D. SMARAGDINA*).

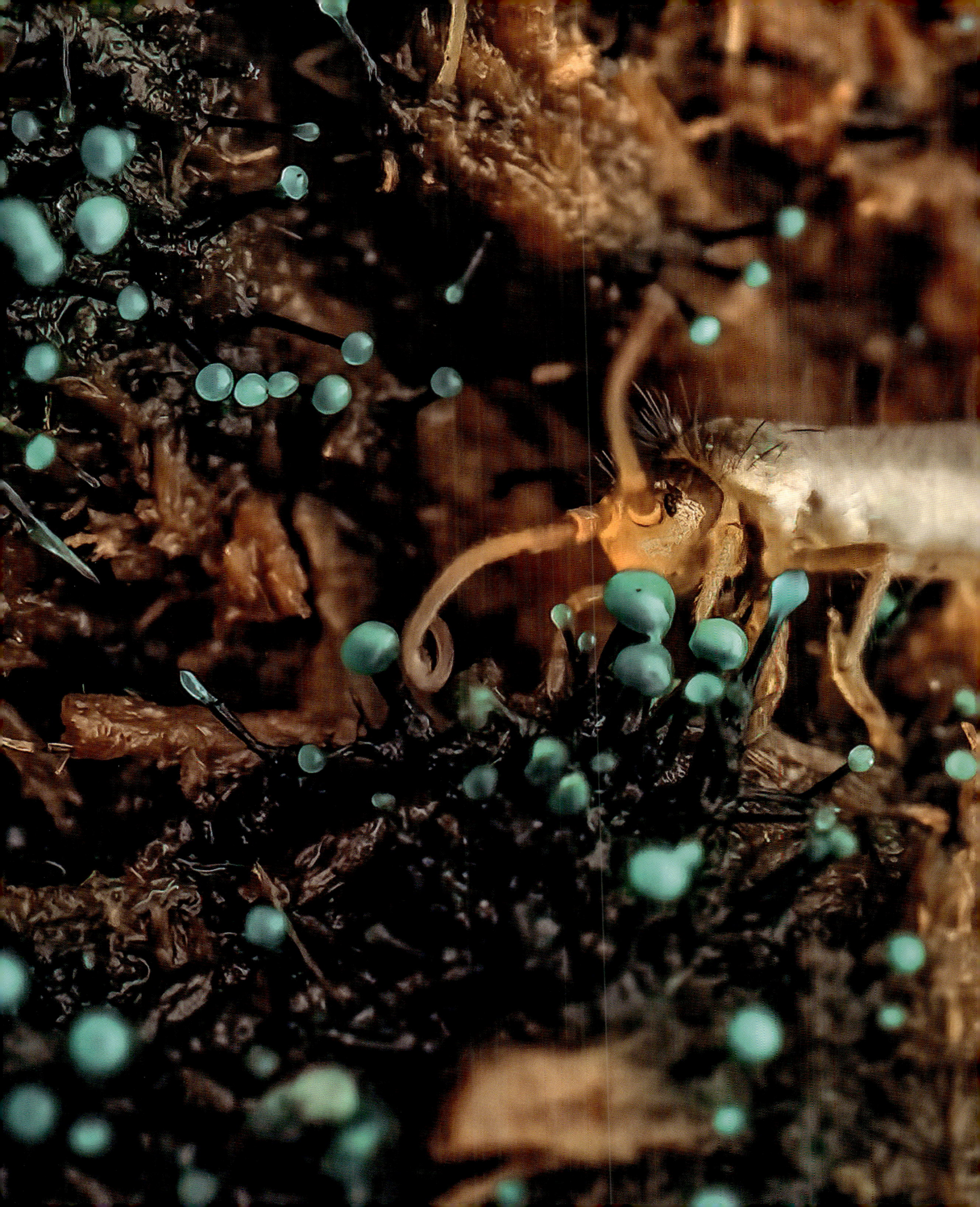

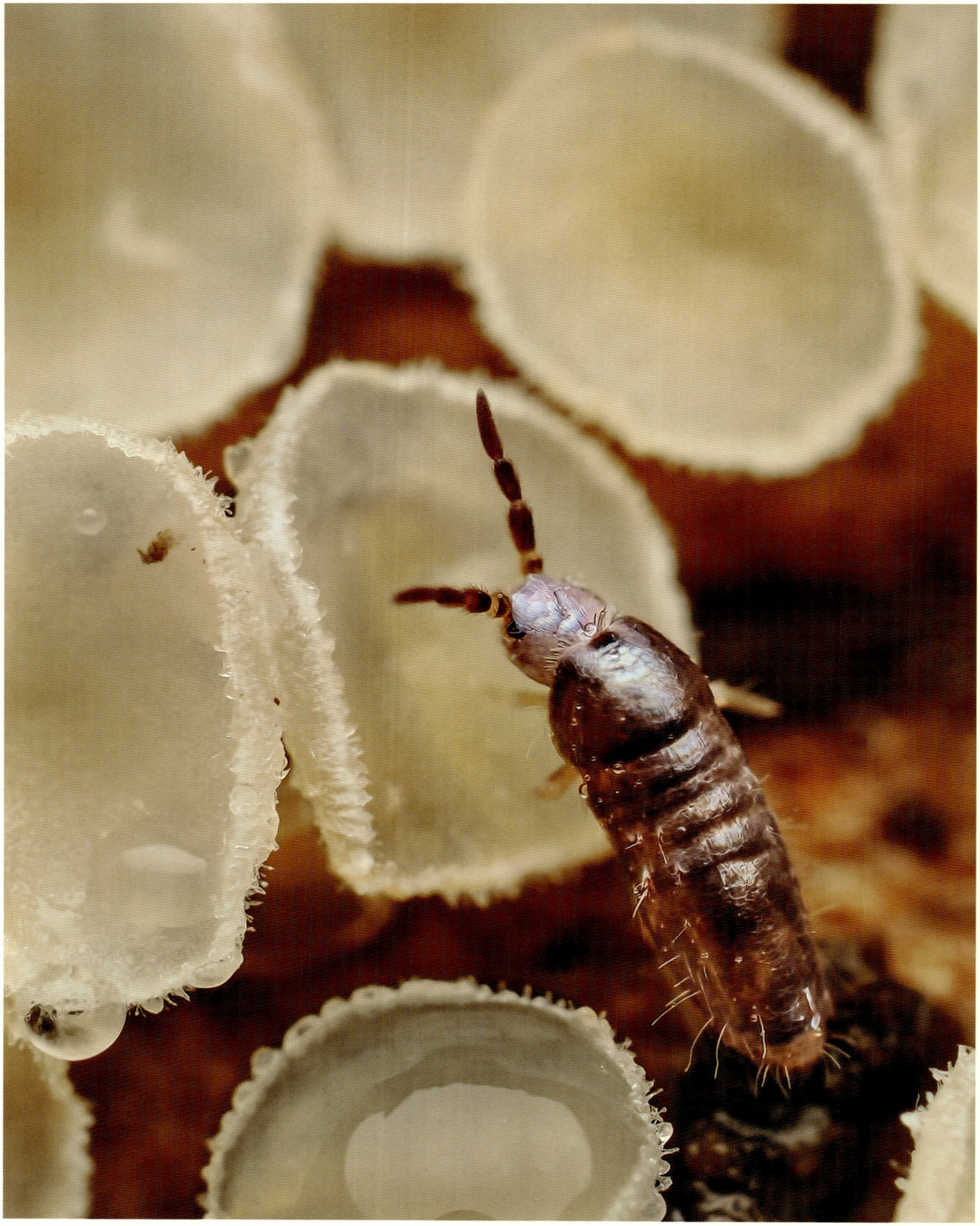

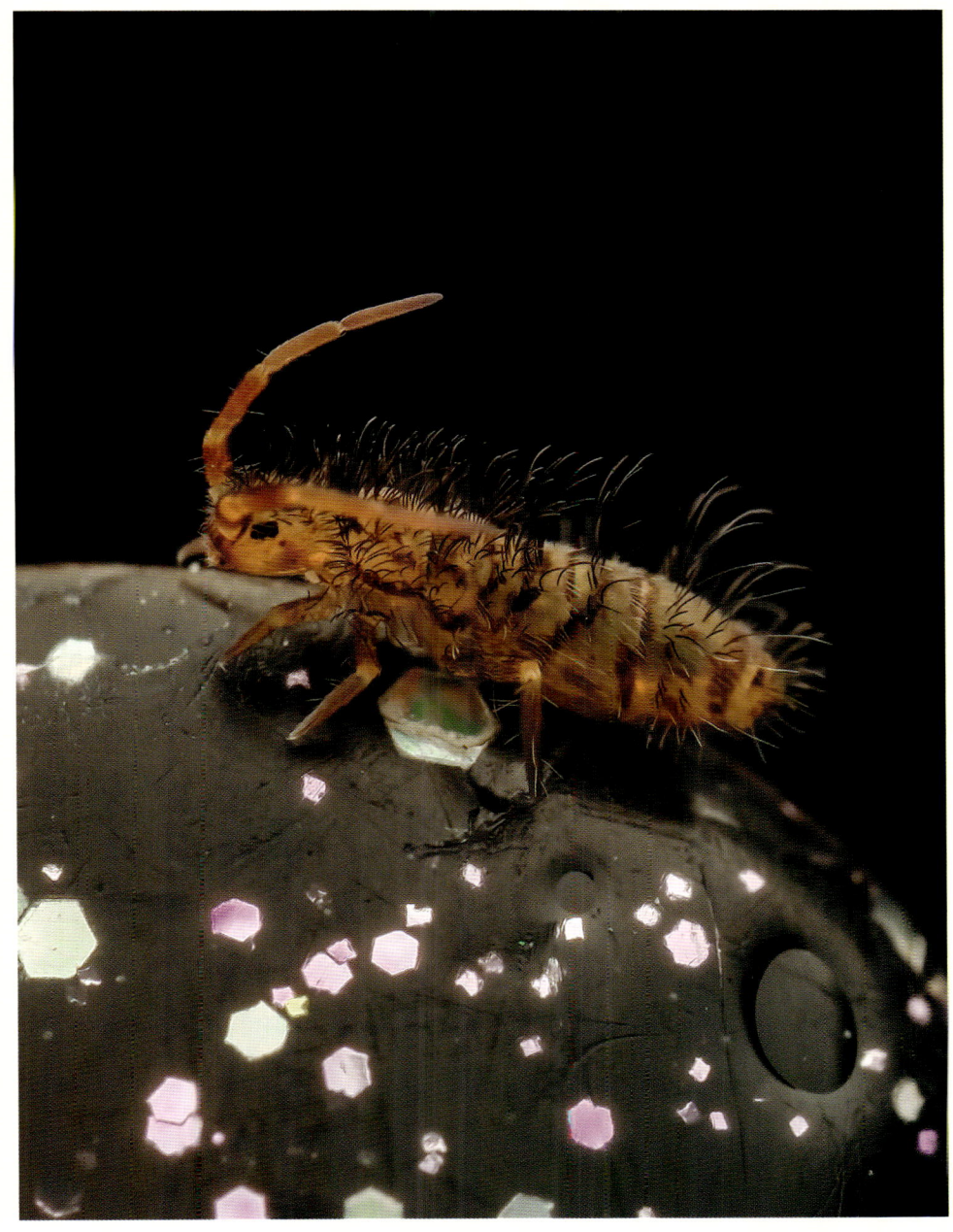

ABOVE: WOOLLY MAMMOTH SPRINGTAIL (*ORCHESELLA VILLOSA*) ON MY FINGERNAIL. OPPOSITE: UNKNOWN SPRINGTAIL AND FUNGI.

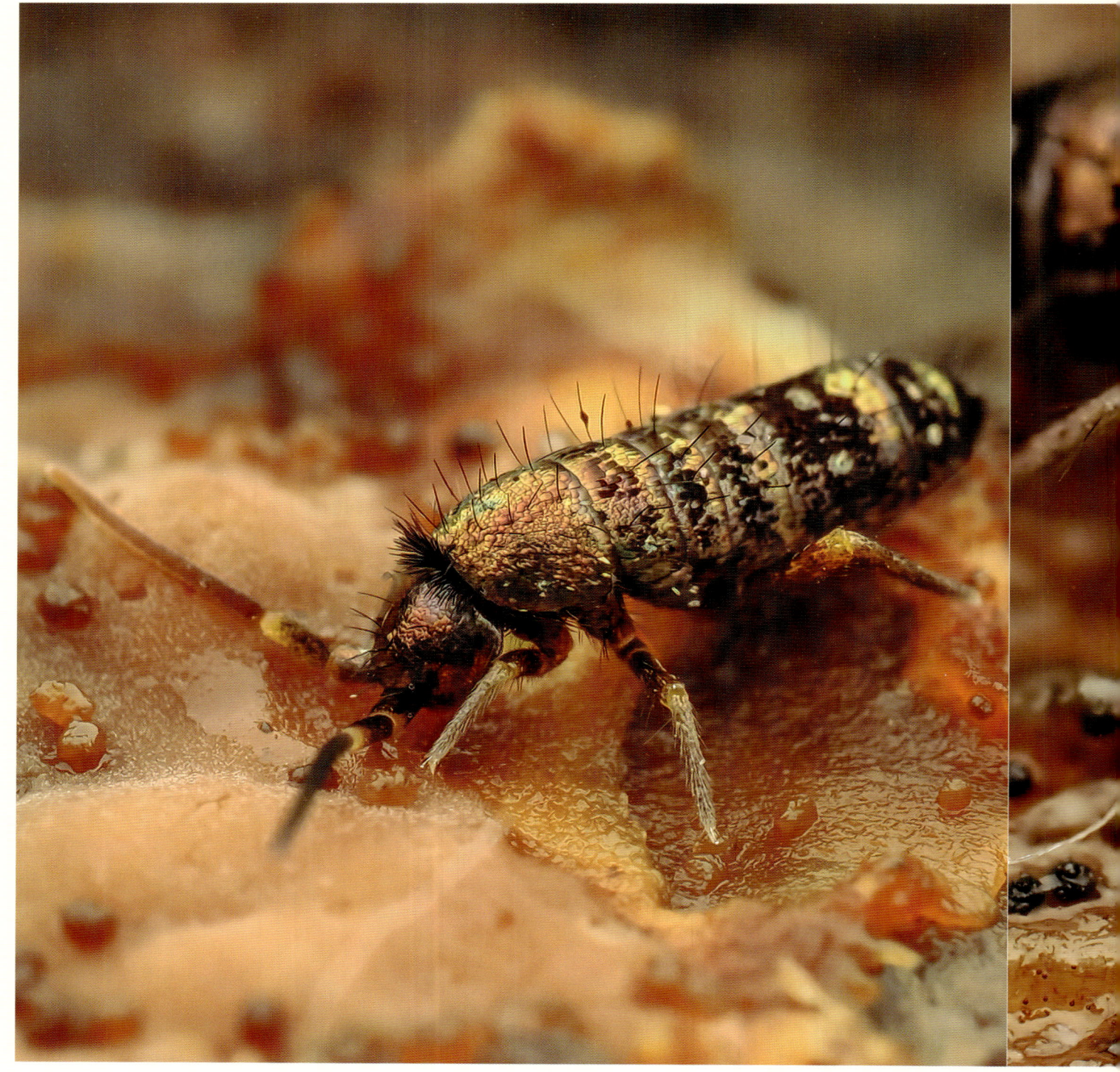

ABOVE AND OPPOSITE: COMMON BULL SPRINGTAIL AND PLASMODIAL SLIME MOLD.

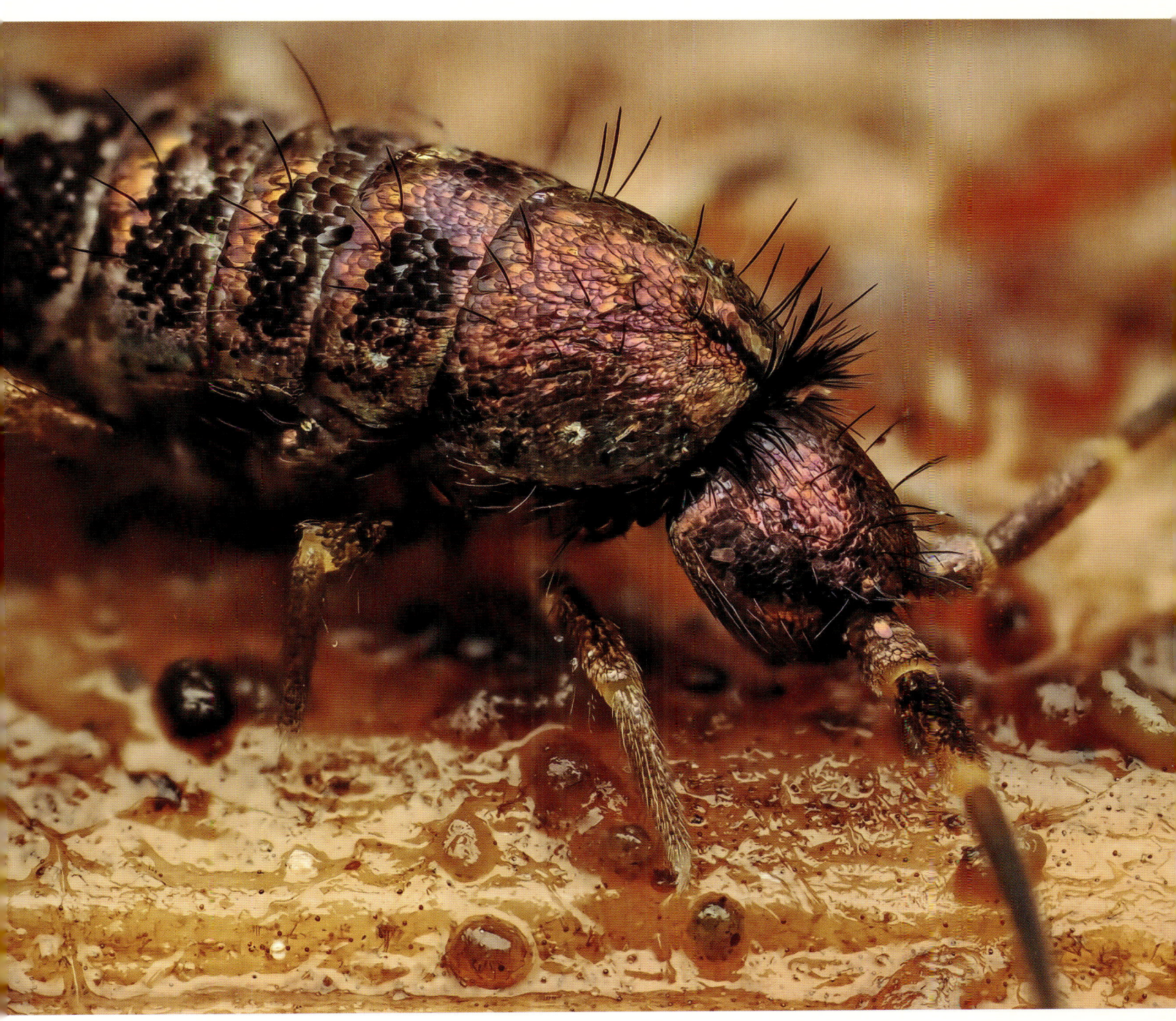

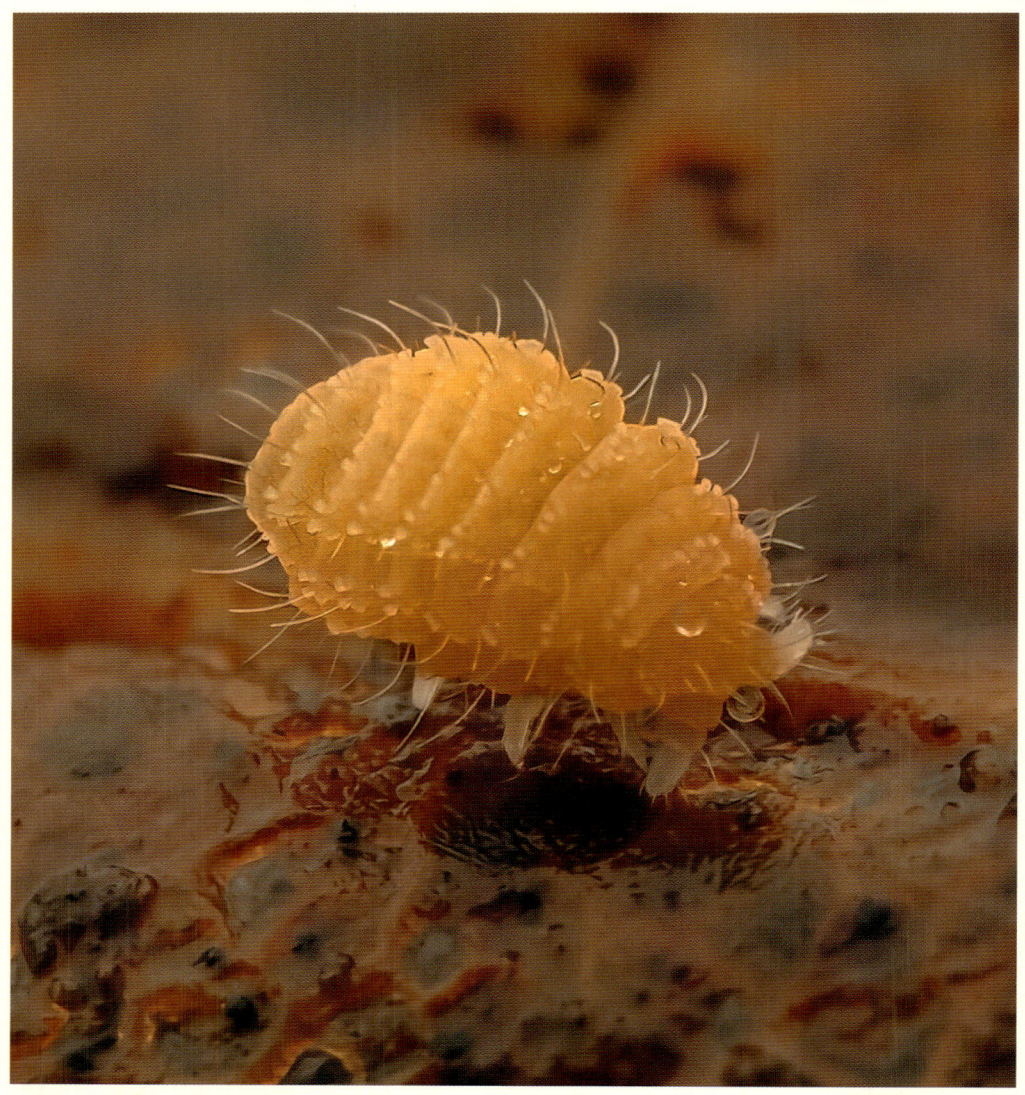

ABOVE AND OPPOSITE: *MONOBELLA* SP.

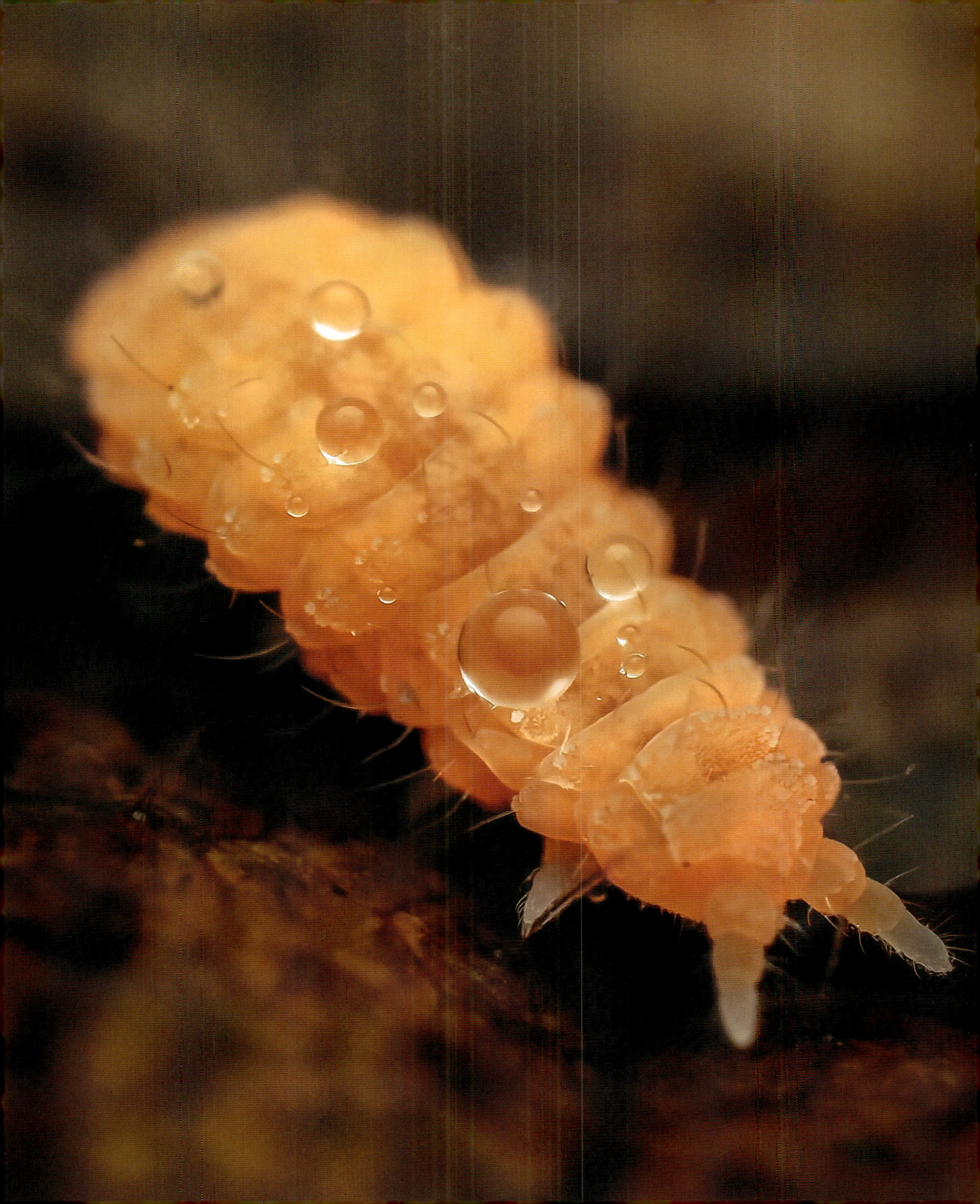

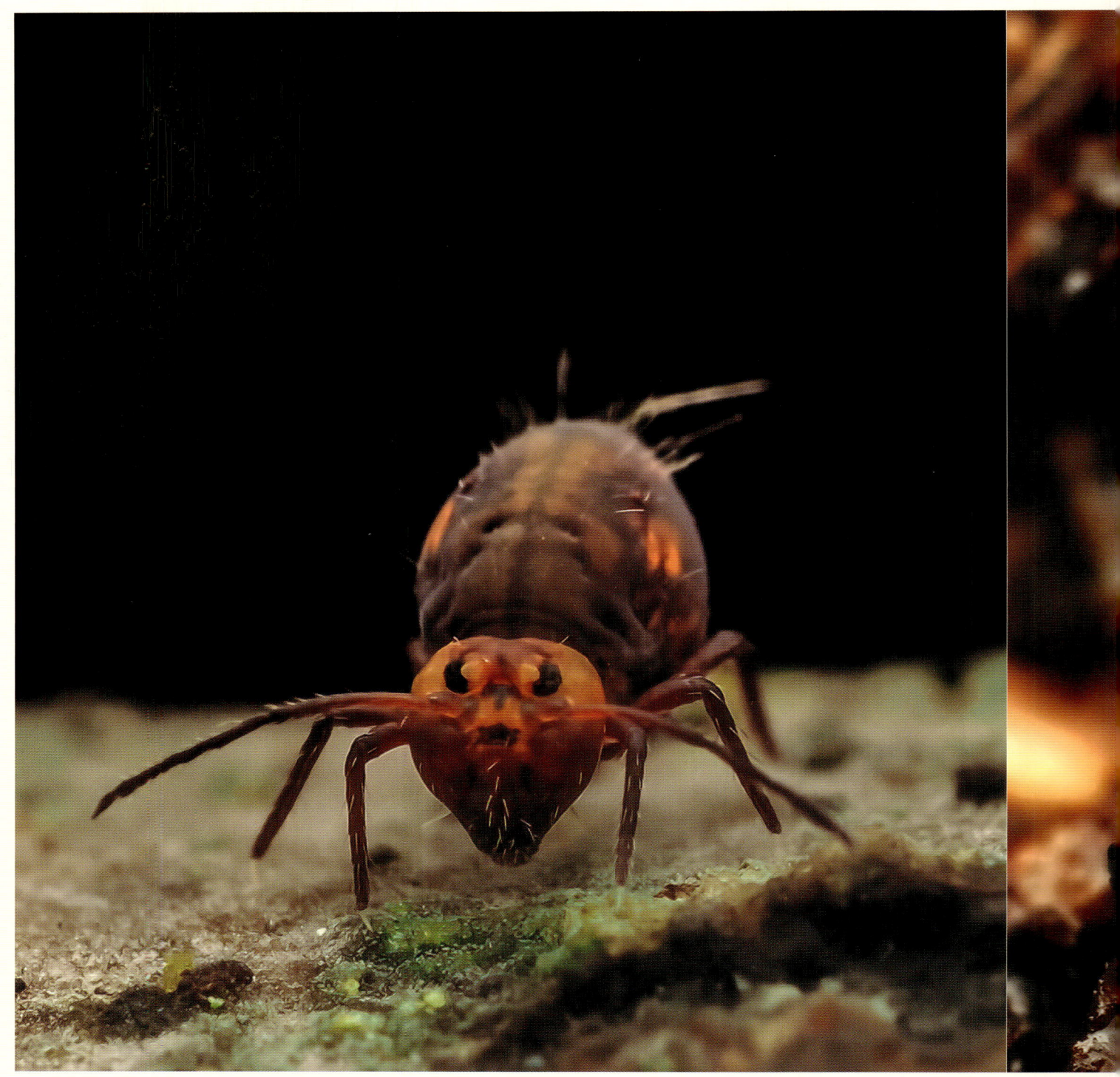

ABOVE: GLOBULAR SPRINGTAIL (*DICYRTOMINA* SP.). OPPOSITE: *ISOTOMURUS* SP.

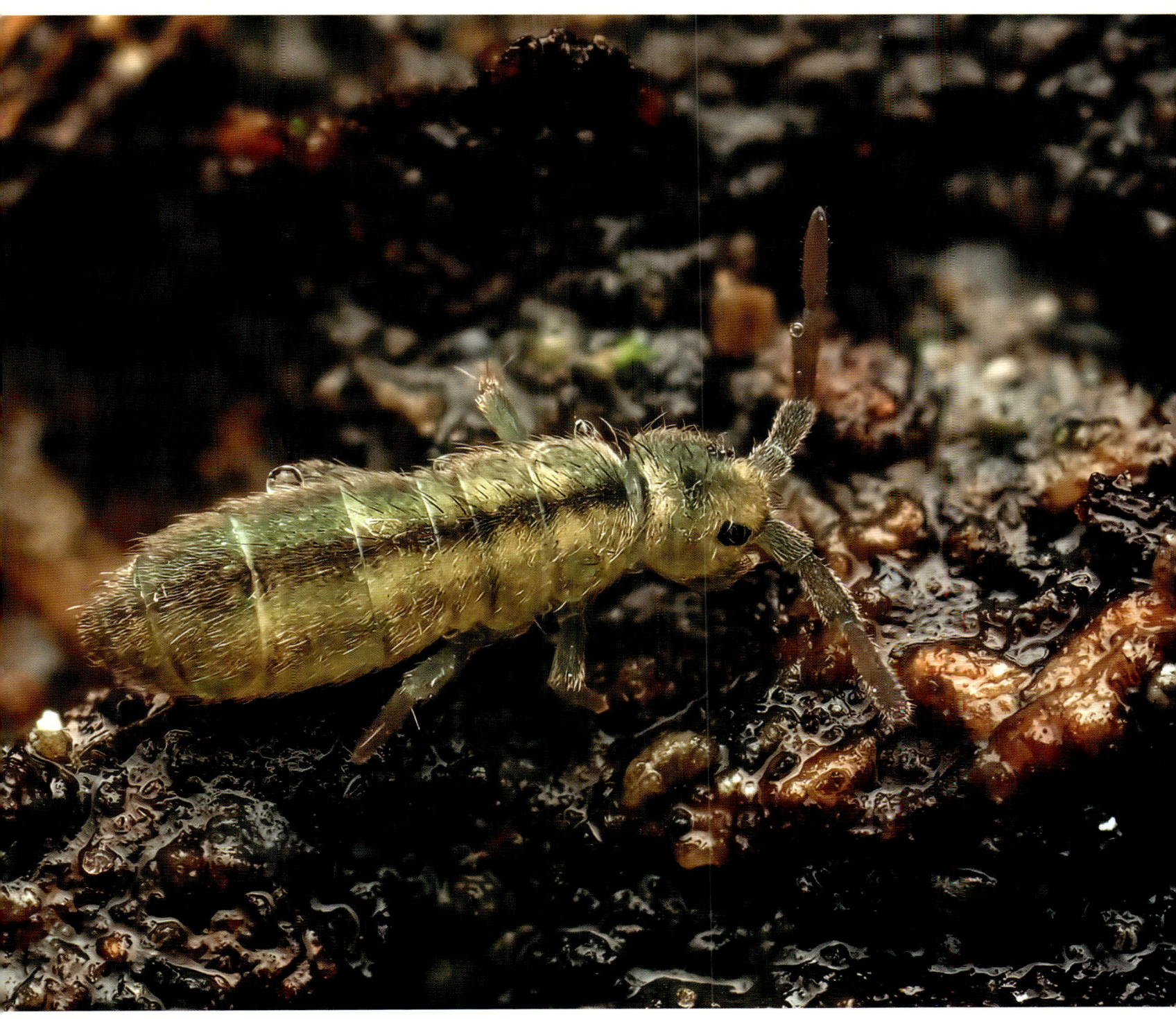

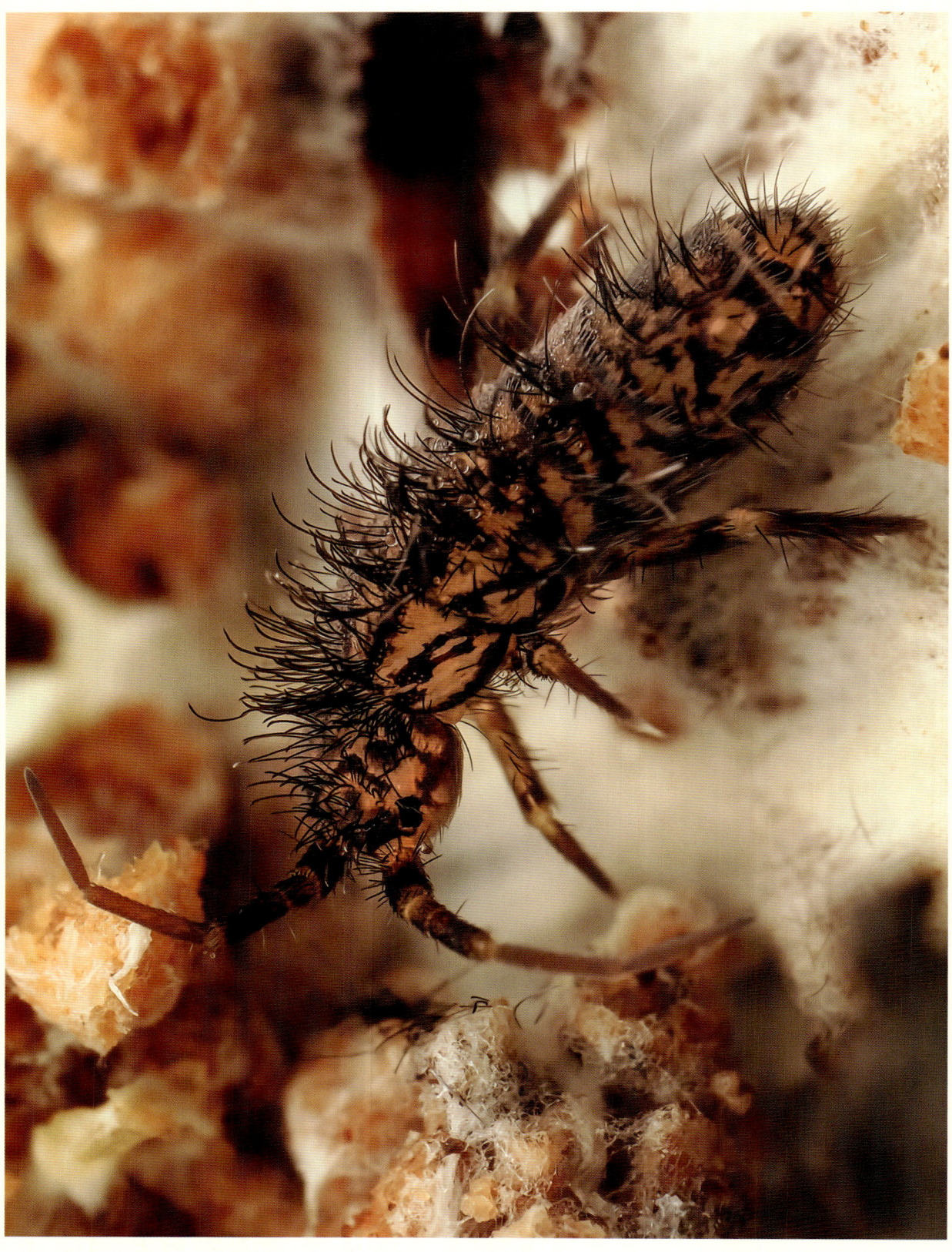

ABOVE: WOOLLY MAMMOTH SPRINGTAIL (*ORCHESELLA VILLOSA*). OPPOSITE: UNKNOWN SPRINGTAIL AND SLIME MOLD.

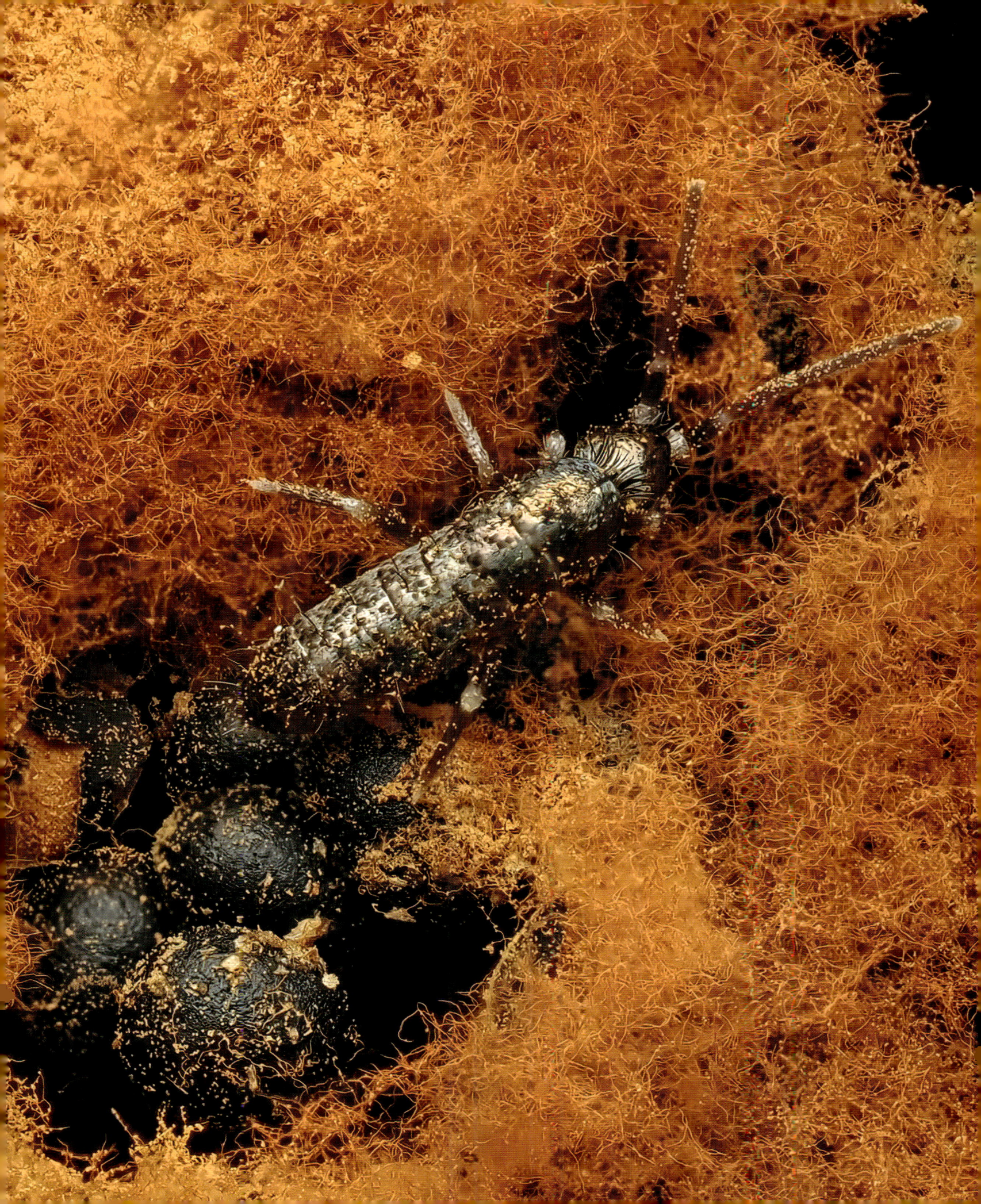

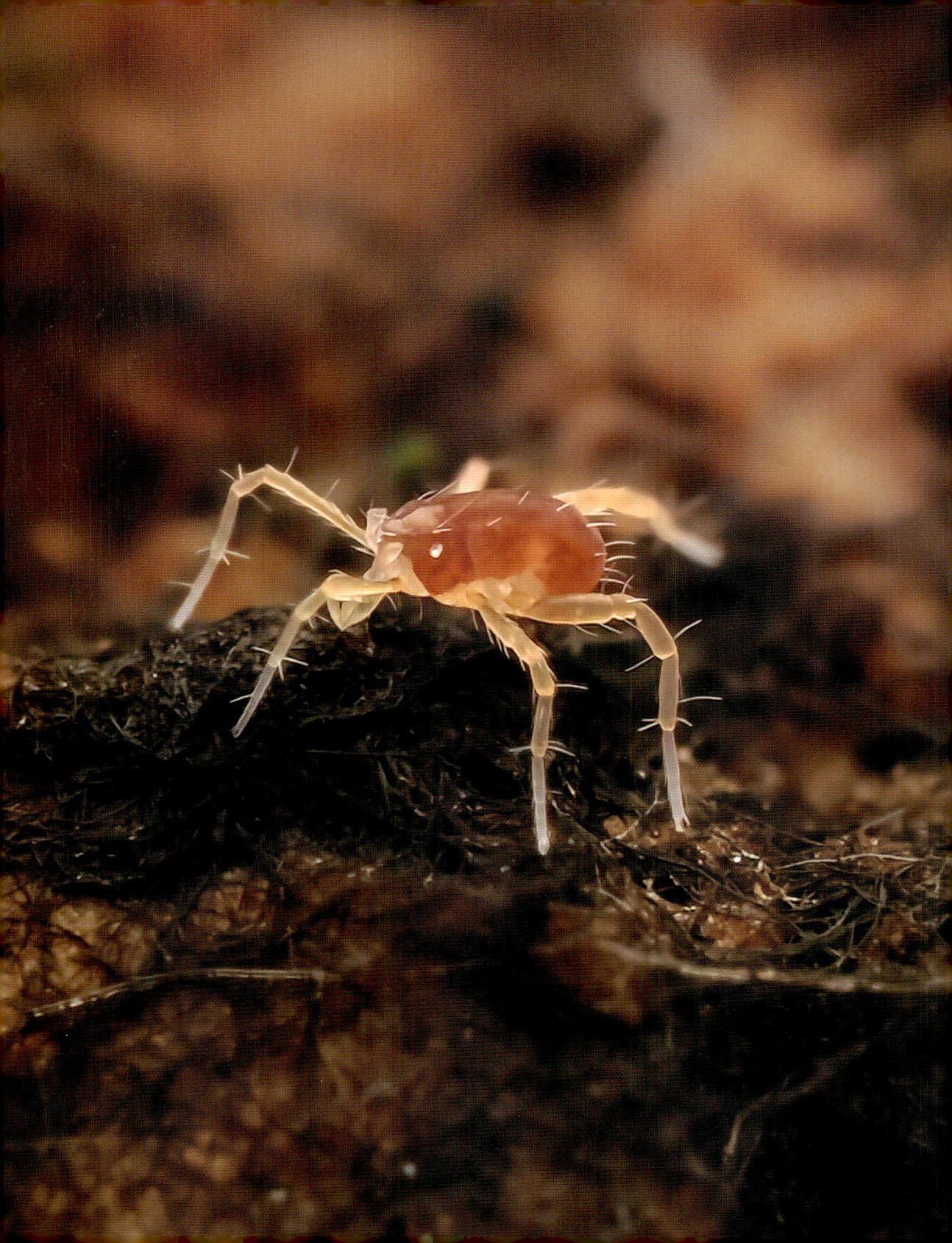

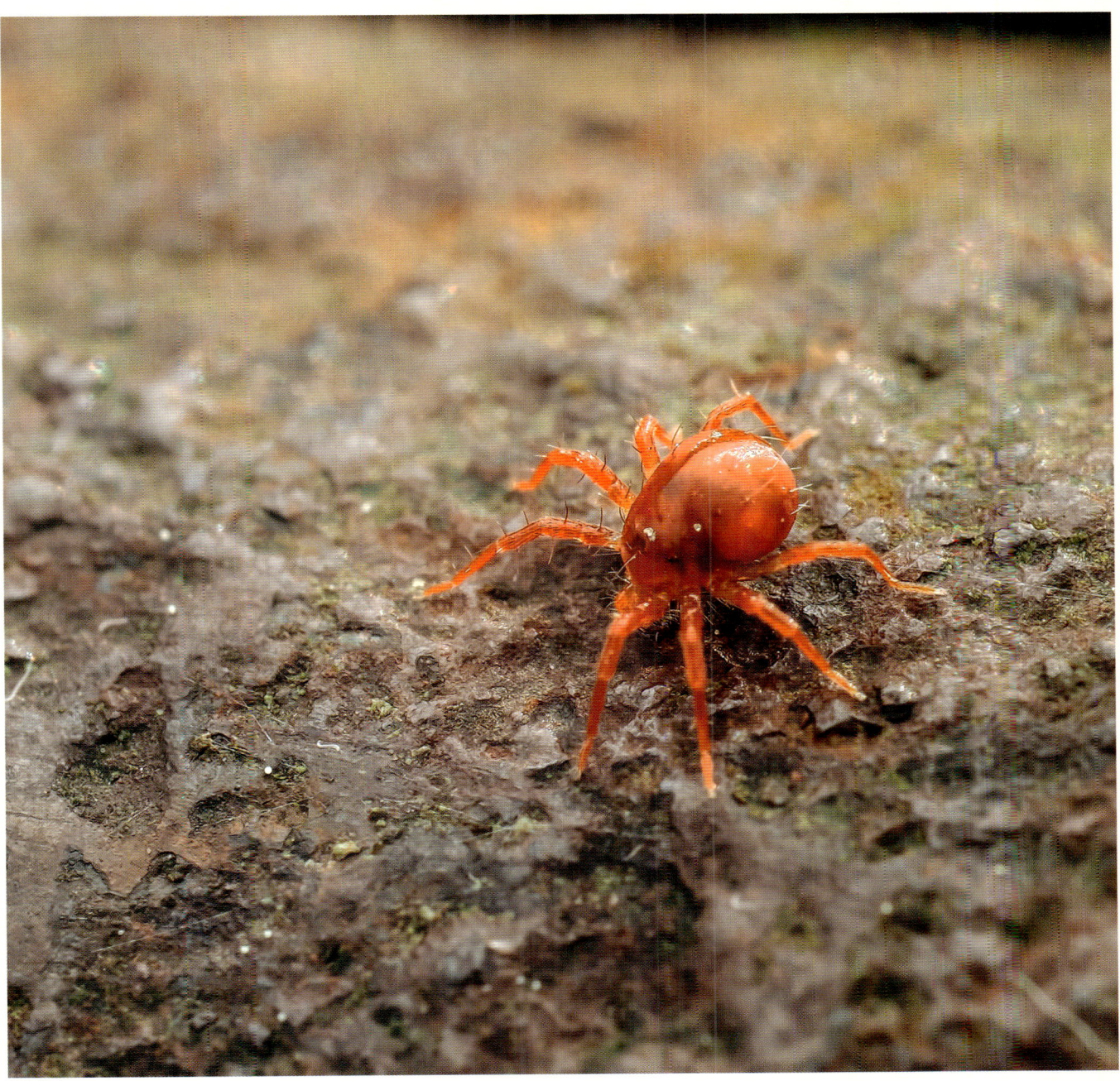

229

Mites are another key group of soil dwellers. Classified as arachnids, and with thousands of species, they come in all shapes and sizes. Some mites are decomposers, while others are predators. The Trombidiidae mite has the outward appearance of a velvety red pincushion, while the *Calyptostoma* mite looks like a walking strawberry. There are even parasitic mites that live off the fluids of animals. (Parasitic mites are my least favorite. I don't find them very visually appealing to photograph thus I generally don't bother.)

ABOVE: WHIRLIGIG MITE (*ANYSTIS* SP.). OPPOSITE: TINY MITE.

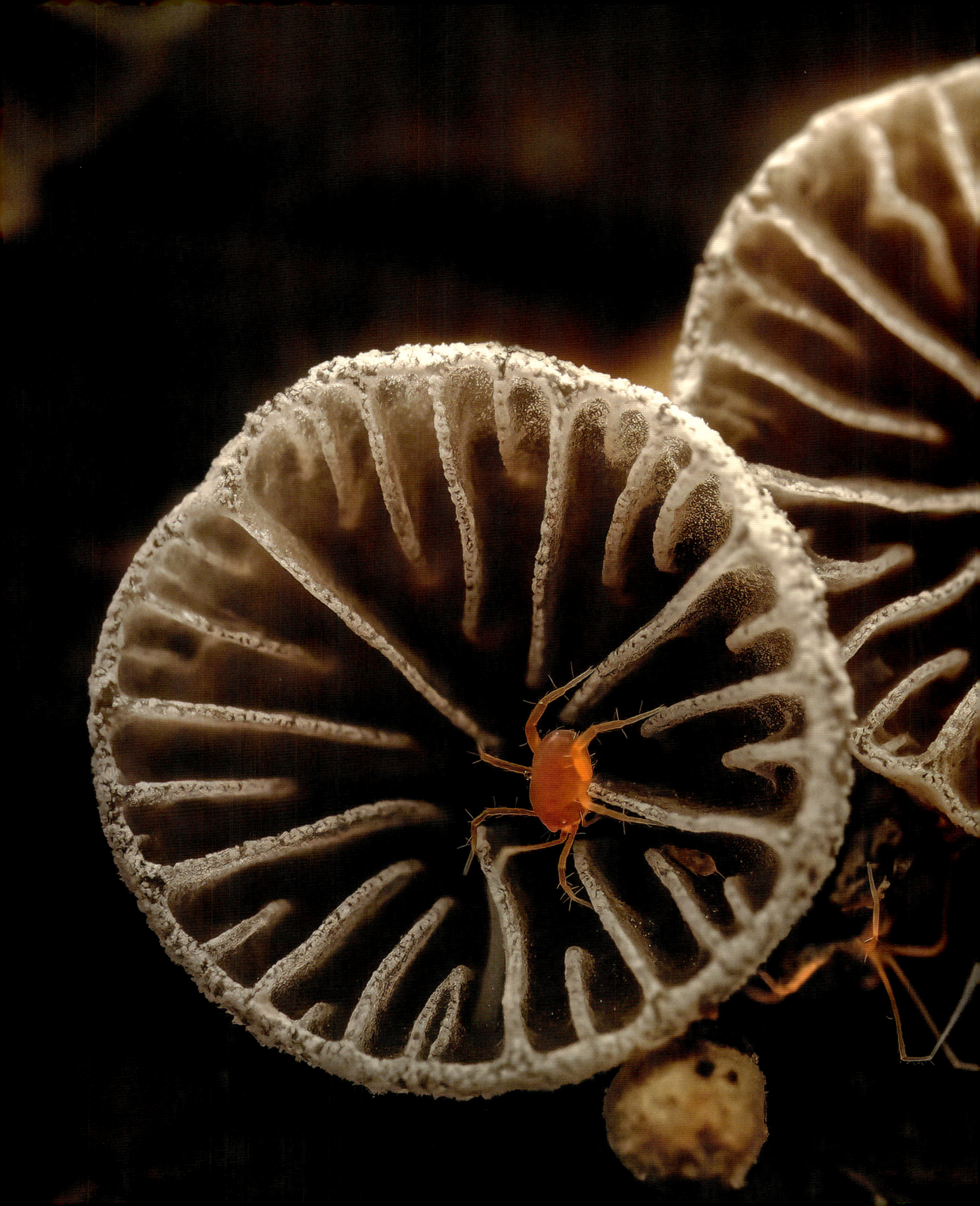

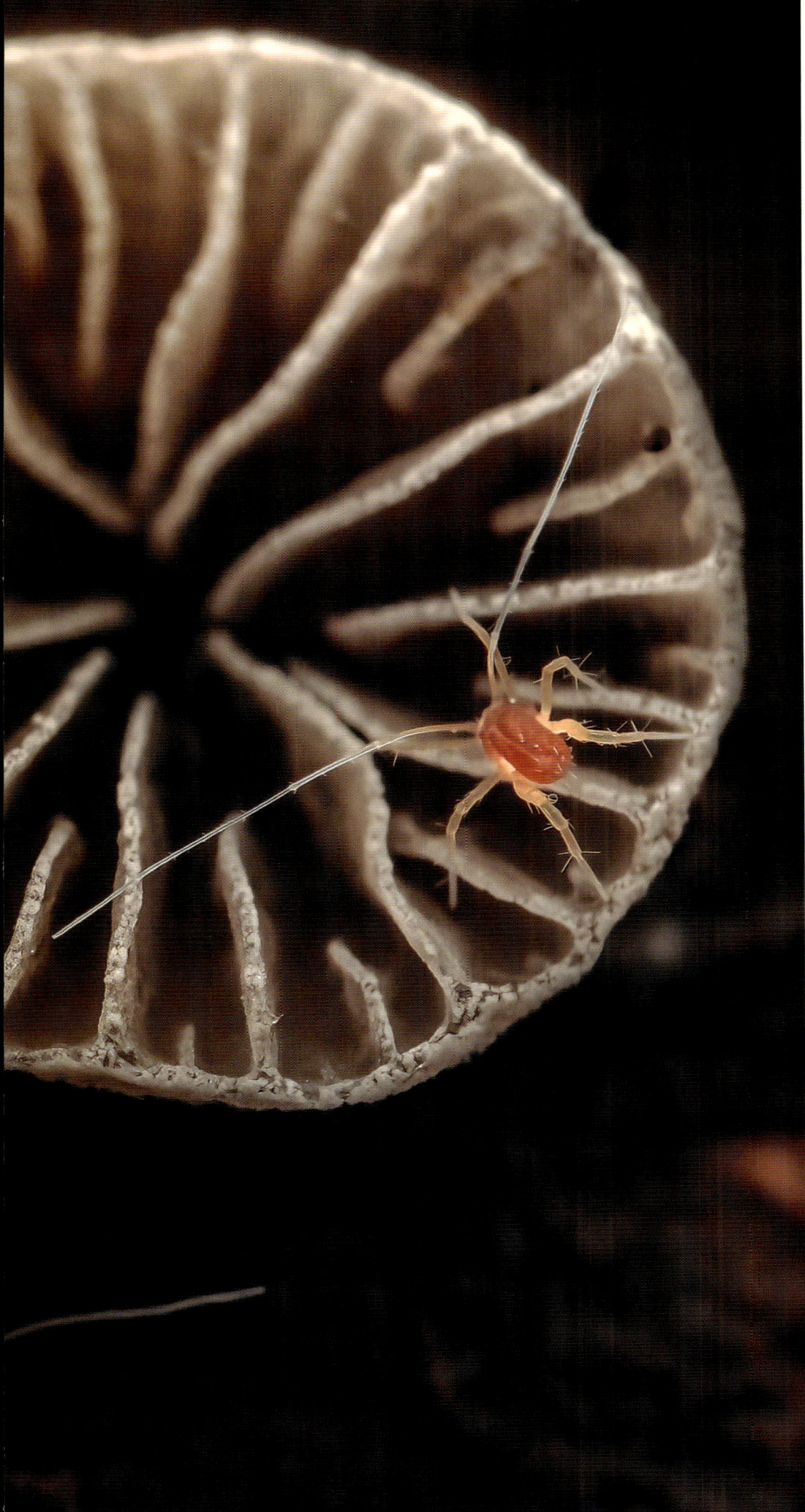

231

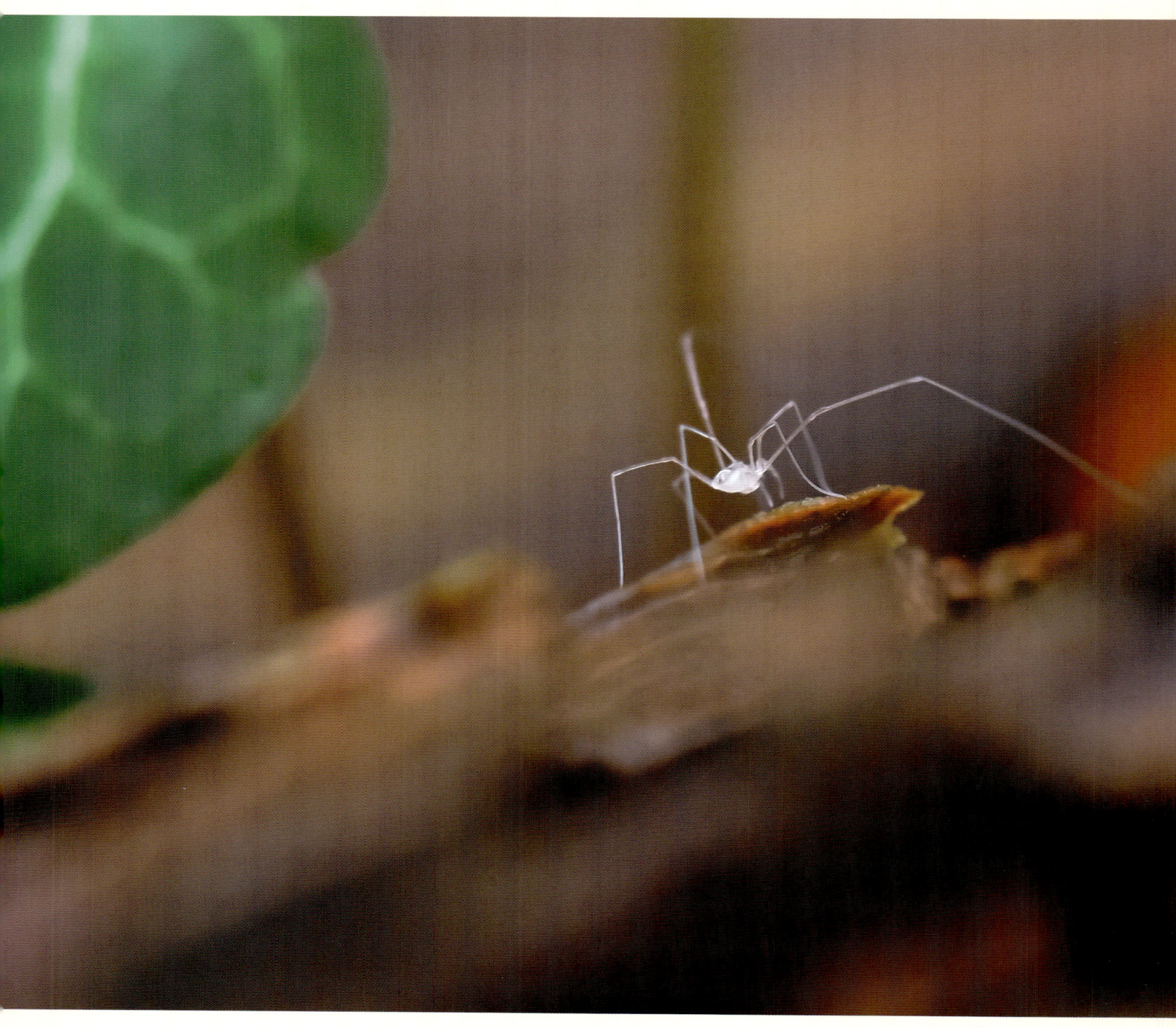

ABOVE: UNKNOWN MITE. OPPOSITE: UNKNOWN MITE ON JELLY FUNGI.

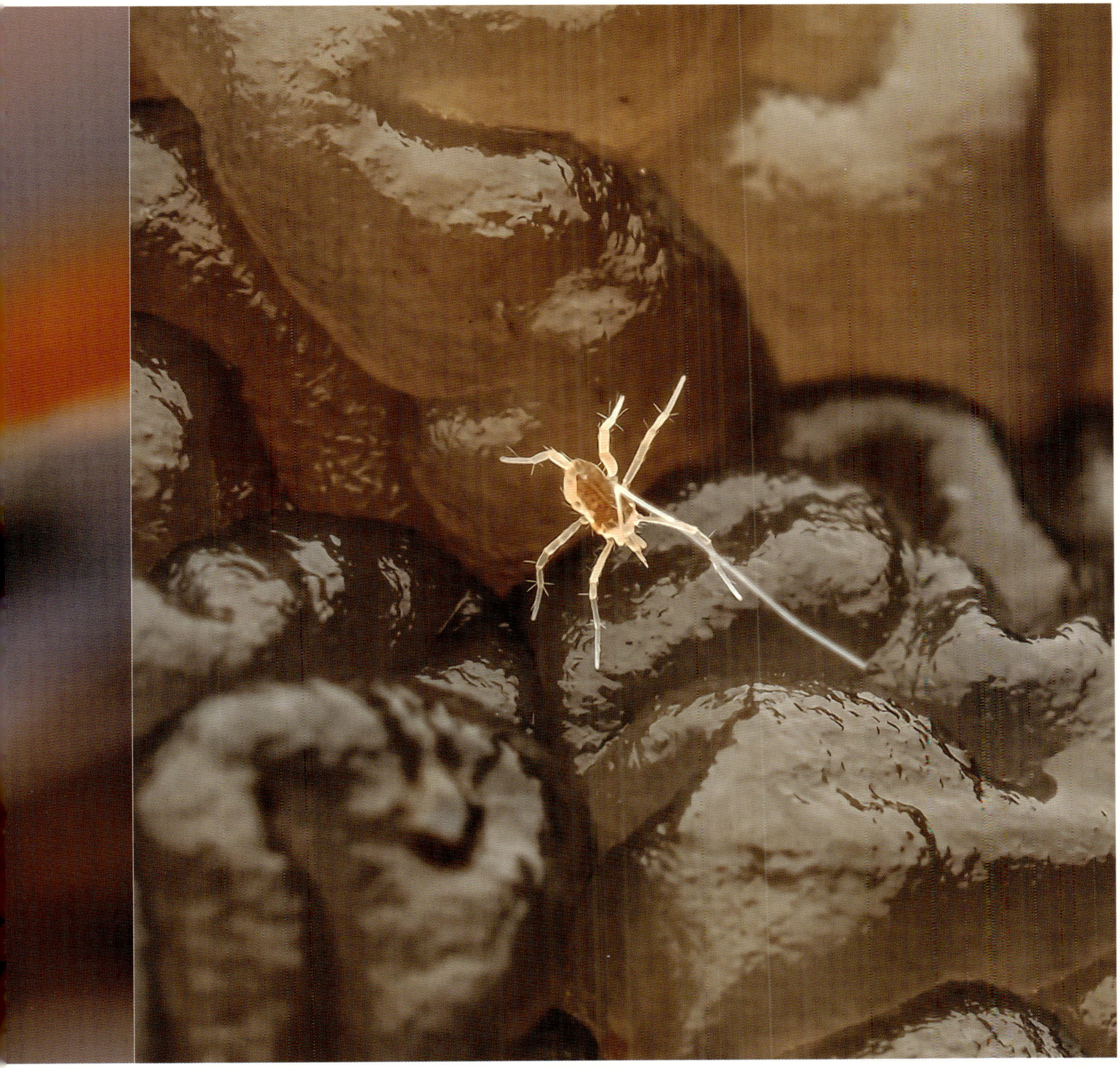

233

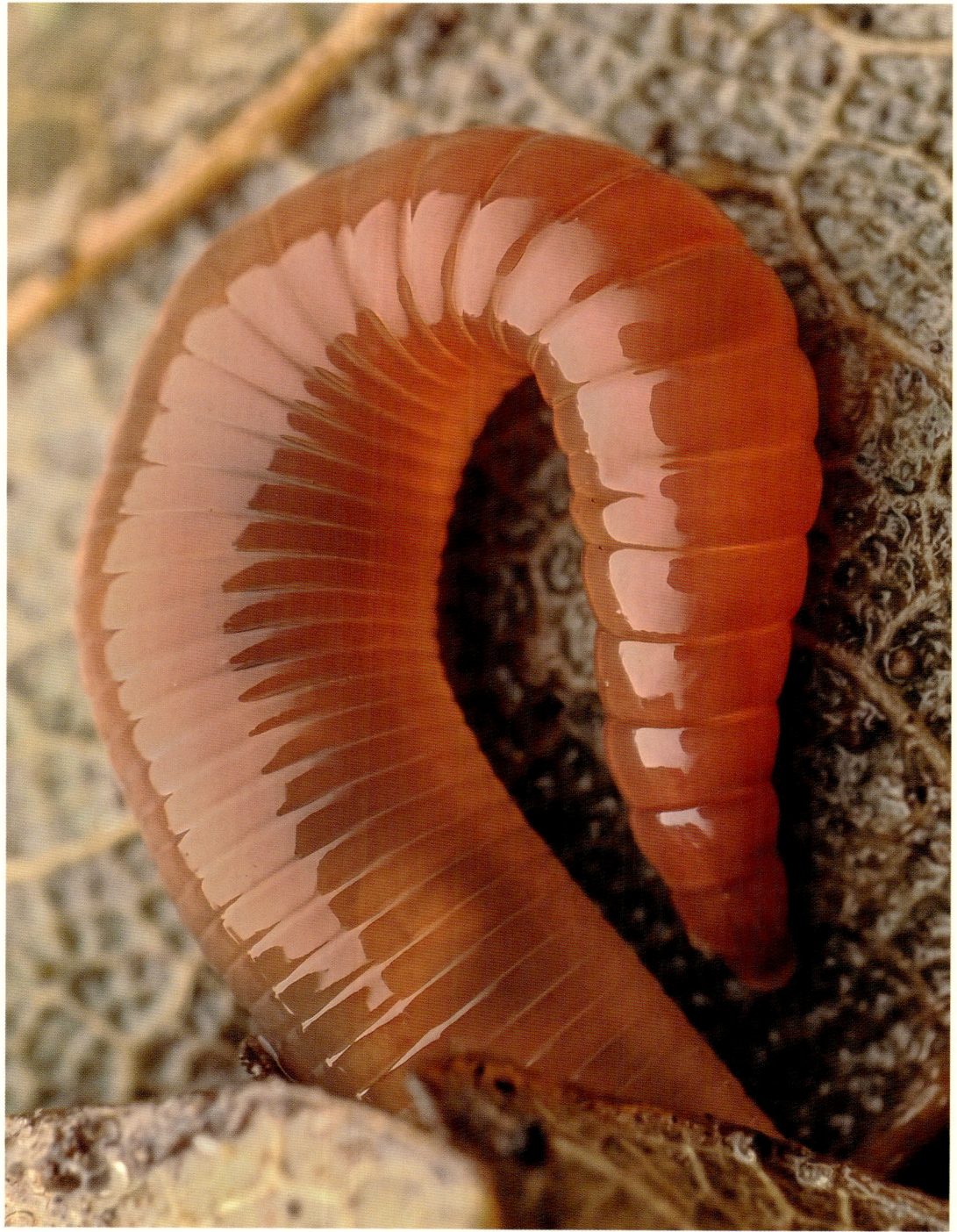

234

Earthworms are probably the most well-known soil animals. They tunnel through the soil, aerating it and improving its structure. Their poop improves soil health, and they breathe through their skin. They tend to rise to the surface of the soil when it begins to rain, much to the delight of birds!

ABOVE AND OPPOSITE: EARTHWORM.

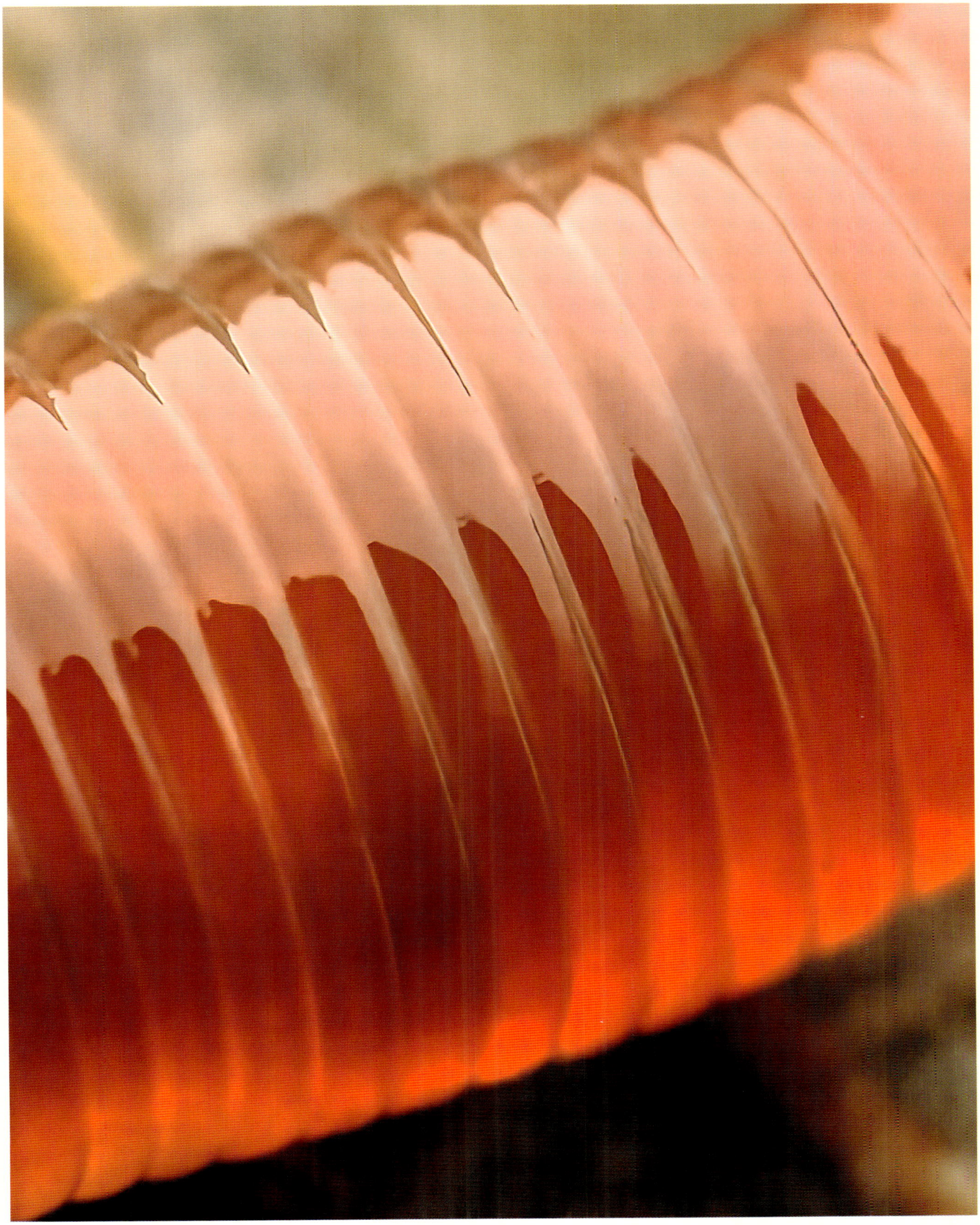

235

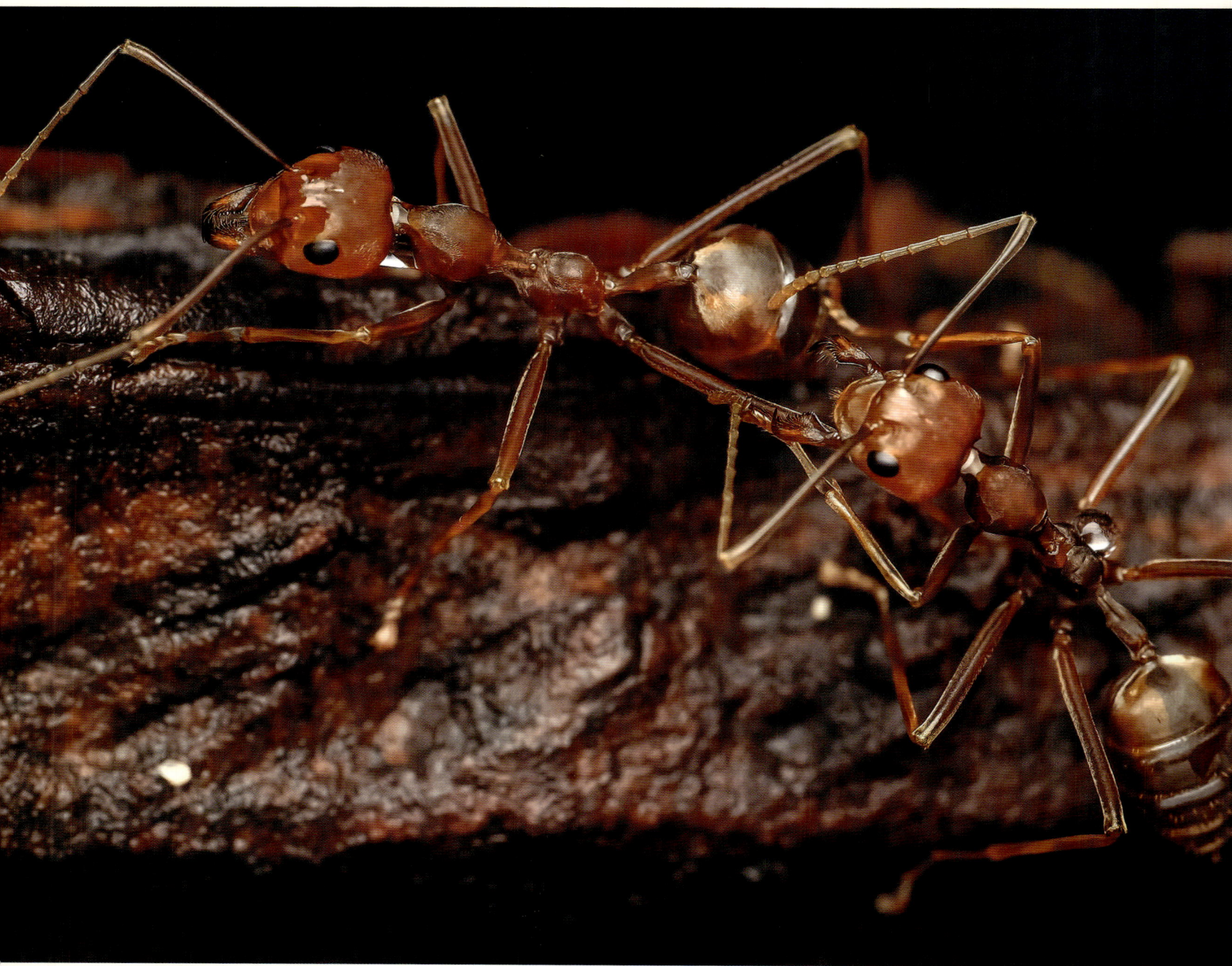

Ants build complex underground colonies and aerate the soil as they tunnel. Ants help with seed dispersal and decomposition. I don't tend to photograph ants as much, but they are amazing creatures.

ABOVE: ANTS. OPPOSITE: ANTS GUARDING APHIDS.

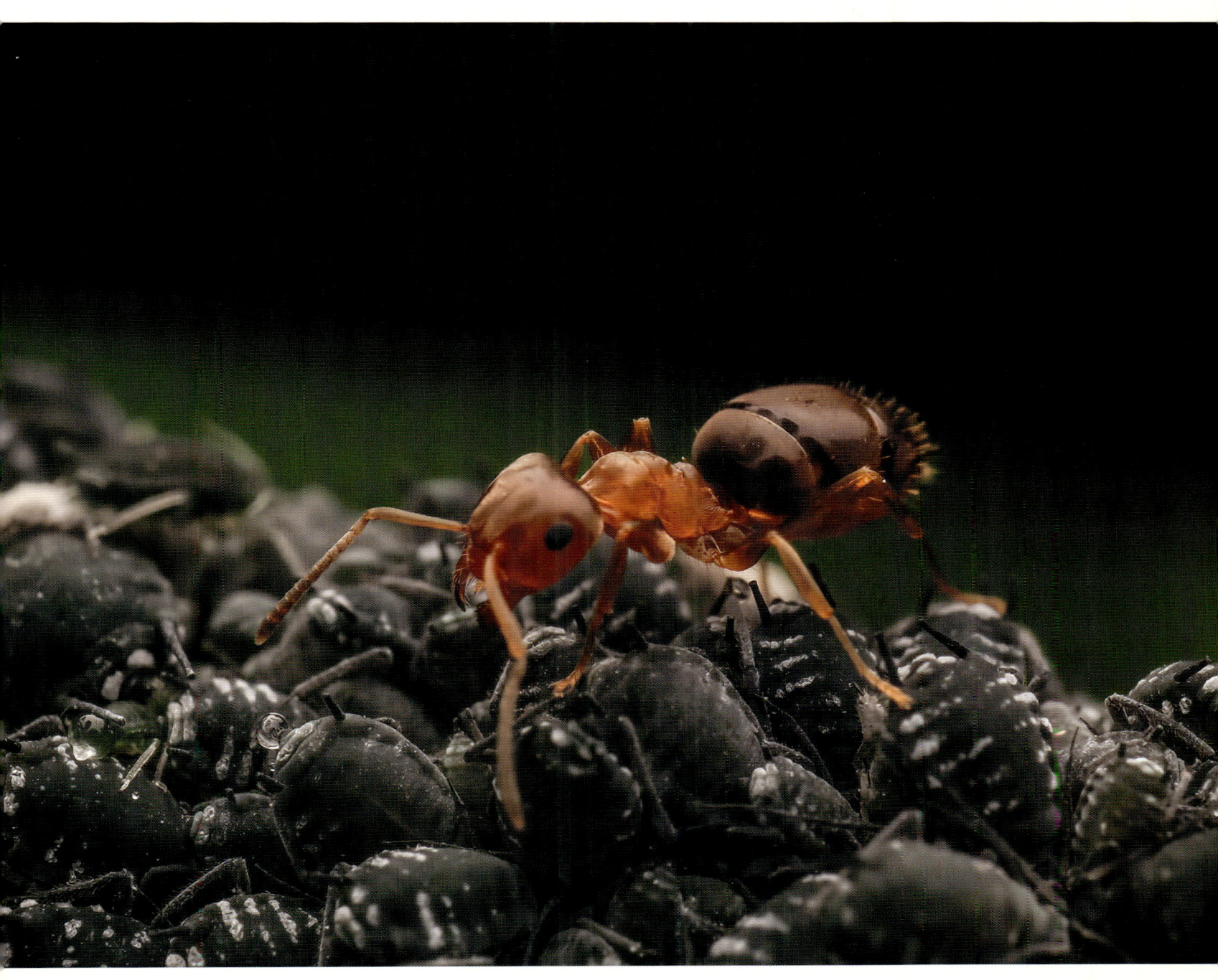

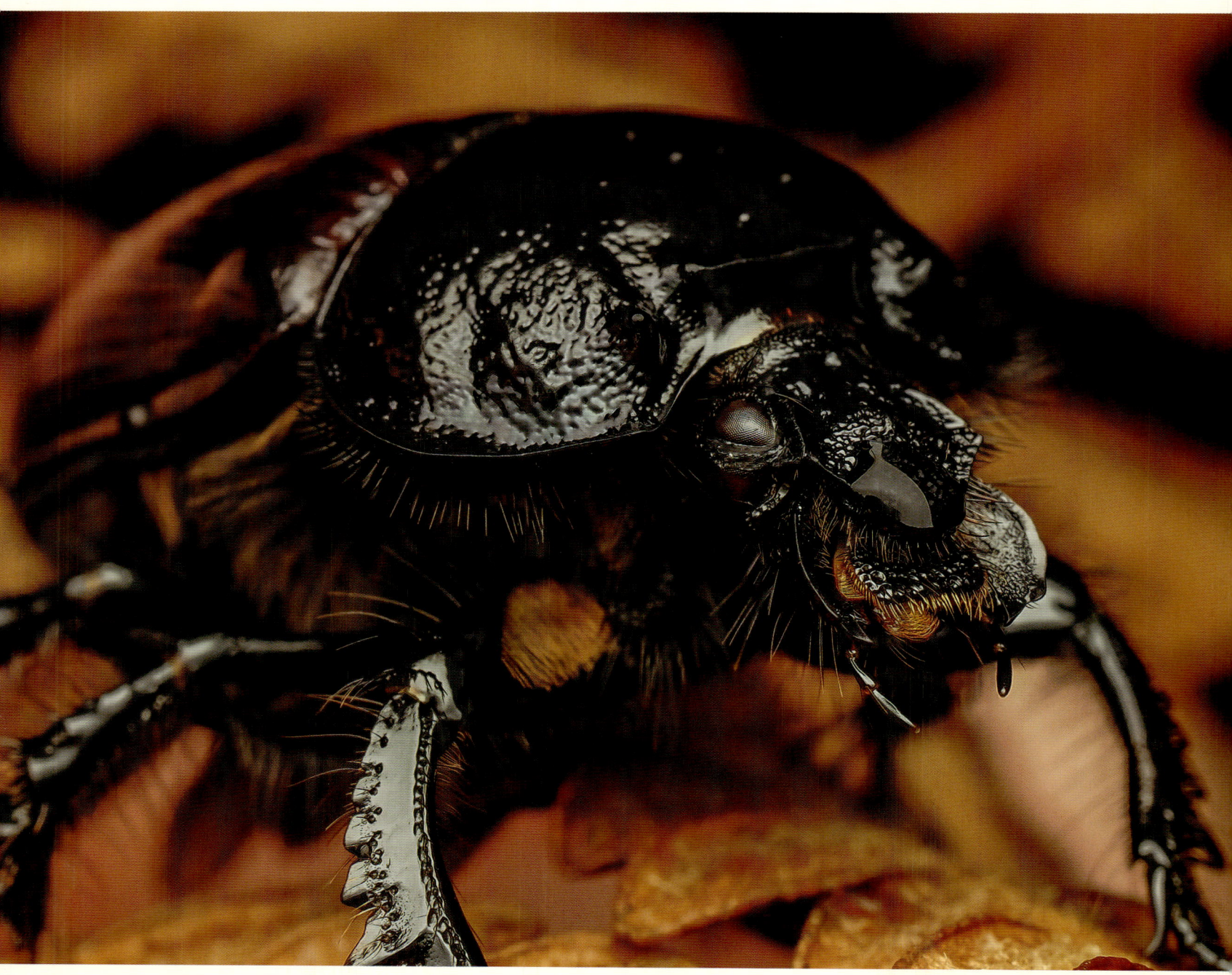

Beetles, including various species of ground beetles, are common soil dwellers. They play multiple roles in the ecosystem, both as decomposers and predators. The beetle world is vast and delightful although ground beetles might appear slightly more menacing. However, beetles

ABOVE AND OPPOSITE: DUNG BEETLE (*SCARABAEIDAE* SP.).

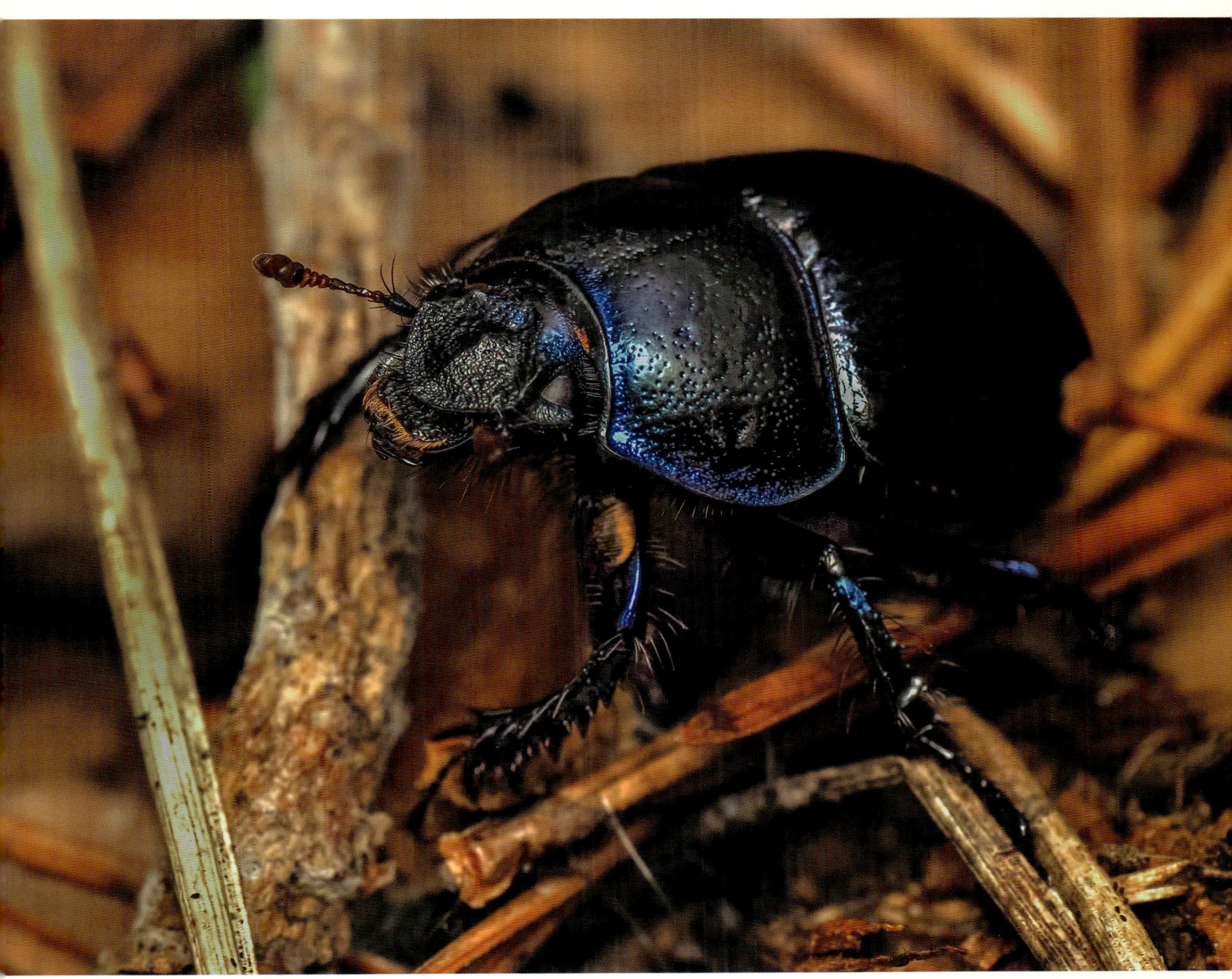

are tolerant subjects to photograph and often
display the most incredible colors.

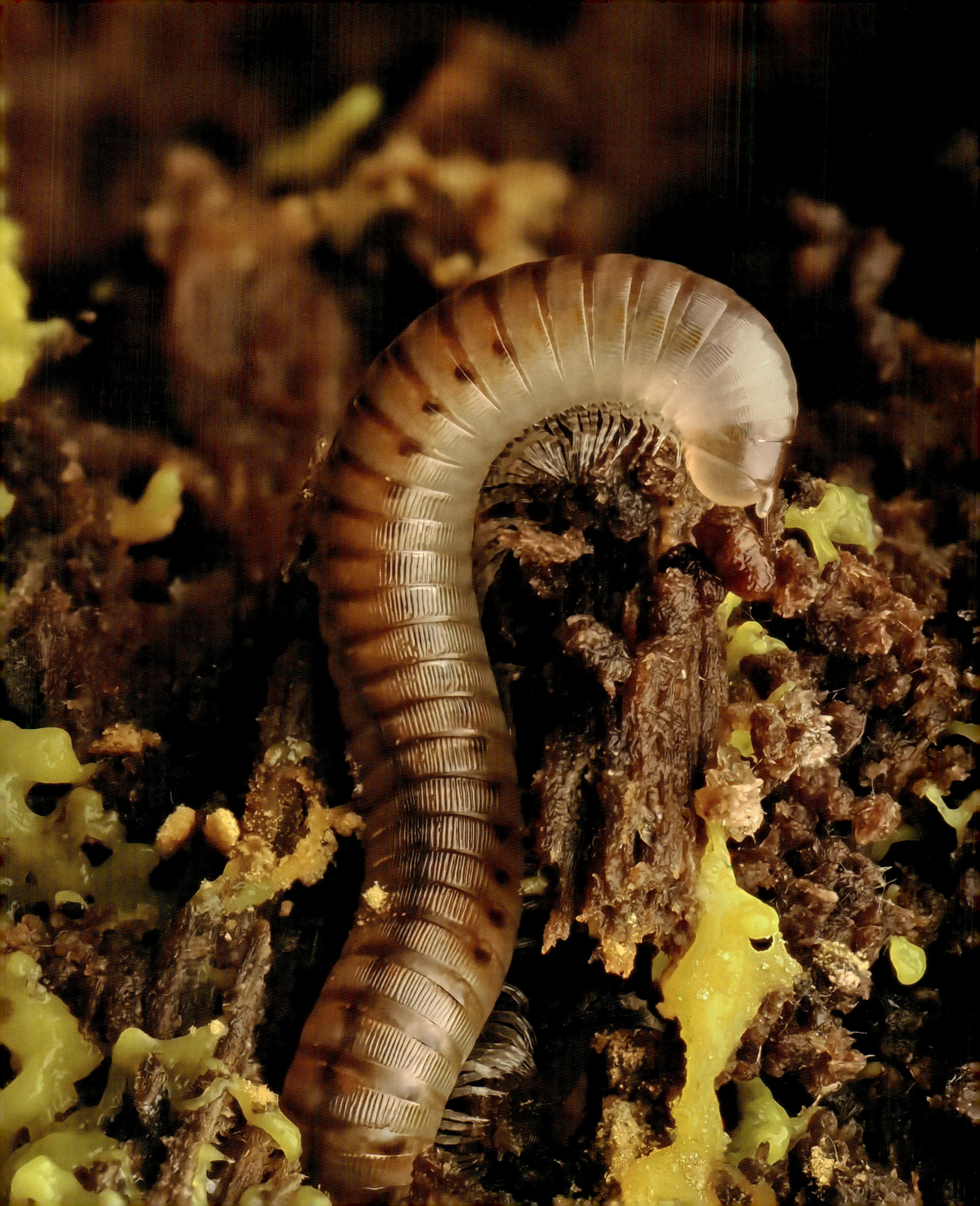

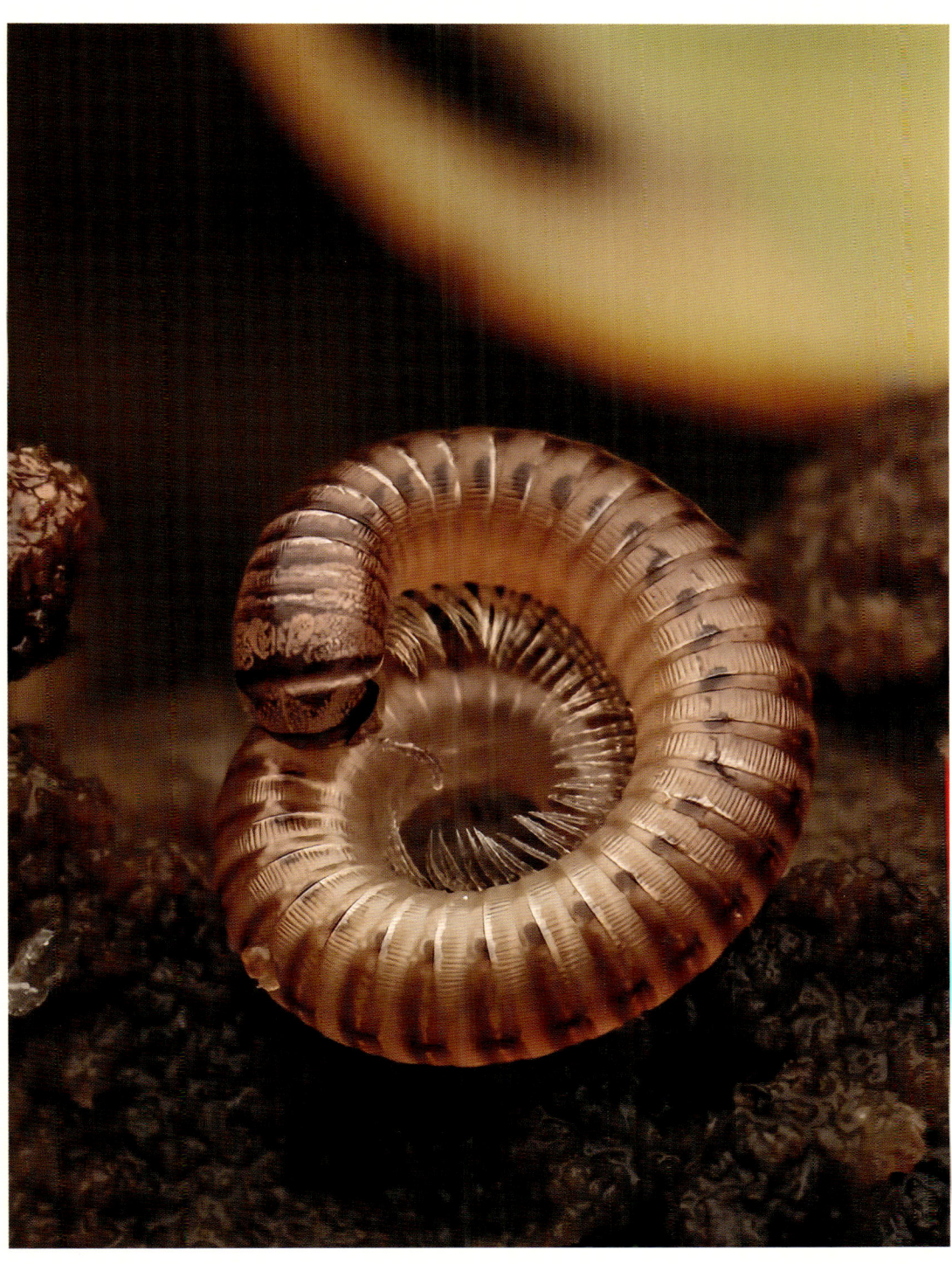

Millipedes and centipedes can be found both in the soil and under bits of wood and logs. Millipedes primarily feed on decaying plant material. Centipedes, on the other hand, are predatory. They are shy and intricately designed, almost architectural with stunning symmetry.

ABOVE: BARREL MILLIPEDE (*CYLINDROIULUS* SP.). OPPOSITE: MILLIPEDE AND PLASMODIAL SLIME MOLD.

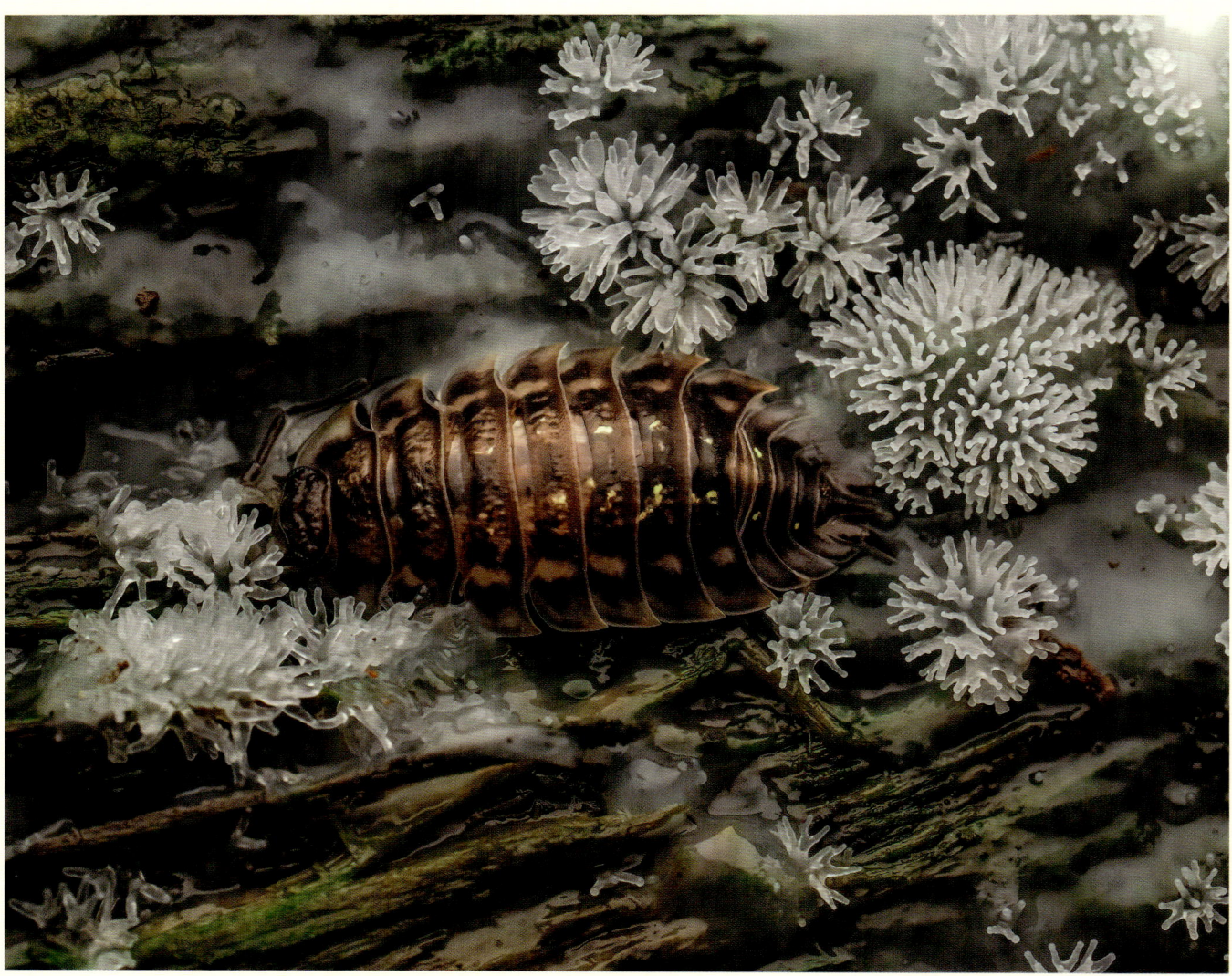

242

Isopods, which include the humble woodlouse and *Armadillidiidae* "Roly-Poly" are actually crustaceans (say what!?) that live in damp soil and leaf litter. Incredibly common and abundant in the UK, you can still get some really interesting behavioral composition shots if you're lucky. They are also herbivores and have become quite popular terrarium pets.

Soil animals' small size and interesting details make for captivating close-up shots but make photographing them more challenging, especially if you are doing natural light photography as much of the year the forest is a pretty dark place. But there are lots of other things that can be found in the soil (such as arachnids, larvae, nematodes, wasp galls, and other insect eggs, to name a few). Soil animals are great subjects for anyone getting into macrophotography for the first time as they are always around, typically slow moving (especially in the winter months), and helpful to practice your macro skills on.

ABOVE: WOODLOUSE (*ONISCUS ASELLUS*) SLEEPING IN COMMON CORAL SLIME MOLD (*CERATIOMYXA FRUTICULOSA*). OPPOSITE: WOODLOUSE AND *ARCYRIA* SPP.

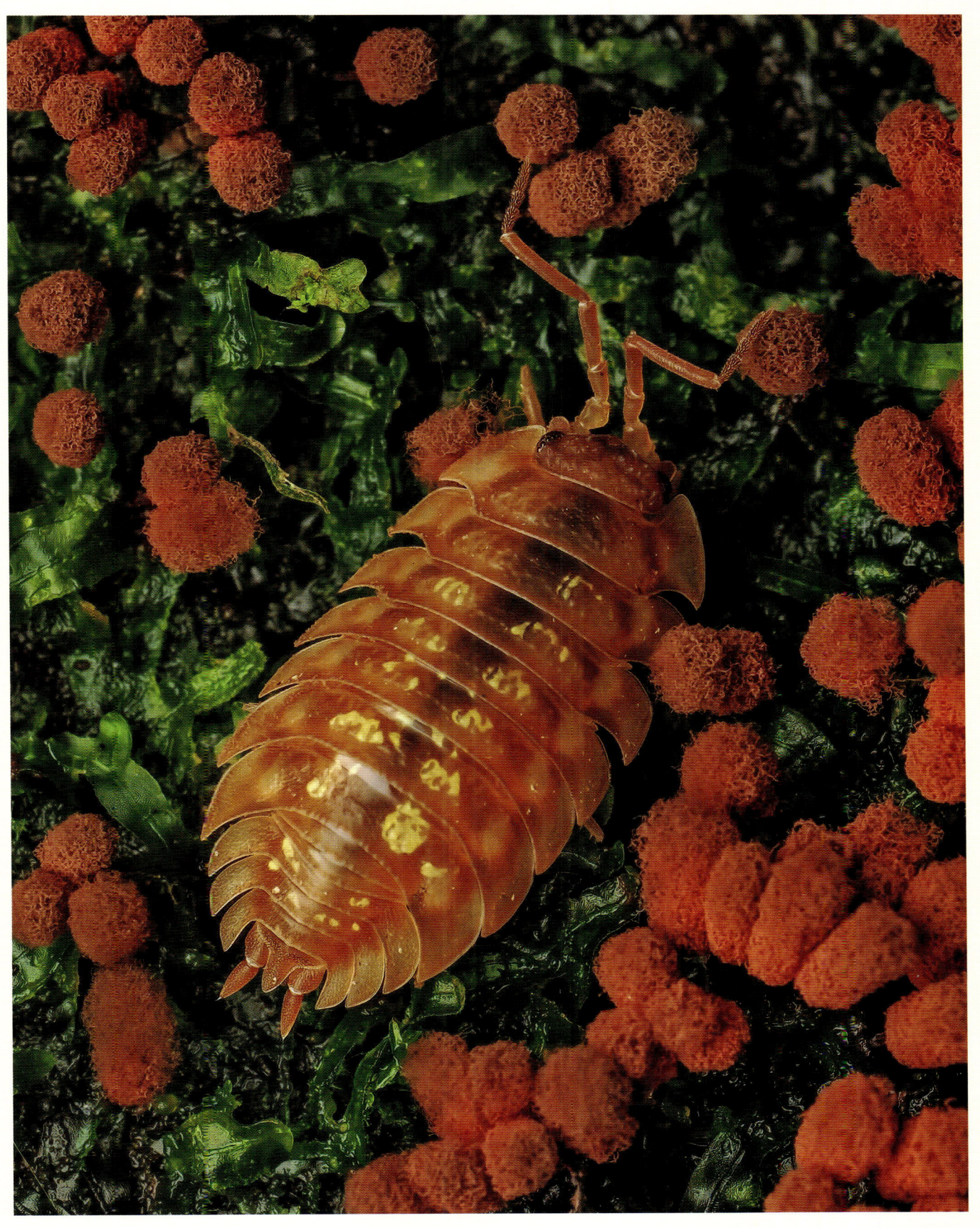

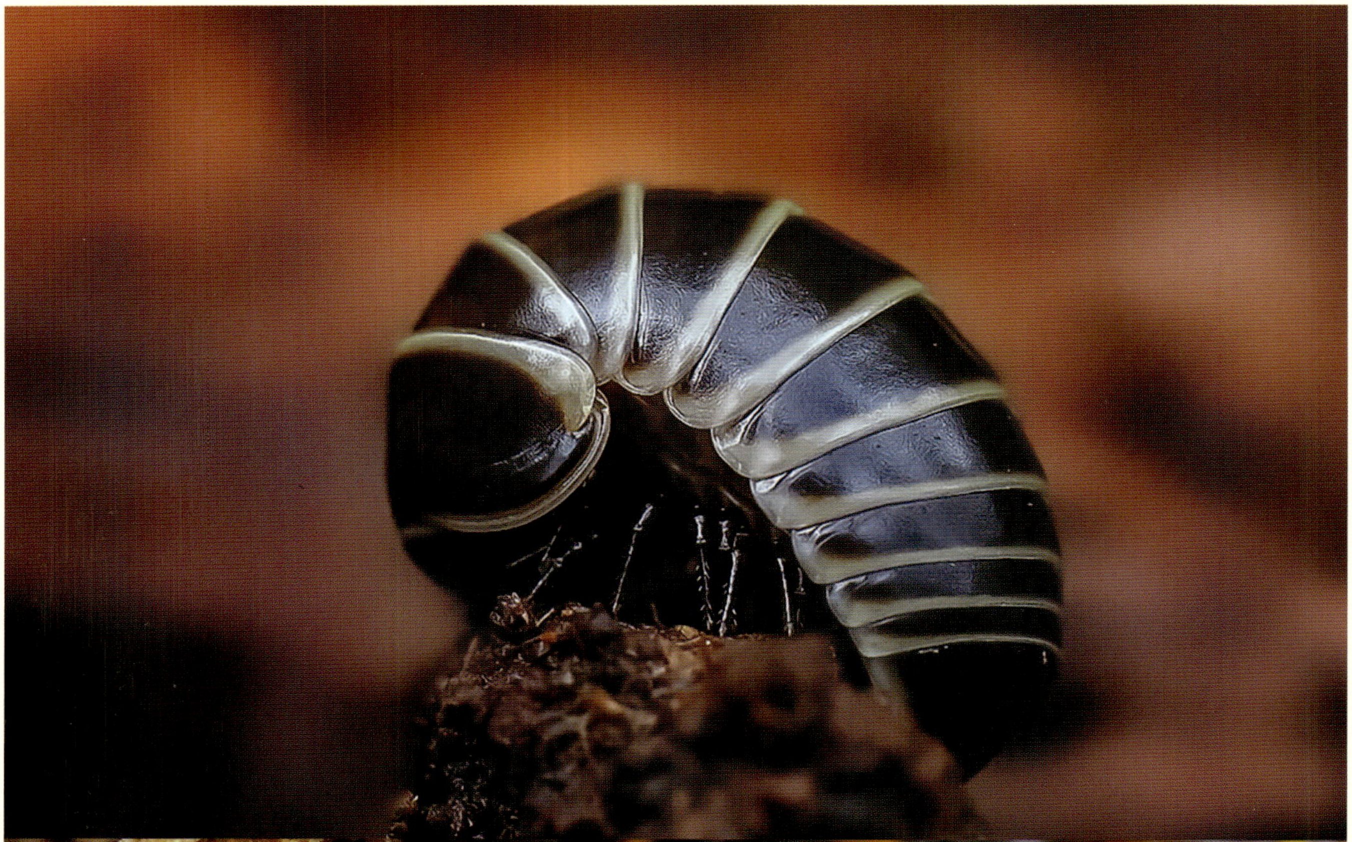

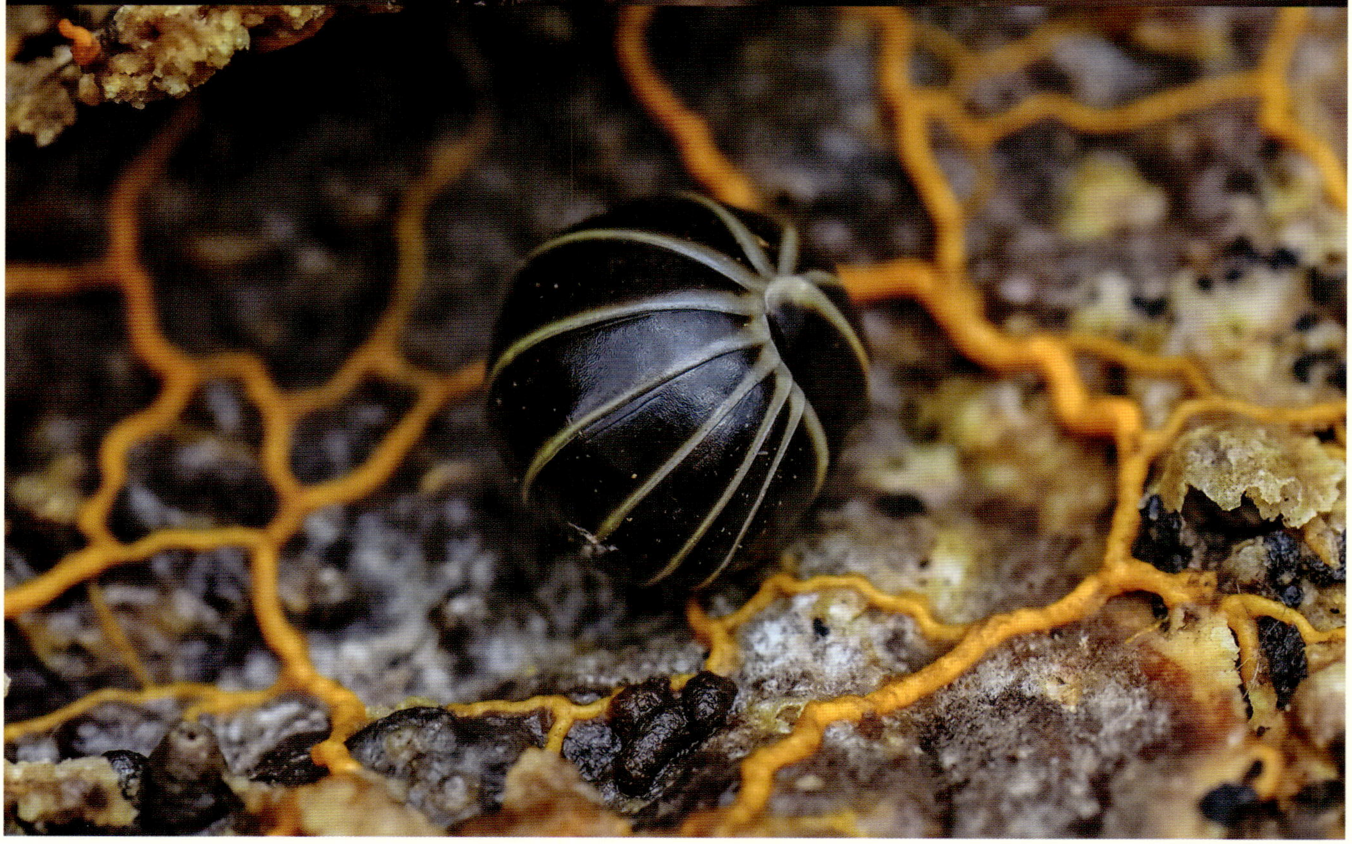

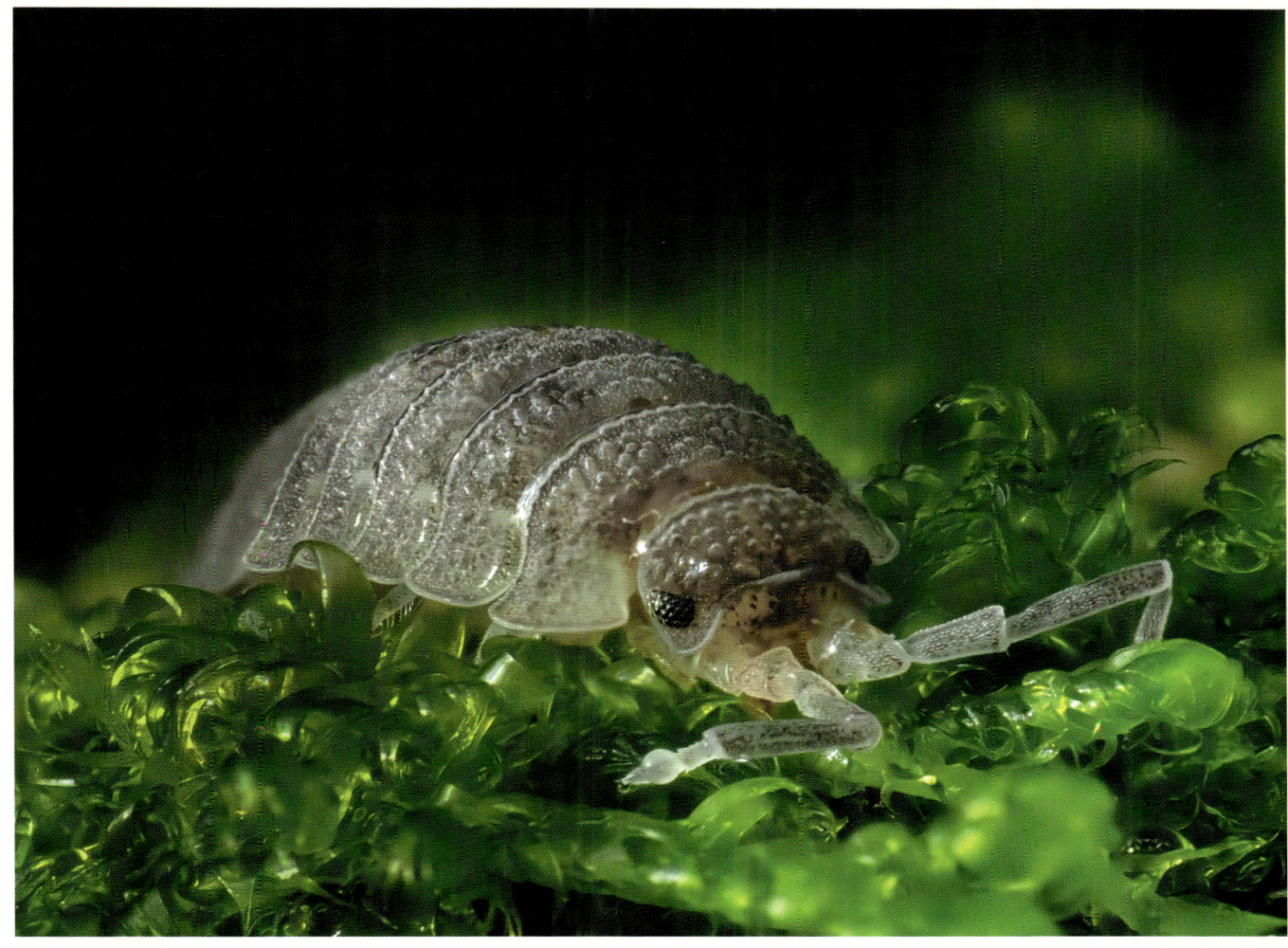

ABOVE: WOODLOUSE (*ONISCUS ASELLUS*) IN MOSS. OPPOSITE ABOVE: PILL MILLIPEDE (*ONISCOMORPHA* SP.).
OPPOSITE BELOW: PILL MILLIPEDE AND SLIME MOLD.

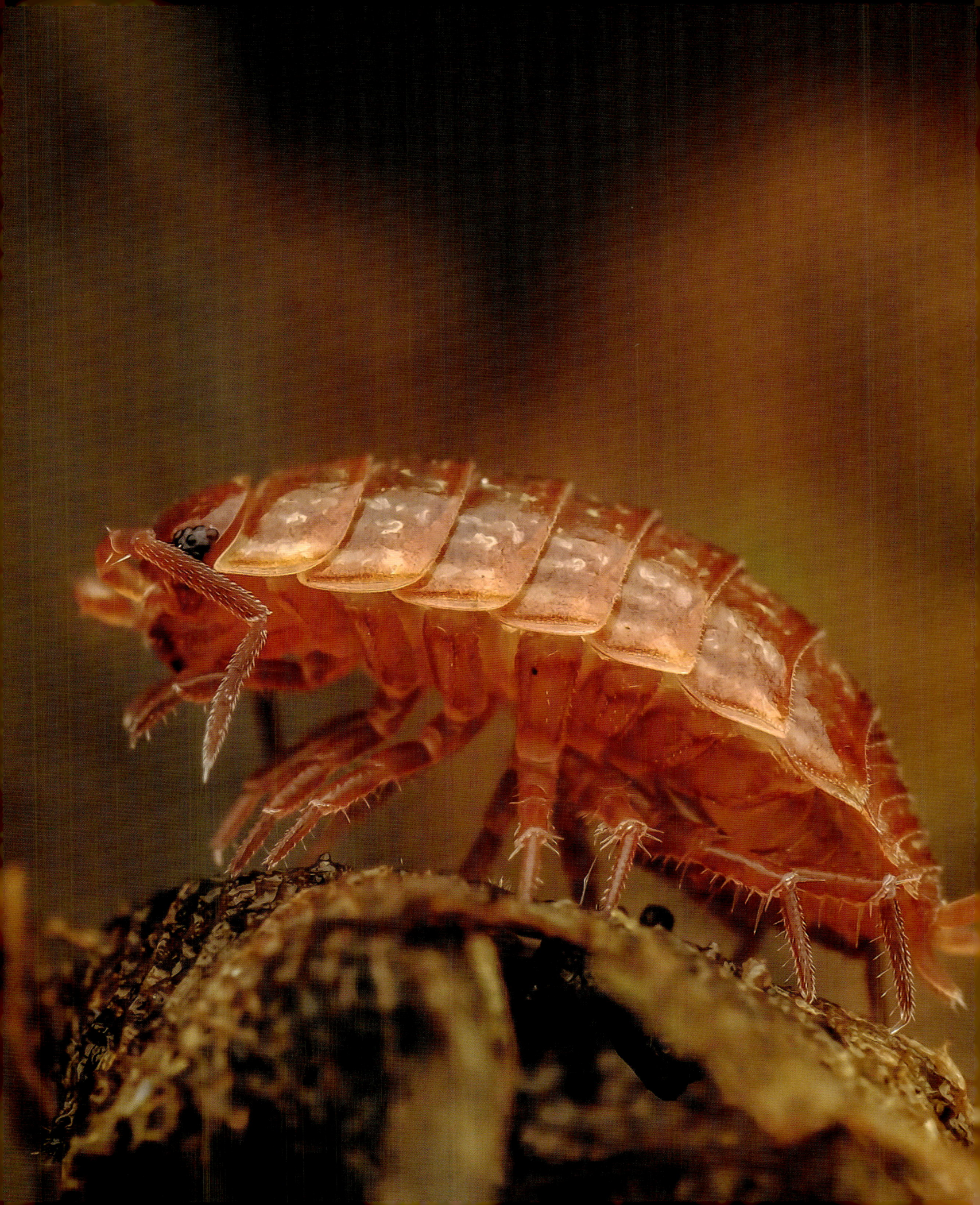

Conclusio

The moral of the story is that there's sor
profoundly humbling and dare I say *hea*
about the forest floor. The forest floor
bustling metropolis of unique organisms, each
crucial role in sustaining the ecosystem. It's a m
overlooked landscape brimming with miniature
many of us never pause to notice.

Please take a walk through any ancient wood
UK—yes, even in urban sprawl like London—an
you can find. London contains over two thous;
woodlands, each one a sanctuary of biodiver:
within the city's chaos. The leaf litter, rotting lo
are a planet on their own where an endless me
survival is taking place. Mushrooms literally pop u
slime molds crawl in creepy slow-motion ball
these beautiful misunderstood tiny critters go
business, oblivious to the giant humans stomping

I can't help but think that macrophotograph
my passport to this kingdom. I am now min
environmental impact, and my love of creating ar
have reawakened. ♋

OPPOSITE: WOODLOUSE (*ONISCUS ASELLUS*).

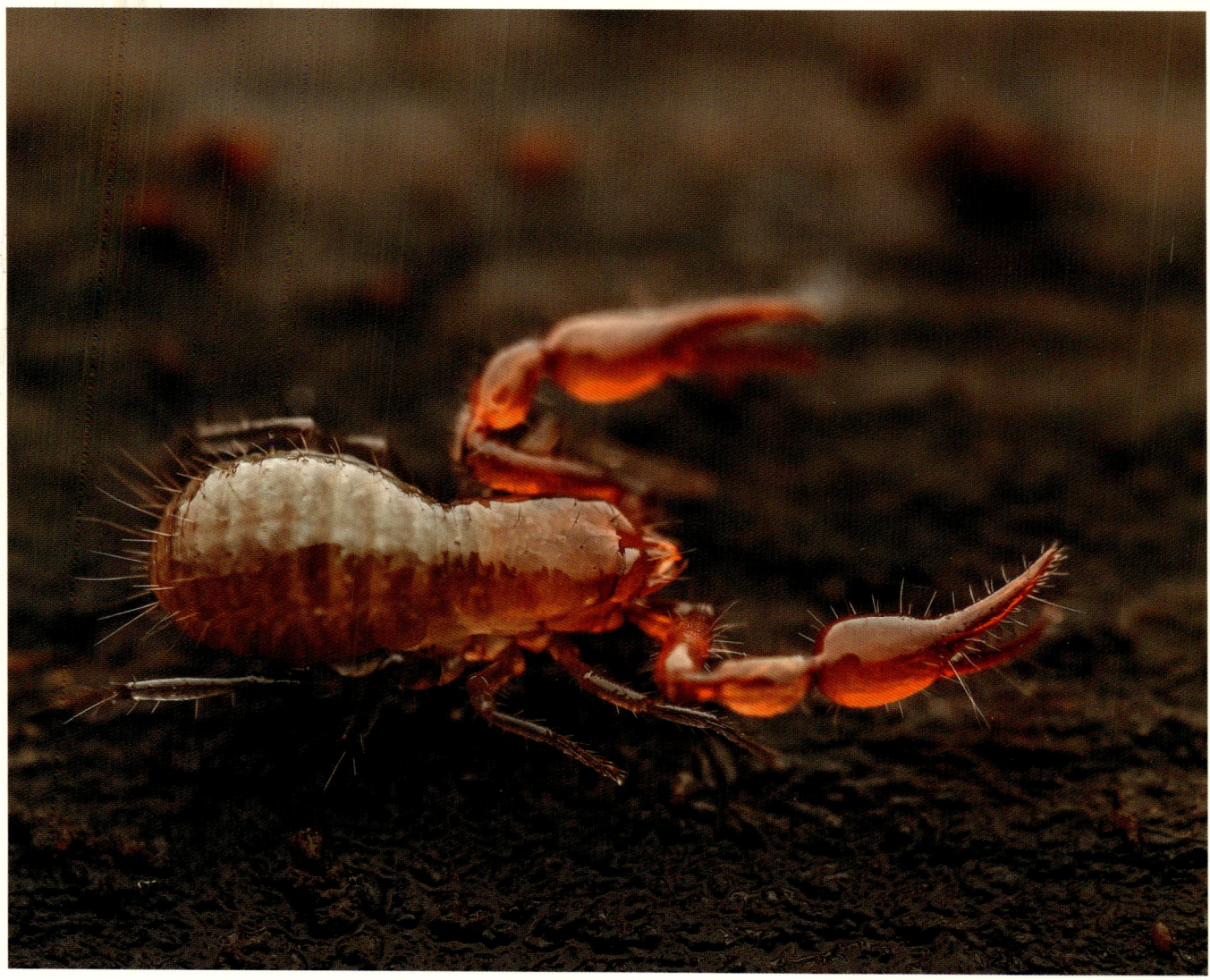

248

I am a better citizen of the Earth. I'm patient, I go out of my comfort zone, and my resolve and determination have strengthened. It's also given me a deeper appreciation for the small things in life, reminding me that every tiny organism has its place and purpose. It's humbled me, frustrated me, and got me through some dark days and sad times. Without this, I'd risk a life of stagnation, losing touch with what really matters and what lights me up as a human.

I hope through my photographs some of this resonates, challenges, and fascinates you. The forest floor isn't just a carpet of leaves and dirt; it's a vital, vibrant world teeming with life that deserves our respect and care. The next time you walk through an urban woodland or leafy nature reserve, slow down a bit, pay closer attention to what you see, and remember: There's an entire tiny universe beneath your feet just waiting to be discovered.

ABOVE: PSEUDOSCORPIONS (*PSEUDOSCORPIONES* SP.). OPPOSITE: INSECT EGG CASINGS IN LEAF LITTER.

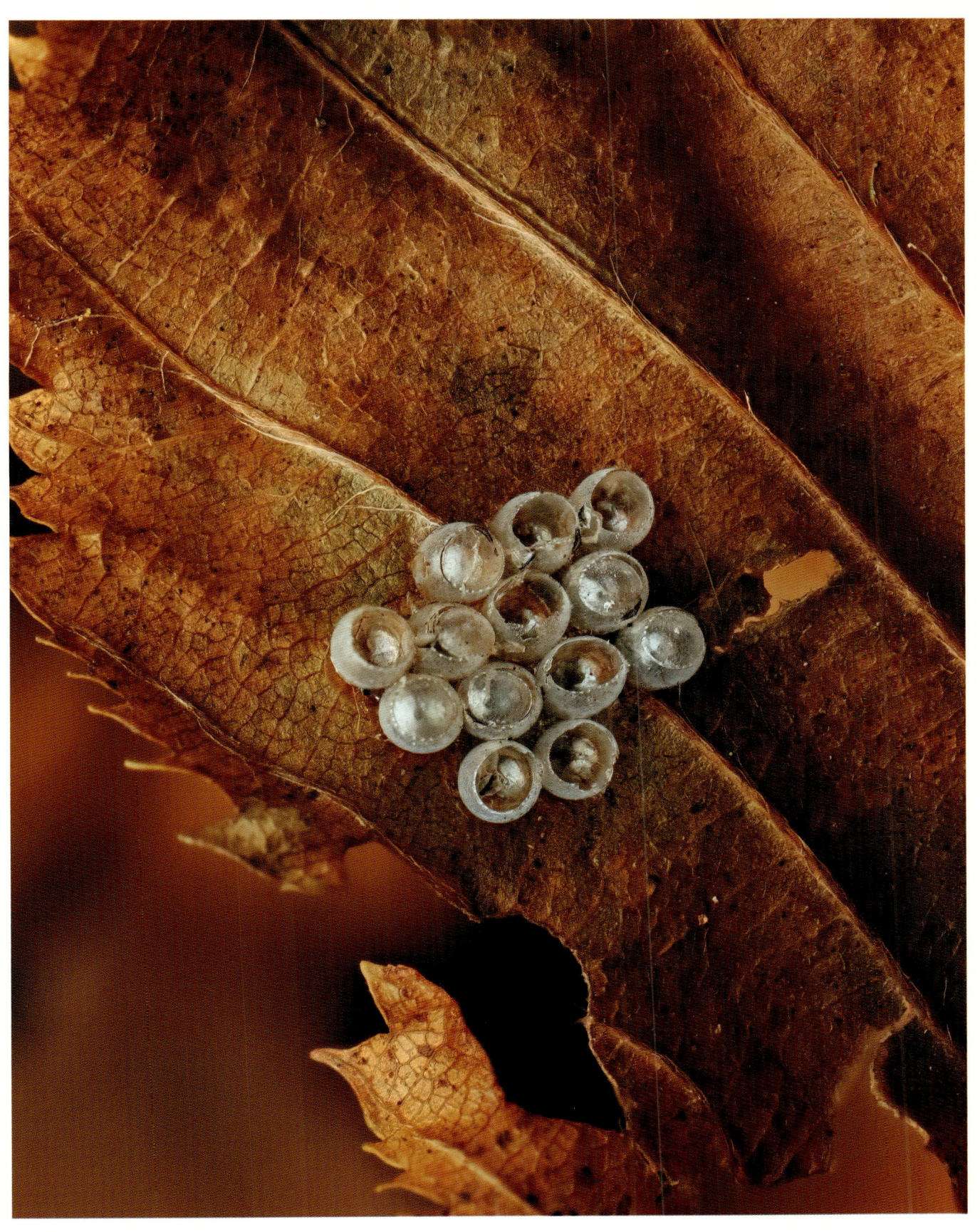

Index

250

Resources

Barker, G.M. *The Biology of Terrestrial Molluscs*. CABI Publishing, 2001.

Bravo, Felipe, Vladimir LeMay, Robert Jandl, and Klaus Gadow, eds. Managing Forest Ecosystems: *The Challenge of Climate Change*. Springer, 2008.

Brock, Paul D. *Britain's Insects: A Field Guide to the Insects of Great Britain and Ireland*. Princeton University Press, 2021.

Chinery, Michael. *Collins Complete Guide to British Insects*. HarperCollins Publishers, 2009.

DellaSala, Dominick A., ed. *Temperate and Boreal Rainforests of the World: Ecology and Conservation*. Island Press, 2011.

Gilbert, O. L., ed. *The Lichen Flora of Great Britain and Ireland*. The British Lichen Society, 2009.

Haskell, David George. *The Forest Unseen: A Year's Watch in Nature*. Viking, 2012.

Kimmins, J. P. *Forest Ecology*. Pearson, 2004.

Lavelle, P., and A. Spain. *Soil Ecology*. Springer, 2001.

Newton, Adrian C., ed. *Forest Ecology and Conservation: A Handbook of Techniques*. Oxford University Press, 2007.

Online Etymology Dictionary. "Macro." 2024. Accessed May 2024. https://www.etymonline.com/word/macro-.

Perry, David A., Ram Oren, and Stephen C. Hart. *Forest Ecosystems*. Johns Hopkins University Press, 2008.

Petersen, Jens H. *The Kingdom of Fungi*. Princeton University Press, February 2013.

Pretzsch, Hans. *Forest Dynamics, Growth and Yield: From Measurement to Mode* l. Springer, 2009.

Rojas, Carlos, and Steven L. Stephenson, eds. *Myxomycetes: Biology, Systematics, Biogeography and Ecology*. Academic Press, September 2021.

Sterling, Phil, and Barry Henwood. *Field Guide to the Caterpillars of Great Britain and Ireland*. Bloomsbury Wildlife, 2020.

Stokland, Jogeir N., Juha Siitonen, and Bengt Gunnar Jonsson. *Biodiversity in Dead Wood*. Cambridge University Press, 2012.

Thomas, Peter, and John Packham. *Ecology of Woodlands and Forests: Description, Dynamics and Diversity*. Cambridge University Press, 2007.

Wohlleben, Peter. *The Hidden Life of Trees: What They Feel, How They Communicate— Discoveries from a Secret World*. Greystone Books, 2016.

Acknowledgments

I must first thank my incredible partner, Christian, who not only tolerated but whole-heartedly supported my sudden and ferocious deep dive into all things forest floor. It has left him a macro-widower most weekends, especially in the autumn months. You are the greatest thing that has ever happened to me—my absolute favorite human—I lubb forever.

Massive thanks to my awesome parents, Lynn and Ken—who have served as the best cheering squad you could ever hope to have and always encouraged my various creative pursuits. Love you guys immensely. My beloved big brother, Dan, who has been there for me through every phase of my crazy life and had my back like a big brother should. I've always known that no matter what life had in store for me, you'd be there to rescue me, celebrate with me, and never judge me. You mean everything to me, loser.

My incredibly gifted Aunt Susan—who I surely must get all my genetic artistic talents from. My incredible Cleveland family—Aunt Joy and Uncle Tony, Aunt Cheri and Uncle Joe—and my incredible cousins who have shown me so much love. I love you all so much.

I have so many spectacular friends and cheerleaders in my life—I'm a lucky gal. Thank you to all of you.

My Macro-A-Team boys, who have been there from the beginning of my macrophotography journey. I respect and admire you all so very much. Thank you for your friendship and all the lolz.

I'm super grateful for all the lovely people I've met on Instagram in the nature nerd and macro-community. And John Plummer and Rosie Foxley-Wood! I wouldn't know what half of the weird things I've found in the forest were without you to investigate for me. Thank you so much for being my number one top ID dudes and beautiful talented artists and people. And thank you Ben Salb for answering all the technical photography questions I've clobbered you with over the years. You are an absolute genius and hero.

A huge thanks to Mark Thackara and OM SYSTEM for believing in me and appreciating my work from the very start. Being an ambassador for a camera company I've idolized my whole life is such an incredible honor, and your support along the way has meant the world to me.

Lastly, my Grandma Jackie (Mimi) who I used to email photos to every week before I knew what the heck I was doing with my camera; she told me I had an eye before I believed it was true. I miss you.

OPPOSITE: WAXCAP FUNGI.

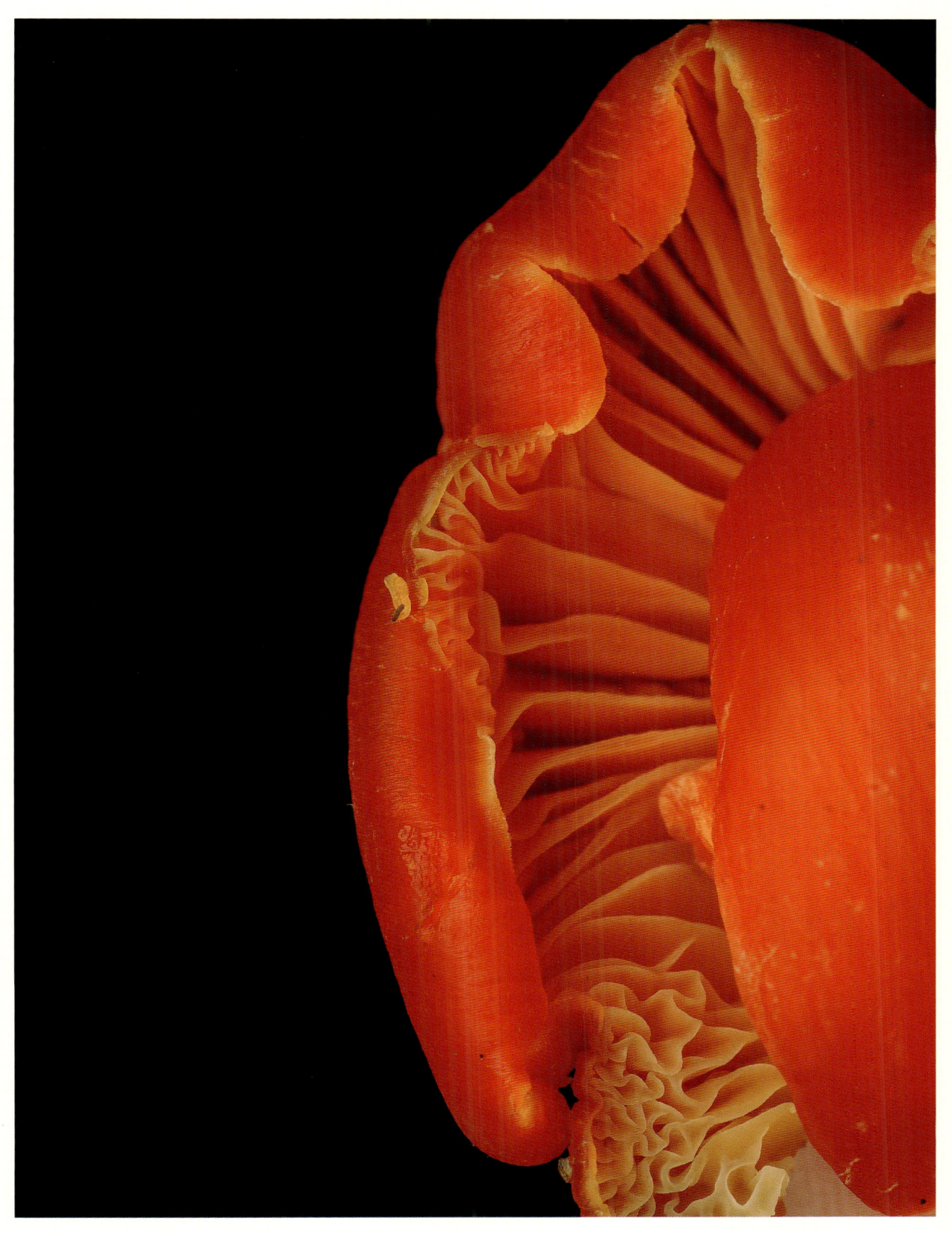

First published in 2025 by Wellfleet Press, an imprint of The Quarto Group,
142 West 36th Street, 4th Floor, New York, NY 10018, USA
(212) 779-4972 www.Quarto.com

Wellfleet Press titles are also available at discount for retail, wholesale, promotional, and bulk purchase. For details, contact the Special Sales Manager by email at specialsales@quarto.com or by mail at The Quarto Group, Attn: Special Sales Manager, 100 Cummings Center Suite 265D, Beverly, MA 01915 USA.

10 9 8 7 6 5 4 3 2 1

ISBN: 978-1-57715-520-1

Digital edition published in 2025
eISBN: 978-1-57715-521-8

Library of Congress Cataloging-in-Publication Data

Names: Rosencrans, Jamie author
Title: Tiny nature : discovering nature's hidden world through the lens of
    macrophotography / Jamie Rosencrans.
Description: New York, NY : Wellfleet Press, 2025. | Includes
    bibliographical references and index. | Summary: "Tiny Nature is a
    journey into the hidden world of the forest floor through the
    captivating lens of macrophotography"--Provided by publisher.
Identifiers: LCCN 2024061807 (print) | LCCN 2024061808 (ebook) | ISBN
    9781577155201 print | ISBN 9781577155218 ebook
Subjects: LCSH: Forest ecology--England--London--Pictorial works | Forest
    fungi--England--London--Pictorial works | Forest
    lichens--England--London--Pictorial works | Soil
    animals--England--London--Pictorial works | Macrophotography
Classification: LCC QH138.L75 R67 2025 (print) | LCC QH138.L75 (ebook) |
    DDC 577.309421--dc23/eng/20250421
LC record available at https://lccn.loc.gov/2024061807
LC ebook record available at https://lccn.loc.gov/2024061808

Group Publisher: Rage Kindelsperger
Editorial Director: Erin Canning
Creative Director: Laura Drew
Managing Editor: Cara Donaldson
Editor: Katelynn Abraham
Cover and Interior Design: Beth Middleworth

Printed in China